Servicing digital circuits in TV receivers

This book is dedicated
to Cheryl and Thomas

Servicing digital circuits in TV receivers

R. Fisher, Tech(CEI), LCG, AMSERT, MRTS
Lecturer in Digital and Television Electronics, Plymouth College of Further Education

Newnes Technical Books

Newnes Technical Books
is an imprint of the Butterworth Group which has principal offices in
London, Boston, Durban, Singapore, Sydney, Toronto, Wellington

First published. 1984

British Library Cataloguing in Publication Data

Fisher, R.
Servicing digital circuits in television receivers.
1. Digital television—Repairing
I. Title
621.388′87 TK6670

ISBN 0-408-01149-1

Library of Congress Cataloguing in Publication Data

Fisher, R. (Robert)
Servicing digital circuits in television receivers.

Bibliography: p.
Includes index
1. Television—Receivers and reception.
2. Television—Repairing 3. Television circuits.
4. Digital electronics. I. Title.
TK6653.F5 1984 621.388′36 83–23776
ISBN 0-408-01149-1

Photoset by Butterworths Litho Preparation Department
Printed and Bound in Great Britain by Thomson Litho Ltd, East Kilbride, Scotland

Preface

During the past five years the application of digital techniques in the domestic television receiver has increased rapidly. The main reason for this increase are two-fold:

Firstly with the advent of the Large Scale Integrated Circuit (LSI) it has been possible to dramatically reduce the cost of large combinational logic circuits to a level that makes their use in domestic equipment commercially viable. Secondly, the introduction of the microprocessor has enabled the broadcast television companies, British Telecom and equipment manufacturers to make available to the general public computer based services that would have been science fiction ten years ago.

Until recently TV engineers have had to repair and install receivers that use analogue techniques throughout, as well as to cope with the very fast changes that have taken place in circuit design. Now that advanced digital circuits are becoming common-place in a large range of TV receivers many engineers are finding the new technology difficult to understand. This situation is aggravated because up until now the receiver has not used digital circuitry at all, hence engineers have never had cause to study the techniques employed. The introduction of facilities such as Teletext and Viewdata has produced many problems for the servicing personnel of television retailers.

This book has been written for the qualified television engineer and for the student who is studying the City and Guilds 224 course 'Digital Information and Reception'. It is assumed that the reader has a sound knowledge of the 625-line television system and understands the operation of all semiconductor devices. Early chapters cover the basic digital techniques that are widely used throughout logic systems, and are designed to give the reader the technical grounding required before describing specialist applications. The remainder of the book contains sections devoted to individual applications of digital techniques within the television receiver. These include Video Display Units, Viewdata, Teletext, Remote Control and Digital Tuning Systems. An introduction to microprocessor techniques is also included for those readers who are working with the latest generation of receivers that have search tuning.

Wherever it has been possible, commercially available systems have been used as examples. A systems approach to the topics has been adopted with the major sections of the systems broken down to component level. Extensive use of block and circuit diagrams has been made. Throughout the book all currents are electron and American logic symbols have been used for illustration.

Technical references are given should the reader require them for further study.

R. Fisher

Contents

Acknowledgements

I would like to express my very sincere thanks to the following companies; without their help this book would not have been possibe.

Mullard Ltd.
Sony U.K.
Texas Instruments
B.B.C. CEEFAX Dept.
I.B.A. ORACLE Dept.
British Telecom PRESTEL
Rank Radio International

Thanks are also due to Edward Arnold & Co. Ltd. for permission to use certain illustrations from their book *Basic Principles and Practice for Microprocessors* (Heffer and King) and McGraw-Hill (UK) Ltd. for permission to use diagrams from *Digital Computer Electronics* (Malvino).

R. Fisher

Chapter 1

Comparison of digital and analogue systems

Since the first regular public television broadcast in 1936 the waveform used to carry the image and sound from the transmitter to the receiver has been a successful marriage of both digital and analogue signals. Picture and sound waveforms are transmitted using analogue techniques while the synchronizing information uses digital line and field synchronizing pulses. In the early 1970s additional digital information was added to the TV signal during the field blanking period in order to provide pages of Teletext at the receiver. The result of this combination of signals is a rather complex waveform which has to be 'deciphered' at the receiver in order to reproduce a high quality colour picture, sound and pages of Teletext. A TV waveform illustrating this combination of signals is shown in *Figure 1.1*.

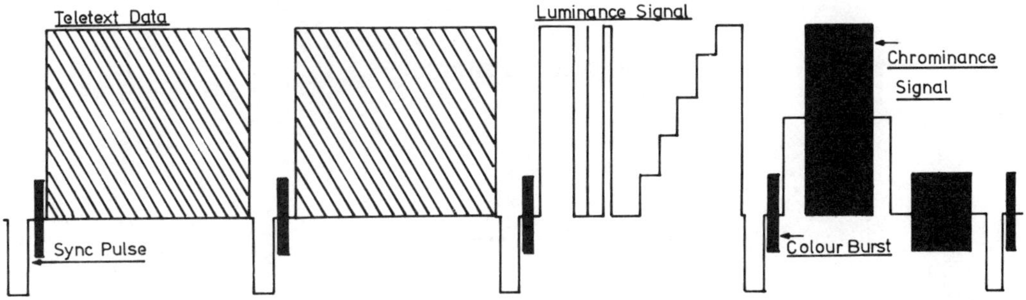

Figure 1.1 TV waveform with Teletext data

Before any attempt can be made to describe the application of digital techniques and circuits in TV receivers, the reader must first understand the major characteristics and merits of both analogue and digital signals. This will provide an introduction to the fundamental differences that exist between the waveforms and the circuits which are used to process them.

1.1 Analogue signals and systems

An analogue signal will vary in amplitude and frequency between maximum and minimum limits. The amplitude of the waveform at any point in time can be anywhere between its high and low limits, while its rate of change can also vary between predetermined maximum and minimum figures.

A system used to process this type of signal is known as an analogue system, and any such system must have a predetermined specification which will enable it to process the analogue signal in the desired way without introducing undue distortion and noise. To illustrate some of the problems involved with analogue systems we will briefly examine the processing of a video signal between the demodulator and the cathode ray tube (CRT) in a monochrome TV receiver.

The frequency response of the receiver's video amplifier will ideally be between d.c. and 5.5 MHz, which means that any input signal, video or otherwise, having a frequency which lies between these two points will be processed by the amplifier. This input signal will be between 2 to 4 V peak to peak, and the amplifier has to increase this amplitude to between 100 and 150 V peak to peak in order to provide sufficient drive for a monochrome CRT. A video amplifier must therefore have a number of characteristics which have to be accurately controlled if it is to provide the required gain to the video signal without affecting its rather complex shape. These characteristics are summarized as follows:

1. It must have the desired frequency response over the frequency range of the input signal.
2. It must not introduce harmonic distortion.
3. It must not introduce noise.
4. It must have a linear gain characteristic.

If the last item in this list is used to illustrate some of the problems encountered in more detail we will find that linearity becomes very critical in television, especially during the conversion of the video waveform into a picture.

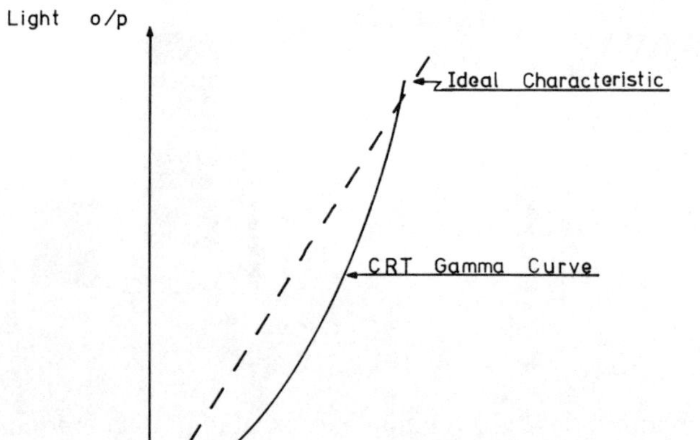

Figure 1.2 Gamma curve of a c.r.t.

When the voltage input versus the light output characteristic of a CRT is plotted, a graph similar to that shown in *Figure 1.2* will be produced, known as the gamma curve. It can be seen that this graph is non-linear and all the technical innovations that have taken place during the history of television have not been able to solve this problem. To overcome this situation the linearity of the video signal is deliberately distorted before transmission using a process known as gamma correction. The non-linearity of the CRT can be effectively removed by introducing an equal and opposite non-linearity into the video signal. Gamma correction therefore ensures that a linear rise in the light input to the camera will produce a linear rise in light output from the CRT. This problem is further complicated because the characteristic of a colour CRT is different to that of a monochrome CRT, consequently a 625-line colour transmission is gamma corrected to compensate for the non-linearity of the colour tube, and any monochrome receiver tuned to the same transmission will display a picture which has an incorrect gamma.

Possibly the greatest merit of the analogue signal is its simplicity. It has the ability to incorporate the two most widely used variables of amplitude and frequency into a single signal. This factor is demonstrated to great effect in the area of audio electronics, where the volume or amplitude of the sound alters the amplitude of the analogue signal and the pitch or frequency of the sound alters the signal's rate of change. Conversion of audio information into an electronic signal is the easiest and was therefore the first application of electronics for use by the general public. As time progressed, the complexity of the analogue signal increased and it is now common practice to transmit several analogue signals simultaneously on the same carrier. Colour television uses this to great advantage where the colour and monochrome signals are transmitted together on the same carrier, and the receiver is able to separate and process these two signals independently to reproduce the colour picture.

It is impossible to cover all the aspects of analogue signals and systems in this very short resumé, but it is hoped that the reader will appreciate the fundamental problems encountered with such systems, and their major disadvantages.

1.2 Digital signals and systems

Digital signals use only two voltage levels; these are either low or high. A low digital voltage is generally known as a logic '0' and a high voltage is known as a logic '1'. These terms 'high' and 'low' are relative since high can be as little as +5 V and low approximately +0.2 V. There are no 'inbetweens' with a digital signal, it is either at logic '0' or at logic '1'. This immediately dismisses all the linearity

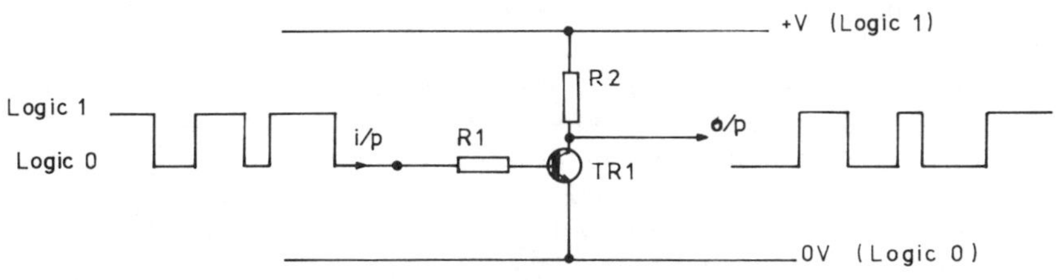

Figure 1.3 Digital signal and inverter

problems encountered with analogue circuits because the transistors or valves are either conducting or non-conducting; they are not required to slowly change from one state to another. *Figure 1.3* shows a digital signal coupled to a simple digital circuit (an inverter). When the input signal is at a logic '1' the transistor will be saturated causing its collector voltage to fall to approximately 0 volts, thus producing a logic '0' at the output. When the input signal falls to a logic '0' there is insufficient voltage on the base of the transistor to maintain conduction. The transistor therefore switches off and its collector voltage will rise to the supply potential producing a logic '1' at the output.

The simplicity of this circuit is immediately apparent, requiring just two resistors and a single transistor, If this is compared with an equivalent analogue inverter shown in *Figure 1.4*, which has a sine wave input and output, it can be seen that the component count and the complexity of the analogue circuit is far greater than the digital example. The example shown in *Figure 1.4* will also be subject to the limitations previously described, whereas the digital inverter only has one major problem to overcome. This is to ensure that the transistor is capable of switching at the speed determined by the rise and fall time of the digital square wave.

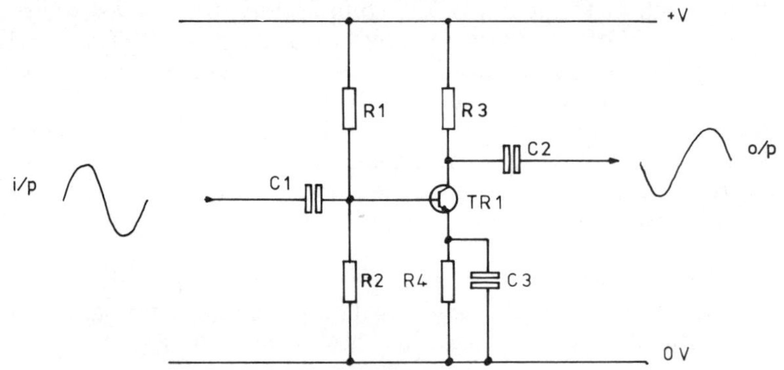

Figure 1.4 Analogue signal and inverter

High semiconductor switching speeds are required for most digital applications, and circuit complexity increases significantly as a result. D.C. coupling is nearly always used between stages of digital circuits. This is essential because the signal consists of two d.c. levels (logic '1' and logic '0') and it is possible that one logic level may be stable for between a few nano-seconds and several hours or more depending on the function of the circuit in question. A.C. coupling may be used if the designer can guarantee that the signal being processed by the circuit will be alternating between one logic level and another on a continuous basis, but in most applications this is not the case.

Nearly all commercially available digital integrated circuits use d.c. coupling between one stage and the next. (A very important point to be remembered when fault finding). At first, d.c. coupling between one circuit and another may seem to be a major limitation as stages cannot be isolated easily during the design phase, or at a later date when fault finding is required. This has, however, been turned to considerable advantage, because it is very difficult to produce capacitors of any significant value using integrated circuit techniques. The use of d.c. coupling together with the introduction of MOS transistors has made the most significant contribution to the rapid increase in component density of digital ICs, and it is now becoming common practice to manufacture single chips containing in excess of 250 000 devices.

Systems which use digital signals in preference to analogue signals are normally complex and 'intelligent', using mathematical as well as logical processing. Obviously the simple representation of the digital signal in *Figure 1.3* is an over simplication of those in common use in computers, TV remote control systems, Teletext and Prestel decoders. It is also apparent that this simple digital signal can only be used to represent two conditions, logic, '1' and logic '0' or circuit on, circuit off. In order to expand this single on-off concept it is necessary to use more than one digital signal to produce a system of any great power.

A large majority of digital circuits use a number of signals in parallel, each signal having only two logic levels. The digital information contained in this group of signals in known collectively as a DATA WORD. The logic level present on a single conductor is known as a BIT, hence the example shown in *Figure 1.5* is known as an eight-bit byte. Each bit of the data is expressed using the figures 0 and 1. 0 = logic '0' and 1 = logic '1'. This further simplification enables an eight-bit byte to be expressed as a combination of 0s and 1s. *Figure 1.5* therefore shows a byte of

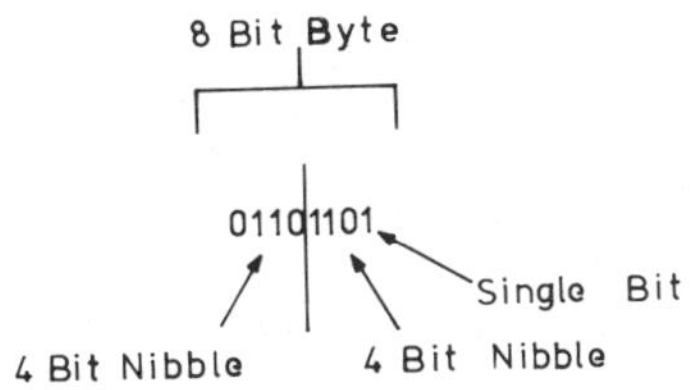

Figure 1.5 8-bit byte

data comprising of eight bits which when used collectively produce a code of '01101101'. Using eight conductors in this way it is possible to produce a total of 256 unique patterns of 0s and 1s and each unique pattern can be used to represent many different types of information. The code shown could be used to represent a character of Teletext, a decimal number, a computer instruction, or a remote control function.

Many digital systems have their own codes which are unique to that system, and Teletext is one such example. A Teletext decoder is a group of circuits which is designed to convert an 8-bit byte of received data into coloured letters and graphics for display on the receivers c.r.t. Each character which appears on the screen is represented by an 8-bit code which is received by the decoder and stored in its memory. A full and detailed description of the Teletext character code is given in chapter 9 but a sample of the character codes is listed below to demonstrate the versatility of a digital code:

A	= 1000001
b	= 1100010
?	= 0111111
FLASH	= 0001000
DOUBLE HEIGHT	= 0001101

Note that these examples only use seven bits. The missing eighth bit is used for detecting errors in the received byte of data so that incorrectly received characters can be prevented from appearing on the screen. Increasing the number of bits used, from say eight to sixteen, increases the number of individual codes that can be produced. In the case of sixteen bits the total number of different codes is 65536.

As the size of the Data Word increases it becomes necessary to sub-divide the byte into smaller groups of bits. This allows the pattern of 0s and 1s to be expressed more clearly. A byte of data is therefore split in half to form two NIBBLES. A nibble is exactly half a byte and therefore in the example shown in *Figure 1.5* the eight-bit byte can be divided into two four-bit nibbles.

At this point it is perhaps necessary to give an explanation for this peculiar terminology before the doubting reader continues.

1. Bit is derived from the two words Binary digIT.
2. Most digital systems process bits of data in parallel bytes, block by block. This processing is analogous to the first stages of the digestive system where food (the input to the system) is processed bite BY biTE.
3. From this point it is easy to see how the word nibble has been derived because any digestive system which takes less than one bite at a time must be nibbling!

1.3 Comparison of digital and analogue communications

Processing data using a number of bits in parallel involves a very large amount of electronics, whereas an analogue circuit normally processes only one, or possibly two, signals. This can be illustrated by examining the application of an emitter follower as a buffer stage which prevents the low input impedance of one circuit from loading the high output impedance of another. If an analogue signal was being coupled via an emitter follower, only a single stage of electronics would be required to provide the necessary isolation between circuits. On the other hand however, if an eight-bit byte was being coupled from one stage to the next via a buffer, eight circuits would be required, one for each bit of the byte. This is especially true where microprocessors or other computing circuits are employed. A comparison of these two circuits is shown in *Figure 1.6*.

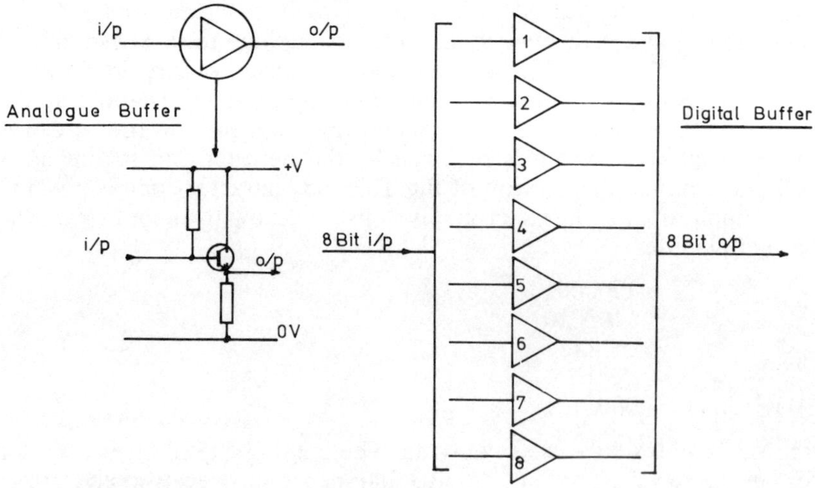

Figure 1.6 Analogue and digital buffer

Communication between one digital circuit and another can also provide several problems for the engineer. If a byte of data is to be coupled from one part of a printed circuit board to another, or from one printed circuit card to another, the easiest solution is to use the required number of conductors between stages.

Although this method is an obvious one it does produce considerable problems in the design of the printed circuit layout, as eight or more tracks take up a considerable amount of space on what will already be a congested printed circuit board. Taking the same situation to chip level, very large problems are encountered because of the amount of chip area this requires.

Coupling data from one piece of electronics to another, say from a Teletext decoder to a printer, presents further complications. Teletext data consists of an eight-bit byte, but only seven bits are used to form the characters which are displayed on the screen. Seven conductors would then be required to couple the data from the Teletext decoder to the printer, and an eighth conductor would have to be included for an 0 V reference or chassis connection. Providing the printer is no great distance from the decoder an eight-way cable could be used with very few problems, but what happens if several miles or hundreds of miles lie between the two digital systems? The British Telecom Prestel system is faced with this problem because it uses cable transmission between the telephone and a British Telecom computer. The telephone system only requires two conductors for its analogue signals so it is impossible to transmit data in parallel form over the telephone lines.

The same situation also arises when data has to be transmitted on a radio frequency carrier. It is obviously a tremendous waste of air space to transmit eight parallel bits of data on eight separate carrier frequencies and then have the same number of receivers to regain the byte of transmitted data. Even if the air space were available the cost of the system would be prohibitive.

An alternative means of transmitting data over long distances is therefore required, not only for communication over land lines but also for radio and satellite communication. The most widely used solution is to transmit the data on a single carrier, or along a single conductor in serial form. Transmission of data in this way

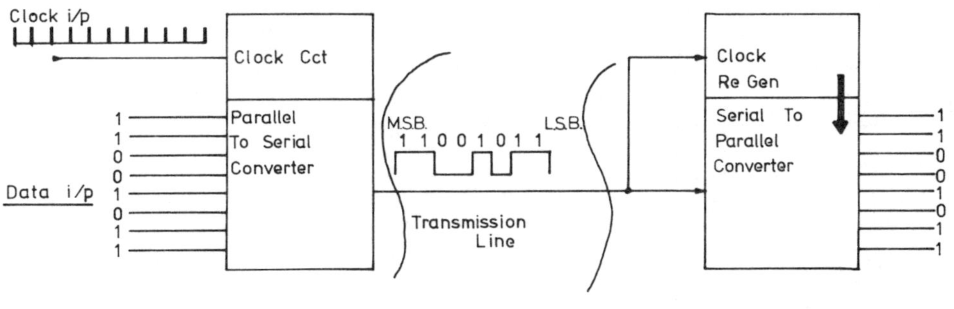

Figure 1.7 Serial transmission and reception of data

is achieved by taking a parallel byte and sending it to the receiver one bit at a time, before taking the next byte of data and repeating the process. A simplified diagram showing how this could be achieved is shown in *Figure 1.7*.

Although the main objective of reducing the number of conductors between one system and another has been achieved, several disadvantages now arise. Possibly the most limiting of these is the reduction in transmission speed. If system bandwidth limitations are ignored then the transmission of an eight-bit byte of data in serial form will take approximately eight times as long as transmitting the same byte of data in parallel form. This is a severe limitation if high data transfer rates are required.

Another but not so serious disadvantage is the increased complexity of the circuitry required to transmit and receive the data. At the transmitter some means of converting a parallel byte of data into serial form is required, and at the receiver the opposite process has to take place where the serial data at the input has to be converted back into a parallel byte before it can be used. In addition the transmitter and receiver must be synchronized so that data is transmitted, received and decoded at the same speed. The receiver must also 'know' which bit is being transmitted at any point in time. It is no good if the receiver expects bit five to arrive when the transmitter is sending bit two!

Serial transmission of data is common practice and there are many systems and standards in use. It is impossible within the limited scope of this book to cover all of the methods of transmitting data in serial form, but those transmission systems used for television orientated applications are fully described in chapter 6.

1.4 Pulse modulation

The descriptions that follow give an introduction to the major types of pulse modulation. The carrier used could be at radio, ultrasonic or light frequencies. Pulse modulation is included here because of its wide use in TV remote control systems, and as an introduction to the serial transmission of data for Teletext. In order to keep this concept as simple as possible an analogue sine wave has been used as the modulating signal in all of the following examples.

Pulse Amplitude Modulation (PAM)

Pulse amplitude modulation is directly analogous to the conventional amplitude modulation used for medium and long wave radio transmissions. The modulating

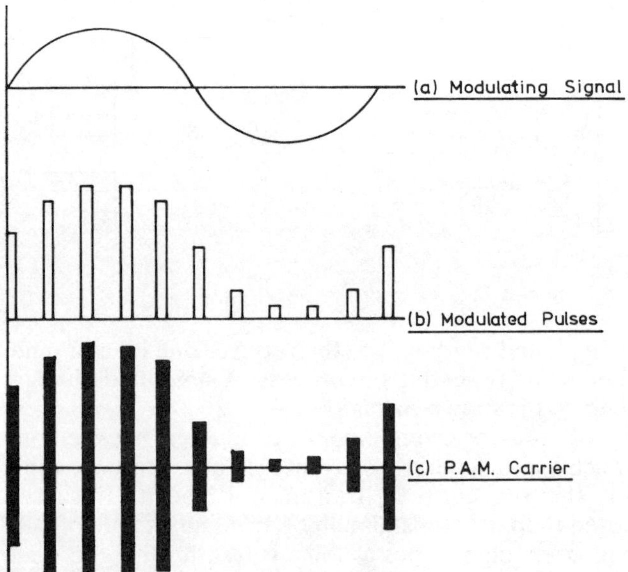

Figure 1.8 PAM (three waveforms)

signal shown in *Figure 1.8 (a)* has its amplitude sampled at a fixed rate, producing a series of pulses which have an amplitude that is proportional to the amplitude of the modulating signal *(Figure 1.8 (b))*. This pulse waveform is then used to amplitude modulate the carrier wave, producing an output from the transmitter as shown in *Figure 1.8 (c)*. Note that the sampling frequency must be considerably higher than the highest modulating frequency if the modulating signal is to be resolved accurately at the receiver.

Pulse amplitude modulation suffers from many of the deficiencies of conventional amplitude modulation, the most significant of these being the introduction of noise. To overcome this problem a constant amplitude carrier is of a considerable advantage, and this concept has led to the introduction of Pulse Width and Pulse Duration Modulation.

Pulse Width Modulation (PWM)

As with the previous example the modulating signal is sampled at regular intervals, but in this case the width of the pulse derived from the sample changes in proportion to the amplitude of the modulating signal. The resulting input signal to the modulator is similar to that shown in *Figure 1.9 (b)*.

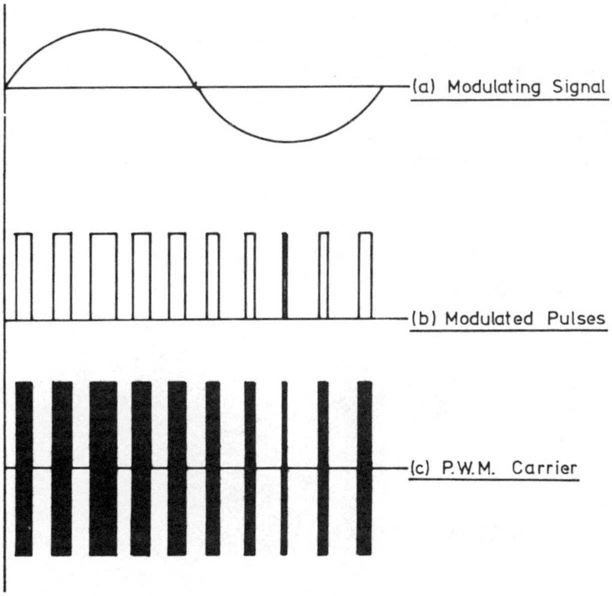

Figure 1.9 PWM (three waveforms)

The modulator used for PWM need only be a simple switch which closes when the modulating pulses are positive and opens when they are zero. This switch can then couple the signal from the carrier oscillator to the output of the transmitter as shown in *Figure 1.10*. Note that both the carrier amplitude and frequency remain fixed. This method of pulse modulation can be used in TV remote control systems because the modulating pulses are applied directly to an infra-red diode which

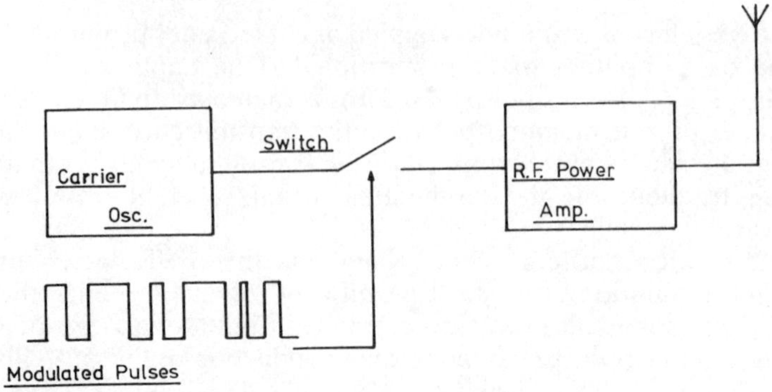

Figure 1.10 PWM using switch as transmitter

converts the pulses into a modulated infra-red light beam. This single light-emitting diode therefore acts as both the modulator and transmitter making it ideal for use within the limited space of a hand held remote control transmitter.

Pulse Position Modulation (PPM)

PPM leaves both the amplitude and the width of the carrier pulses unchanged, but alters the timing intervals at which they occur. Under conditions of no modulation bursts of carrier frequency will be transmitted at the nominal sampling rate. If the modulating signal rises in a positive direction the position of the sampling pulses is delayed, and if the modulating signal falls in a negative direction the sampling pulses are advanced. This is illustrated in *Figure 1.11*. PPM is virtually immune from interference but does require complex electronics to process the information.

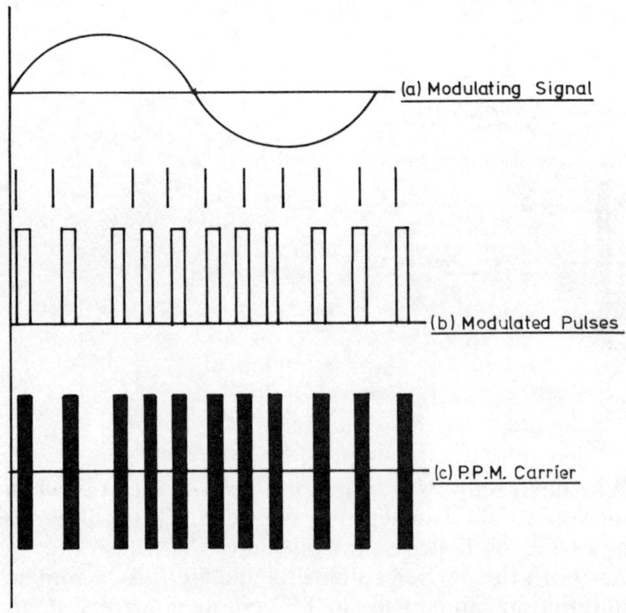

Figure 1.11 PPM (three waveforms)

1.5 Analogue to digital conversion

There are many situations where a device producing an analogue signal is required to interface directly with a digital system. For example a joystick control in a TV games unit will produce an analogue signal in the form of a variable d.c. voltage. If the TV games unit is microprocessor based it will have to convert this voltage in to a digital signal before it can be used by the microprocessor. To achieve this conversion circuits known as analogue to digital convertors (ADC) have been developed.

An ADC accepts an analogue input, samples its amplitude, and produces a correspondingly coded digital output. Low cost ADCs are available in a single integrated circuit package and normally produce an eight-bit byte at their output. A simplified schematic of an ADC is shown in *Figure 1.12*. In order to convert a

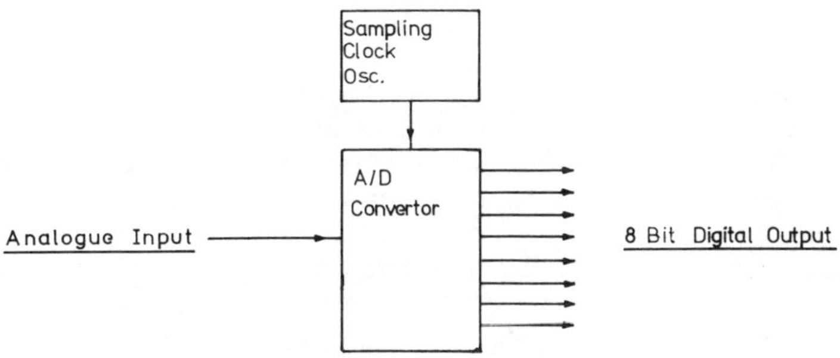

Figure 1.12 ADC schematic

continuously changing analogue input into a digital output, the analogue waveform must be sampled at regular intervals, each sample producing a digital code at the output of the convertor. The sampling rate used must be at least twice the highest frequency contained in the analogue signal if an accurate digital replica is to be produced.

The accuracy to which the amplitude of the input voltage is resolved is determined by the number of bits used in the digital output. If the convertor produces an eight-bit byte an accuracy of 1 in 256 can be obtained; e.g. the input waveform can be subdivided into 256 discrete levels which can be identified by the 256 unique codes present at the output of the ADC. If greater accuracy in the conversion process is required then a larger number of bits has to be used. A twelve-bit word would allow the input voltage to be subdivided into 4095 sampling levels resulting in an extremely accurate analogue to digital conversion.

The input voltage to an ADC need not be a.c. but could be any analogue voltage from say a strain gauge, d.c. generator or photo-electric device. The ADC will, however, produce a digitally coded output with a value that is proportional to the amplitude of the input voltage. Hence a digital signal can not only be used to represent mathematical or character data but it can indicate the amplitude of a voltage, temperature, position or pressure etc, depending on the device which produces the analogue input to the ADC.

1.6 Digital to analogue conversion

Processing analogue signals which have been converted into digital form is of little use unless a conversion in the opposite direction can be achieved so that the digital signal can be converted back into an analogue waveform. A device known as a digital to analogue convertor (DAC) is used to achieve this and a simplified diagram is shown in *Figure 1.13*. The coded digital input waveform is presented to the convertor in parallel form, in this example using eight bits.

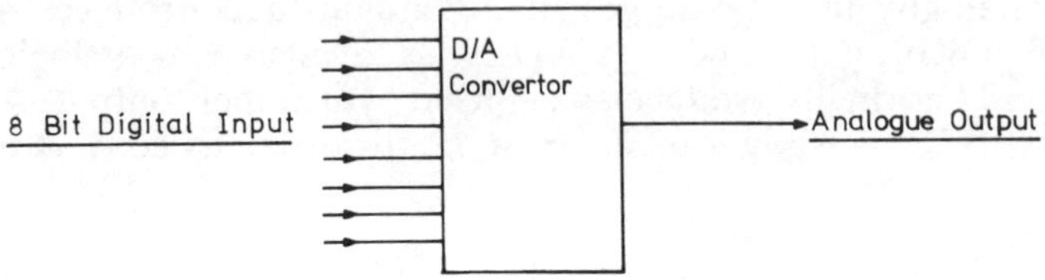

Figure 1.13 DAC schematic

The convertor uses the digital input to produce an analogue output voltage which has an amplitude that is proportional to the value of the coded data at the input. If the digital input to the convertor is made to change quickly enough, the convertor can produce what appears to be a continuous a.c. output. It will in fact be a d.c. voltage which changes rapidly.

The first part of this chapter has described and compared some of the major characteristics, advantages and disadvantages of both analogue and digital signals and systems. Digital circuits have made a significant impact in TV mainly because they are ideally suited in applications which require some form of mathematical or logical execution. The remainder of this chapter will outline the binary and hexadecimal number systems which form the foundation of all the digital theory covered in the following chapters. It is therefore imperative that the reader fully understands the remaining pages of this chapter before continuing.

1.7 The binary number system

Digital circuits can only operate using the two logic levels previously described, hence with one bit of data it is only possible to count to one, (nought to one). In order to produce digital circuits that carry out mathematical processes it is necessary to devise a counting system with a number base of two, rather than the decimal system which has a number base of ten. Such a system is known as the

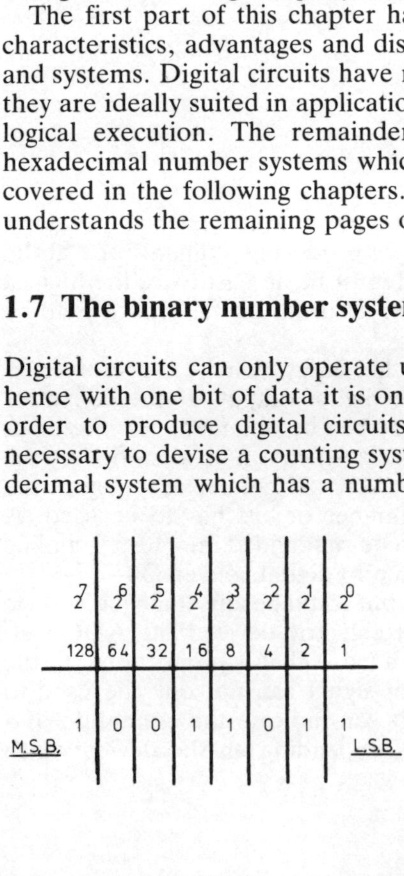

Figure 1.14 8-bit byte (10011011) weighted with MSB and LSB

Binary Number System. Each bit of data can be given a binary weighting which is expressed as 2^x. If an eight-bit byte of (10011011) is used as an example then each bit of the byte is weighted as shown in *Figure 1.14*. The right-hand bit is weighted 2^0 and makes the least significant contribution to the total value of the binary number and is therefore known as the Least Significant Bit (LSB).

The left-hand bit has a weighting of 2^7 and will make the biggest contribution to the total value of the number and is therefore known as the Most Significant Bit (MSB). If the byte of data is now given the binary weightings shown, the equivalent decimal number can be calculated. This process is shown in *Figure 1.15*. Converting the binary number to decimal is an easy process. Each bit which is set to a logic '1' is referred to its binary weighting, and the weighting is noted. All the weighted values are then added together to produce the equivalent decimal number.

2^7	2^6	2^5	2^4	2^3	2^2	2^1	2^0
1	0	0	1	1	0	1	1

128 + 0 + 0 + 16 + 8 + 0 + 2 + 1 = 155

Figure 1.15 Figure 1.14 converted to decimal

If a decimal number is to be represented in binary form then some means of conversion from decimal to binary is required. *Figure 1.16* shows how the decimal integer 155 is converted into binary.

2	155		
	77	R1	LSB
	38	R1	
	19	R0	
	9	R1	
	4	R1	
	2	R0	
	1	R0	
	0	R1	MSB

Figure 1.16 Decimal to binary conversion

This is achieved as follows. The decimal number is divided by two producing an integer result and a remainder. The remainder will be either a '0' or a '1' and it is this part of the process that produces the binary digit. Division is repeated until the integer is reduced to zero and the binary equivalent of the original integer is obtained by reading the remainders in reverse order; e.g. The last remainder produced is the most significant binary digit, and the first is the least significant binary digit.

It is possible to convert numbers of less than one into binary form and *vice versa*. This process is of course vital for accurate calculations in computing systems, but it is rarely used in digitally based TV receiver systems. It is therefore beyond the scope of this book.

1.8 The hexadecimal number system

Bytes of data containing a large number of bits can become rather cumbersome when it is required to print or display them on a CRT. In addition, if bytes of data are to be entered into a digital system manually, e.g. via a keyboard, some means of simplifying the representation of the data is required. This simplification is achieved by using the Hexadecimal number system which is widely used by microprocessors and other computing systems using 8-bit data processing techniques.

This coding system enables up to sixteen decimal numbers to be represented using a single character. Four binary bits also represent a maximum of sixteen different codes from '0000' to '1111', consequently a single hexadecimal character can be used to represent a decimal number up to the value of 15 or a four-bit binary code.

Hexadecimal	Binary	Decimal
0	0000	0
1	0001	1
2	0010	2
3	0011	3
4	0100	4
5	0101	5
6	0110	6
7	0111	7
8	1000	8
9	1001	9
A	1010	10
B	1011	11
C	1100	12
D	1101	13
E	1110	14
F	1111	15

Figure 1.17 Hexadecimal binary and decimal conversion chart

Decimal and binary numbers from 0 to 9 are represented by their decimal equivalent, as shown in *Figure 1.17*. Decimal and binary numbers from 10 to 15 are represented by the alphabetical characters A, B, C, D, E, F. Many keyboards which are used to input data to microprocessor systems are labelled using hexadecimal numbers, because each key entry will input four bits of data to the system, and two key entries will input an eight-bit byte.

Binary and hexadecimal codes are the two most commonly used coding systems in digital electronics. Many others exist such as Octal, Binary Coded Decimal, and the Gray code.

Chapter 2

Logic circuits

Logic networks fall into two main groups known as Combinational Logic and Sequential Logic. Combinational logic systems generate an output signal when a given combination or combinations of input signals exist. Sequential logic circuits depend on a sequence of events (which have already occurred) within the circuit in order to produce one or more outputs. Such systems include counters, shift registers and the manipulation of serial data.

In this chapter we will examine the design, application, and operation of the common combinational logic circuits.

2.1 Positive and negative logic systems

A positive logic system is one in which the higher of the two voltages used in the digital signal represents a logic '1' and the lower of the two voltages represents a logic '0'. In a negative logic system the lower of the two logic levels represents a logic '1' and the higher a logic '0'. The designer of a circuit is free to choose which logic system he is to use, but it is important that the choice of positive or negative logic is clearly stated before describing any aspects of the system, or an individual circuit. Throughout the description of digital signals in chapter 1 positive logic was used, where a logic '1' is represented by +5 V and a logic '0' is represented by 0 V. Unless specifically stated all circuits and descriptions in this book will use positive logic.

The voltage levels used to represent logic '1' and logic '0' can also vary considerably depending on the system in use and its application. So far we have used voltage levels of 0 and +5 V, but there are applications where these voltages may differ considerably. A system may use −12 V for logic '1' and +12 V for logic '0'. Such a system uses negative logic but provides an overall peak to peak voltage of 24 instead of 5 volts.

2.2 Tri-state logic

In the descriptions of digital signals used so far only two logic levels have been used. There is however, a third logic 'condition' which is commonly used in microprocessor based systems, known as the Tri-State output. The term Tri-State is a little misleading because digital circuits are not capable of producing a third logic state or level. They either produce a logic '1' or a logic '0'. It is possible however to open circuit the signal feed between the output of one logic circuit and the input to the next simply by inserting a series switch. In the example shown in *Figure 2.1* an inverter formed by the MOS transistor TR1 has a second MOS transistor TR2 connected in series with its output.

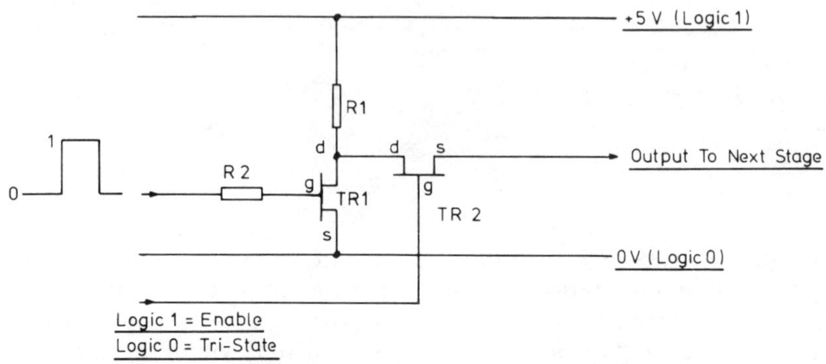

Figure 2.1 MOS Inverter stage

Assume the system uses positive logic where a logic '0' = 0 V and a logic '1' = +5 V. TR2 will conduct when a logic '1' is placed on its gate, causing a low resistance to be present between its source and drain. This low resistance allows the output from TR1 to be coupled to the next stage of the circuit. If the signal on the gate of TR2 is changed to a logic '0' then the transistor will switch off resulting in a high resistance between its source and drain. An effective open circuit now exists between the output to the next stage and the drain of TR1, preventing the signal from leaving the circuit.

Logic circuits that use Tri-state conditions are found where the outputs of several circuits have to be switched to form a single input to one or more stages.

2.3 Digital pulse parameters

Ideally any digital signal should be a perfect pulse waveform in which the rise or fall from one logic level to another is instantaneous. However this is not possible because most digital signals are derived from an electronic circuit which will suffer from limitations such as the transistor switching speed.

In practice a typical digital pulse will appear as shown in *Figure 2.2*. This type of pulse could be produced by practically any digital circuit such as the output stage shown in *Figure 2.3 (a)*. When this transistor produces a logic '0' at its output the input waveform to its base will be a logic '1', causing the transistor to saturate. Under these conditions the output circuit of the stage can be represented by the potential divider shown in *Figure 2.3 (b)*, where RL is the value of the transistor collector load and RT is the internal saturation resistance of the transistor between collector and emitter. Although this saturation resistance is quite low it will produce a very small voltage drop across it. Consequently the logic '0' produced by

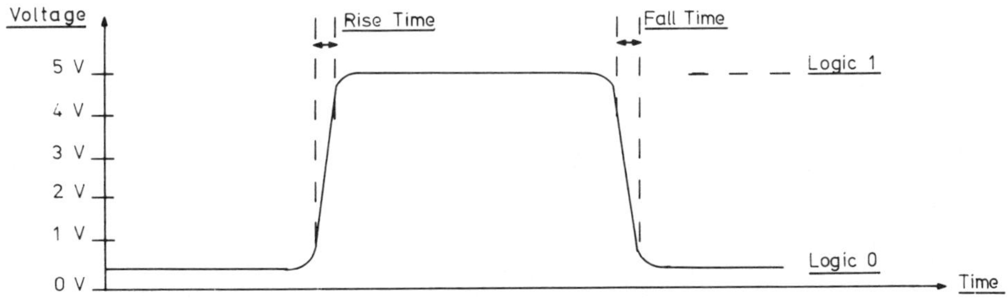

Figure 2.2 Typical digital pulse

this stage will never fall to exactly 0 V but will always be a small positive value above 0 V. (Typically 0.2 V).

When the circuit produces a logic '1' at its output, the transistor will be cut off resulting in a very high resistance between its collector and emitter. The current flow through the potential divider formed by the transistor's internal resistance and the load resistor will now be equal to the leakage current of the transistor, which is only a few micro amps. The voltage drop across RL is now very small indeed, and the output voltage from the circuit will rise to the supply potential, see *Figure 2.3 (c)*.

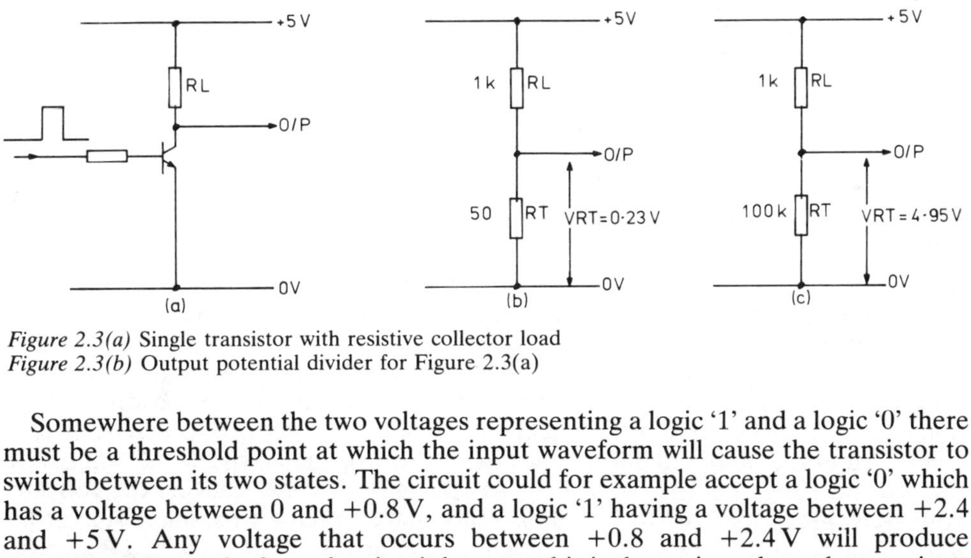

Figure 2.3(a) Single transistor with resistive collector load
Figure 2.3(b) Output potential divider for Figure 2.3(a)

Somewhere between the two voltages representing a logic '1' and a logic '0' there must be a threshold point at which the input waveform will cause the transistor to switch between its two states. The circuit could for example accept a logic '0' which has a voltage between 0 and +0.8 V, and a logic '1' having a voltage between +2.4 and +5 V. Any voltage that occurs between +0.8 and +2.4 V will produce unpredictable results from the circuit because this is the region where the transistor will change from one state to another. It is imperative that any digital system works well within the specified tolerances of the logic signal, so that reliable and predictable results can be produced.

If a logic circuit has no external load, such as the one shown in *Figure 2.3 (a)*, there is no danger that the resulting output voltage from the circuit will lie outside the amplitude parameters of the pulse. In practice there will be one or more circuits connected to this output and the input impedance of these circuits will effect the parameters of the digital signal.

2.4 Fan out

The fan out capability of a circuit is its ability to drive a number of input circuits simultaneously without causing either of the two logic levels to fall outside their specified range. Examine the situation shown in *Figure 2.4*. TR1 forms the output stage of a digital circuit, and its output signal is feeding a number of circuits all of which have an input stage identical to that used by TR2. When TR1 is conducting its output voltage will fall to approximately +0.2 V to produce a logic '0'. The input diode D1 of each driven circuit will now be reversed bias and consequently very little current will flow in the output circuit of TR1, leaving its output voltage unaltered. When TR1 is not conducting its collector voltage will rise to the supply potential causing the input diode of each driven circuit to conduct.

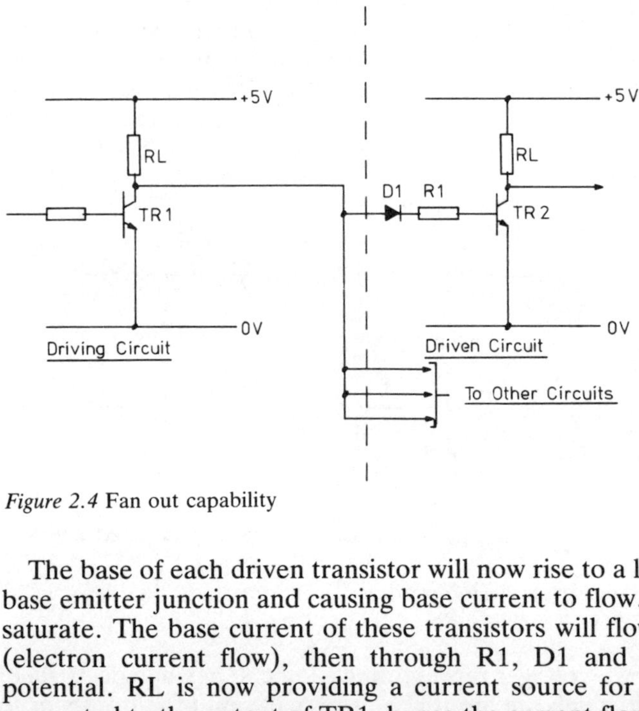

Figure 2.4 Fan out capability

The base of each driven transistor will now rise to a logic '1', forward biasing the base emitter junction and causing base current to flow. The driven transistors now saturate. The base current of these transistors will flow from the emitter to base, (electron current flow), then through R1, D1 and RL to the positive supply potential. RL is now providing a current source for each of the driven circuits connected to the output of TR1, hence the current flowing through RL will rise. If this current rises then the voltage across RL must also rise, causing the collector voltage of TR1 to fall below the supply potential and reduce the amplitude of the output pulse. It can be seen that as the number of driven circuits is increased a point will be reached where the output voltage of TR1 will fall below the threshold point causing unpredictable operation of the driven circuits.

Alternatively the input circuit to the driven stages may appear as shown in *Figure 2.5* causing a different situation to arise. With TR1 non conducting a logic '1' results at its output, reverse biasing the input diode D1 in each of the driven circuits. The only current that now flows through RL is

(a) The leakage current from TR1;
(b) The leakage from the input diode D1.

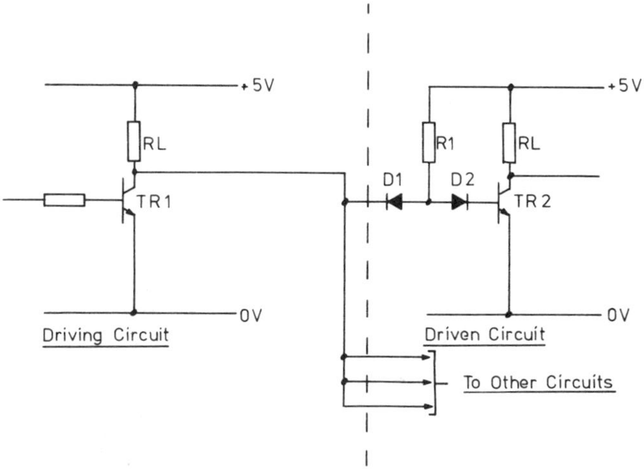

Figure 2.5 Input circuit to driven stages

The sum of these currents is very small indeed and this has little effect on the amplitude of the output pulse.

When TR1 is conducting, producing a logic '0' at its output, then the input diode D1 on each driven circuit will be conducting. Current flows from emitter to collector of TR1 via D1 and R1 to the positive supply potential. This current is in addition to the current flowing from TR1 emitter collector via RL to the supply. TR1 is therefore carrying:

(a) Its normal conduction current;
(b) The input current for each driven stage.

As a result of this situation the voltage between TR1 collector and emitter will rise causing the output voltage representing a logic '0' to rise towards the threshold point. Secondly the power dissipation in TR2 will also rise, and if this became excessive the output stage of the driving circuit could be damaged.

Every logic circuit will therefore have a specified fan out figure. This indicates the maximum number of circuits that can be driven from the output without causing the amplitude of the digital signal to fall to a point where the operation of the system becomes unreliable. It is therefore important to note that while servicing digital equipment some of the signals encountered may not be exactly the correct amplitude, but providing the voltages measured are within the specified parameters the system will operate reliably.

2.5 Rise and fall times

It was mentioned earlier in this chapter that an ideal digital signal will change from one logic state to another instantaneously. But this ideal situation is never achieved because of the switching delay that takes place in the transistors used in the circuits. When examining rise and fall times it is important to remind ourselves that bipolar transistors are current-operated devices, consequently input and output current waveforms have to be considered before rise and fall times in output voltage can be calculated. Typical switching waveforms of a transistor are shown in *Figure 2.6*.

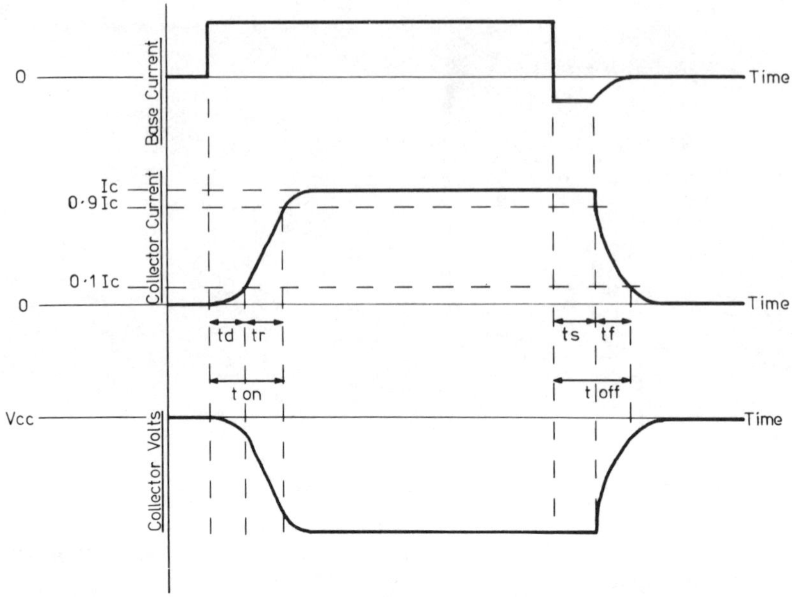

Figure 2.6 Transistor switching waveforms

Assume that an *n-p-n* transistor is operating in the common emitter mode as an inverter and it is non conducting. If leakage current is ignored we can say that there is no base or collector current, and the collector voltage will be at the supply potential. When a positive drive voltage is applied to the base, the transistor will remain off for a very short period of time and the collector current will still be zero. This is known as the delay time (*td*) and it is the result of the time it takes for the base current to propagate through the base region before it brings the transistor into conduction.

The delay time finishes once the collector current rises to 10% of its maximum value. From this point the collector current rises rapidly, but not instantaneously, towards its maximum value. The time taken for the collector current to rise from 10% to 90% of its maximum value is known as the rise time (*tr*). This rise time is determined by several factors including:

(a) The magnitude of the base drive;
(b) The current gain of the transistor;
(c) The frequency response of the transistor.

During the rise time period the collector voltage of the transistor will fall from 90% to 10% of the supply potential. By examining the leading edge of the collector current waveform it can be seen that the total period of time it takes for the transistor to switch on (*t* on) is the sum of the delay and rise times.

When the base voltage is reduced to zero, a short time is required to remove the negative base charge before the transistor can begin to switch off. This time period is known as the storage time (*ts*), and during this time the collector current will remain constant. Once the base charge has been removed the collector current will start to fall, and the time it takes this current to fall from 90% to 10% of its maximum value is known as the fall time (*tf*). During this period the collector voltage will rise again from 10% to 90% of its maximum voltage. This is known as the turn off time (t off) and it can be calculated using the formula

$$t \text{ off} = ts + tf$$

If the base current and collector voltage waveforms are compared the reader will notice that the whole signal has been delayed by the transistor and this becomes a very important factor in the design of all logic circuits. This delay is known as the propagation delay (t pd) and it is specified at the 50% voltage levels of the input and output pulses.

$$t \text{ pd} = \frac{t1 + t2}{2}$$

An example is shown in *Figure 2.7*

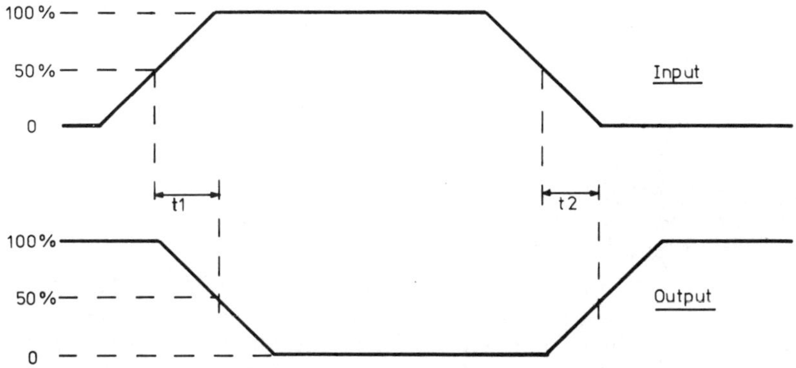

Figure 2.7 Propagation delay

A digital circuit may consist of several transistor stages connected in cascade consequently the propagation delay of each transistor must be added before the total delay through the circuit is known. The amount of delay depends on several factors including the circuit design, stray capacitance on the circuit inputs and outputs, power consumption and supply voltage.

2.6 Logic gates

Logic gates form the foundation of all digital circuits and they are so named because their behaviour can be likened to a gate that can be either opened or closed depending on a set of predetermined laws of the circuit. The fundamentals of each gate are relatively simple to understand and by themselves should not produce any problems for the reader. Unfortunately logic systems uses many gates, ranging from a few to several hundreds depending on the type of system. It is when circuits get to this complexity that they become difficult to understand and fault find. As a gentle introduction we will start off with the simplest of logic gates.

The NOT gate

The NOT gate is a simple digital inverter, hence the output will be the inverse of the input. In its simplest form the NOT gate could be constructed using a common emitter *n-p-n* transmittor but in practice the circuit is more complex in order to improve propagation delay, fan out etc.

As all logic gates are contained in integrated circuits there is no need to examine the circuit diagram of an individual gate. Logic symbols have been developed to represent the complete combination of components which make up each gate, and the logic symbol for a NOT gate is shown in *Figure 2.8*. As digital signals only have two states we can summarize the input and output conditions that will exist for this gate in what is known as a truth table. See *Figure 2.9*. A truth table will specify all

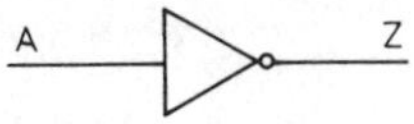

Figure 2.8 NOT gate cct symbol

A	Z
0	1
1	0

Figure 2.9 NOT gate truth table

the possible input conditions to the gate and the resulting output condition for each combination of inputs. Mathematical equations have been derived for the function of all logic gates using what is known as a Boolean expression. For the NOT gate this is:

$$Z = \overline{A}$$

where $Z=$ the output of the gate;
A = the input to the gate.
NOT = an inversion process

It is not within the scope of this book to go into the mathematics of logic circuits in any great detail because this is primarily the concern of the design engineer. But it is useful for the service engineer to understand the basic expressions because they are used extensively in manufacturers data on integrated circuits and can be of considerable help when dealing with the more complex logic elements.

The buffer

A buffer simply provides isolation between elements of a logic circuit. It may for example be required where a logic gate has a fan out of ten circuits and it is required to connect twenty circuits to this single output. If this sort of problem arises the answer is to feed the output of the gate to the inputs of two separate buffers and then use each buffer to feed 10 circuits.

A logic buffer simply provides isolation between its input and output without affecting any of the characteristics of the input signal. Its logic symbol is similar to the NOT gate shown in *Figure 2.8*, but it does not have the circle at the output. This circle indicates an inversion process. The logic symbol for the buffer is shown in *Figure 2.10*, and the truth table in *Figure 2.11*. The Boolean expression is:

$$Z = A$$

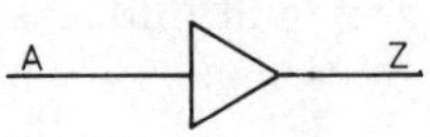

Figure 2.10 Buffer cct symbol

A	Z
0	0
1	1

Figure 2.11 Buffer truth table

The AND gate

Figure 2.12 shows the logic symbol of a two input AND gate. The gate is so named because the output Z will only rise to a logic '1' when inputs A AND B are both at logic '1', under any other input condition the output Z will be a logic '0'. A two input AND gate has been used in this example so that the description can be kept as simple as possible. AND gates are available with several inputs. The truth table of a two input AND gate is shown in *Figure 2.13* and is naturally a little more complex than those used for the buffer and NOT gates.

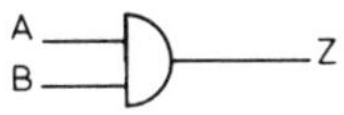

Figure 2.12 AND gate cct symbol

A	B	Z
0	0	0
0	1	0
1	0	0
1	1	1

Figure 2.13 AND gate truth table

Two input columns are required for the inputs A and B, and one column for the output Z. Again the truth table contains all the possible combinations of input signals, and as the number of inputs to a gate increases so does the complexity of the truth table. Note that the inputs A and B progress through a binary count from 00 to 11 (0 to 3 in decimal). Consequently there is a maximum of four different input combinations to this gate and the truth table states the output condition Z for each of the four input conditions. The Boolean expression for the AND gate is:

$$Z = A.B$$

where, Z = the output;
A and B = the two inputs;
. = the logic AND function

If the number of inputs to the gate were increased to four, the Boolean expression would read:

$$Z = A.B.C.D$$

Sometimes a situation will arise where a designer requires a four-input AND gate, but only has two-input AND gates available for use in the circuit. In this situation it is possible to use a combination of two-input AND gates to produce the

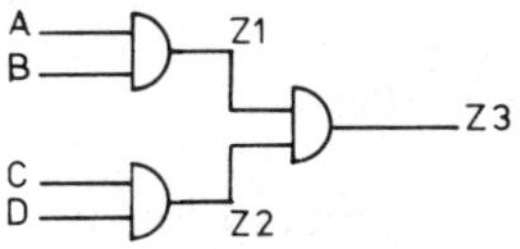

Figure 2.14 4 i/p AND gate using 2 i/p AND gates

four-input gate required. Consider *Figure 2.14*. Three, two-input AND gates are used to make up the required four-input gate. Mathematically the circuit produces:

$$Z1 = A.B \qquad Z2 = C.D. \qquad Z3 = Z1.Z2$$

Substituting, for Z1 and Z2

$$Z3 = A.B.C.D.$$

The NAND gate

NAND stands for NOT AND, hence this gate is equivalent to an AND gate followed by a NOT gate. Its circuit and truth table is shown in *Figures 2.15 and 2.16* respectively. The Boolean expression for the gate is

$$Z = \overline{A.B}$$

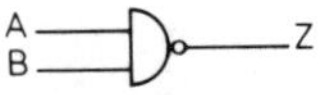

Figure 2.15 2 i/p NAND cct symbol

A	B	Z
0	0	1
0	1	1
1	0	1
1	1	0

Figure 2.16 NAND gate truth table

As the NAND gate is effectively two gates in one it is quite easy to derive the truth table by using those already examined for the AND and NOT gates. The truth table for the AND gate is shown in *Figure 2.13*, all that is required is to invert the output Z to produce the truth table for the NAND gate.

The two-input NAND gate is the most versatile of logic circuits, and it can be used to produce any other logic function. A NAND gate can be used as a NOT gate simply by connecting the two inputs together to form a single input. Examination of the NAND truth table shows that if the two inputs are at a logic '0', Z will be at logic '1' and if the two inputs are at logic '1', Z will be at logic '0'. No other input combination is possible because the two inputs are joined.

The flexibility of this particular arrangement can be further illustrated by producing a two-input AND gate by using two NAND gates. The first of these gates which are shown in *Figure 2.17* is the NAND gate followed by the inverter connected NAND gate. This circuit can be redrawn as shown in *Figure 2.18*. If the output from the AND gate is inverted twice it will return to its original state hence the last two stages effectively do nothing at all leaving just the two-input AND function.

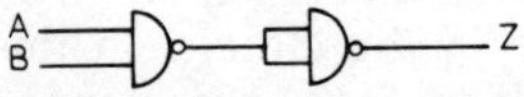

Figure 2.17 2 i/p AND using NAND gates

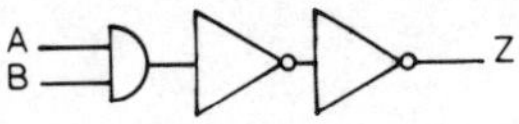

Figure 2.18 2 i/p AND followed by 2 NOT gates

Although logic circuits can be custom-made to carry out any number of functions it is important to remember various tricks of the trade such as those described above. It has become common practice to use these techniques where several logic gates are unused in an integrated circuit and the designer requires the use of a gate which is not contained on the circuit board. It is more economical to use existing unused logic gates to produce the required function rather than add extra circuitry at an additional cost. The reader is almost certain to meet this type of situation when working on logic circuits.

The OR gate

The output of an OR gate will rise to a logic '1' when A OR B OR both inputs are at logic '1'. The logic symbol and truth table are shown in *Figures 2.19 and 2.20* respectively. The Boolean expression is

Z = A + B

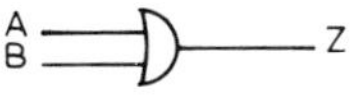

Figure 2.19 2 i/p OR cct symbol

A	B	Z
0	0	0
0	1	1
1	0	1
1	1	1

Figure 2.20 OR gate truth table

In this example the + sign is used as a LOGIC OR function. The OR gate can be used for 'combining' two or more signals to produce a single output. The example shown in *Figure 2.21* uses a single OR gate to combine two sets of pulses labelled (a), and (b). Waveform (a) is a logic version of inverted line sync. pulses, each positive pulse being 4.7 micro-seconds wide with the distance between leading edges being 64 micro-seconds. Waveform (b) is a positive field pulse. The output from the OR gate will be a positive version of a closed circuit TV sync. signal which can then be added to a non-composite video waveform to produce the final TV signal. This method of combining TV pulse waveforms is quite common in TV games units, TV cameras, test pattern generators, etc.

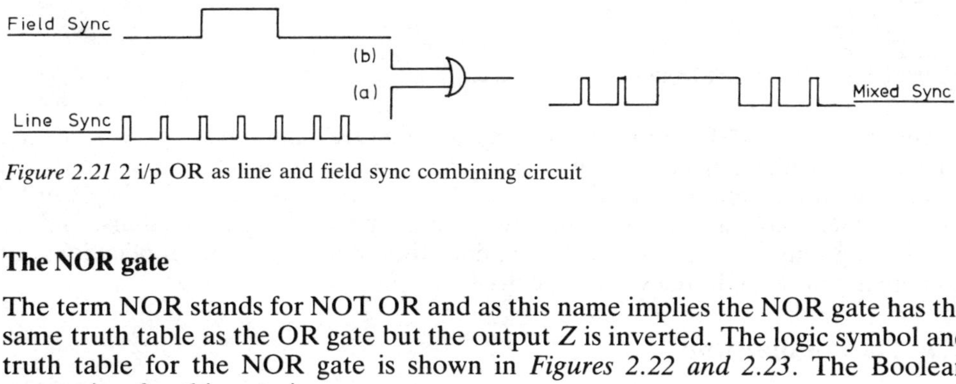

Figure 2.21 2 i/p OR as line and field sync combining circuit

The NOR gate

The term NOR stands for NOT OR and as this name implies the NOR gate has the same truth table as the OR gate but the output *Z* is inverted. The logic symbol and truth table for the NOR gate is shown in *Figures 2.22 and 2.23*. The Boolean expression for this gate is

$$Z = \overline{A + B}$$

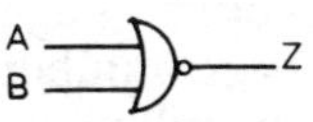

Figure 2.22 2 i/p NOR cct symbol

A	B	Z
0	0	1
0	1	0
1	0	0
1	1	0

Figure 2.23 NOR gate truth table

The EXCLUSIVE OR and EXCLUSIVE NOR gates

This is the last in the series of logic gates we will be examining. The logic symbol of the XOR gate is shown in *Figure 2.24* and the truth table in *Figure 2.25*. The Boolean expression is

$$Z = A \oplus B$$

Figure 2.24 2 i/p XOR cct symbol

A	B	Z
0	0	0
0	1	1
1	0	1
1	1	0

Figure 2.25 2 i/p XOR truth table

The output of the XOR gate will only rise to a logic '1' if the inputs are dissimilar. If all the inputs are at a logic '1' or a logic '0' the gate will produce a logic '0' at the output.

As with the previous gates an inverter can be used at the output of the XOR gate producing a XNOR gate producing the logic symbol shown in *Figure 2.26* and the truth table shown in *Figure 2.27*.

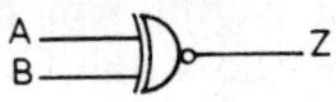

Figure 2.26 Logic symbol of XNOR gate

A	B	Z
0	0	1
0	1	0
1	0	0
1	1	1

Figure 2.27 Truth table XNOR gate

The AND, NAND, OR, NOR, XOR, XNOR, and NOT gates are the elementary building blocks for any digital system and they can all be interconnected to form highly complex logic elements such as Teletext decoders, or they can be used in their ones and twos to produce the simplest of logic functions. Before proceeding any further we will examine the more relevant applications of combinational logic in television circuits.

2.7 The comparator

Comparison of a single bit or byte with another is a fundamental exercise which is used in many digital systems. In a Teletext decoder for example the page number selected by the user has to be compared with the page number being received, and when a match occurs the correctly received page is stored in the decoder's page memory. A circuit which compares the page number received and the page number selected is therefore required. A true binary comparator will indicate the following three conditions:

1. input A = input B
2. input A > input B
3. input A < input B

If the two bits are fed to a two input XNOR gate, the gate will give a logic '1' out when the two inputs match. Refer to the truth table shown in *Figure 2.27*. Hence with a simple two input XNOR gate the condition A=B can be detected.

To find if A > B a NOT gate with a 2 input AND gate is connected as shown in *Figure 2.28*. Input B has been inverted and applied to the AND gate together with input A. The AND gate will only produce a logic '1' at its output when A AND $\overline{B}$ are at logic '1'. This is the same as saying that a logic '1' will only be present at the output when A is at logic '1' and B is at logic '0'. Hence A > B.

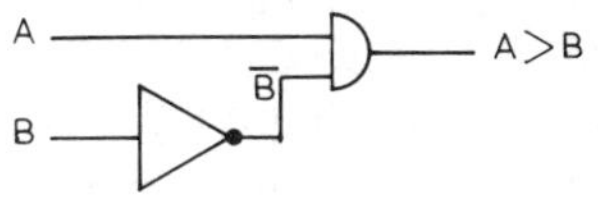

Figure 2.28

To detect if A < B, then A is inverted and the B goes directly to the AND gate. In this mode the circuit will only produce a logic '1' at the output when A is at logic '0' and B is at logic '1' i.e. A < B (see *Figure 2.29*).

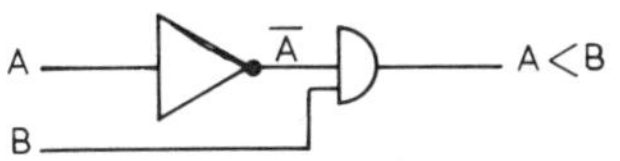

Figure 2.29

The three circuits under discussion can be combined to produce the full binary comparator which is illustrated in *Figure 2.30*. The principles of the previous three circuits have been incorporated into a single circuit.

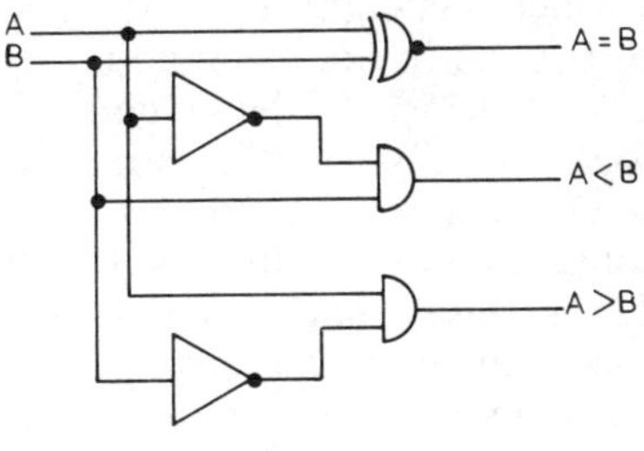

Figure 2.30 Full binary comparator

It is possible to produce binary comparators in a single IC package thus saving considerable space and expense. Manufacturers' data would include a full logic diagram to indicate the operation of the circuit, but the whole of the electronics would normally be represented by a single block with just the inputs and outputs labelled. An example of this form of circuit symbol is shown in *Figure 2.31*.

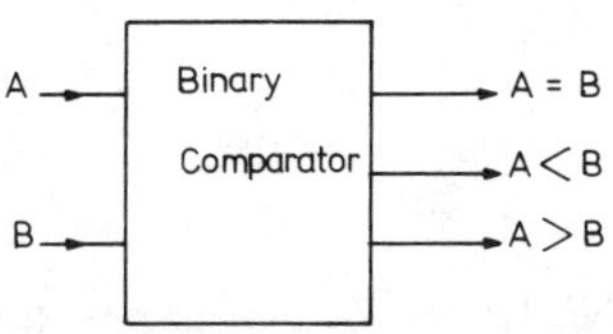

Figure 2.31 Block of binary comp. and truth table

2.8 Decoding binary numbers

Many digital systems have to identify just one combination of 0s and is within a byte of data. Circuits have been produced which interrogate each byte of data and produce an output every time the predetermined number is detected. Consider a system which operates using an eight-bit byte and it is required to detect the presence of the decimal number 125 within each byte of data. The choice of number in this example has been made at random and in a practical system there would be a specific reason for its choice.

Before the number 125 can be detected it is first of all necessary to calculate the binary equivalent of the decimal number, as shown in *Figure 2.32*. The mathematics for this type of calculation was fully described in chapter 1. The solution calculated in *Figure 2.32* only produces seven bits, and as we are using an eight-bit byte it is necessary to add the eight most significant bit to the left hand side of the answer to complete the byte (a logic '0').

```
2|125
   62   R1   LSB
   31   R0
   15   R1
    7   R1
    3   R1
    1   R1
    0   R1   MSB

  Ans = 1111101
```

Figure 2.32 Conversion of 125 to binary

Having derived the correct eight bit binary code we have to design a circuit which will produce a logic '1' at its output every time this pattern of bits is present within the byte. The simplest way to achieve this is to invert the logic '0s' contained in the required pattern to produce a second byte of data which has eight logic '1s', and then feed all eight bits to an eight input AND gate. Hence, in the example under discussion, it will be necessary to invert bit 7 and bit 1 using two NOT gates to produce the second byte of data containing eight logic '1s'. *Figure 2.33* shows the complete circuit of the detector. The output of the AND gate will only rise to a logic '1' when all eight of its inputs are at logic '1', consequently the circuit will only produce an output when the binary coded number 125 is present in the byte. Under any other input condition the AND gate will produce a logic '0' at the output.

Depending on the application of the system just described, the circuit used may not be accurate enough for the desired requirements. If it was being used in a situation where the data at the eight inputs to the circuit was changing at high speed

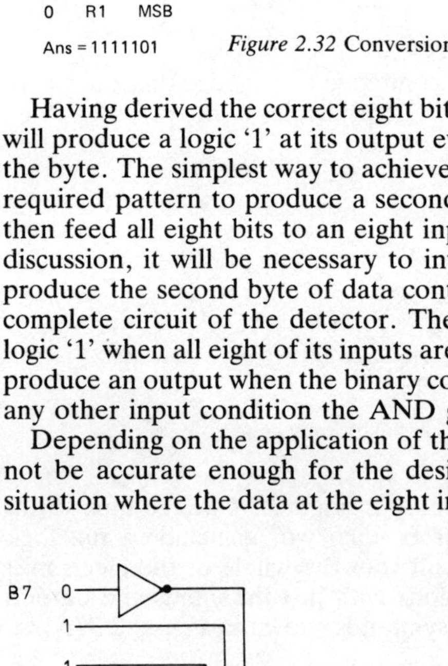

Figure 2.33 Decoder for binary number 125 plus output enable gate

it is possible for the output to rise to a logic '1' as the data on the inputs is changing. This will cause the circuit to identify the desired pattern under false conditions because the data at the input is changing from one logic level to another.

This problem can be overcome by feeding the output from the detector to one input of an additional two input AND gate. The second input of this AND gate is fed with an enable signal, see *Figure 2.34*. When this signal is at a logic '0' the final output of the circuit will be at logic '0' irrespective of the input conditions. The circuit is therefore disabled while the input data changes. Once the input data is stable the AND gate is enabled by a logic '1' and the result from the decoder will be passed via the AND gate to the output. The final AND gate is being used as a single pole single throw switch. The astute reader will realize that this circuit can be simplified down to a single nine input AND gate.

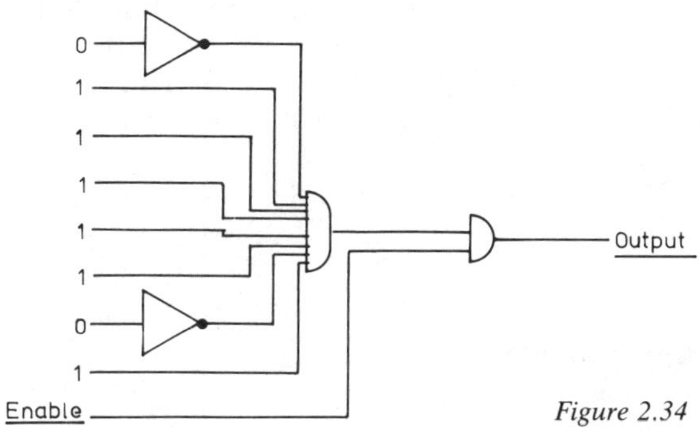

Figure 2.34

There are many situations where decoders are required, not only in combinational logic circuits such as the example under review, but in sequential logic circuits where we have to identify specific conditions in a counting system. This concept is further explained in chapter 3.

The two examples of combinational logic just described provide a small introduction to the versatility of logic systems and gates. It is possible for a designer to produce any number of logic operations with the gates available and we have chosen the two examples of comparison and decoding as an introduction to the use of gates in TV logic.

2.9 Binary addition

It was stated in chapter 1 that digital electronic circuits are capable of carrying out mathematical as well as logical operations. So far only logic operations have been considered and, to introduce the concept of mathematical circuits, we will examine the addition of two binary numbers in both theory and in practice.

If any two bits of data are added together four possible bit combinations arise. They are as follows:

(a) $0 + 0 = 0$
(b) $0 + 1 = 1$
(c) $1 + 0 = 1$
(d) $1 + 1 = 0$ (carry 1 to next column)

Situation (a)
Here it is stated that if both bits are set at logic '0' then nothing is added to nothing resulting in zero.
Situation (b) and (c)
Here binary 1 is added to binary 0 resulting in a binary 1.
Situation (d)
Here a binary 1 is added to a binary 1 producing a sum of 0 and a carry of 1 to the next column of the addition sum. For example 1 + 1 = 2 in the number base of ten hence binary 1 + binary 1 = 0 and a carry to the two's column.

All possible combinations of inputs and outputs for this simple binary addition are summarized in the truth table shown in *Figure 2.35*. Here it is shown that any circuit which will add two binary digits, will have two inputs and two outputs. A combinational logic circuit which satisfies this truth table, is known as a half adder, see *Figure 2.36*. The sum output of this half adder is an XOR function of the A and B inputs, thus satisfying the sum column of the truth table. The carry output for the next column is produced by an AND gate whose inputs are A and B. Thus a carry to the next column is only produced when A AND B are both at logic '1'.

A	B	S	C
0	0	0	0
0	1	1	0
1	0	1	0
1	1	0	1

Figure 2.35 Half adder truth table

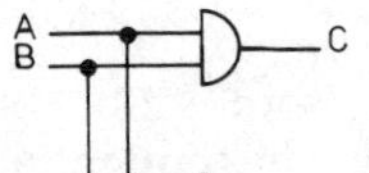

Figure 2.36 Half adder circuit

This example of a half adder is just one of many possible circuits which will fulfil the truth table. Many others are capable of achieving the same result. An elementary circuit such as this makes the first step towards solving more complex mathematical problems.

Extending this theory further we find that there are many situations where a carry is passed from one column of the problem to another during binary addition. For example:

$$\begin{array}{r} 0111 \\ +\ 0101 \\ \hline =\ 1100 \end{array}$$

When adding these two numbers a carry is produced every time during the addition of the three LSB bits. To execute such an addition electronically a three-input adder is required, rather than the two-input adder previously discussed. These three inputs will be A B and CARRY. Such a circuit is known as a full adder and it can be produced using two half adder circuits and an OR gate as shown in *Figure 2.37*. To simplify the circuit a single box is used for each half adder but the reader must remember that this half adder is in fact the circuit used in *Figure 2.36*.

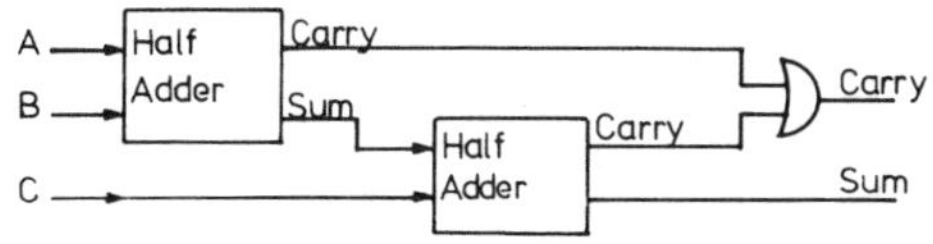

Figure 2.37 Full adder circuit

Assume that A = 1, B = 1, and CARRY = 0. *Figure 2.38* shows a full adder with this combination of inputs. The first half adder with the A and the B inputs has a sum of 0 and a carry of 1. The second half adder produces a sum and a carry of 0. These two circuits will thus produce a final output from the full adder of SUM = 0, CARRY = 1.

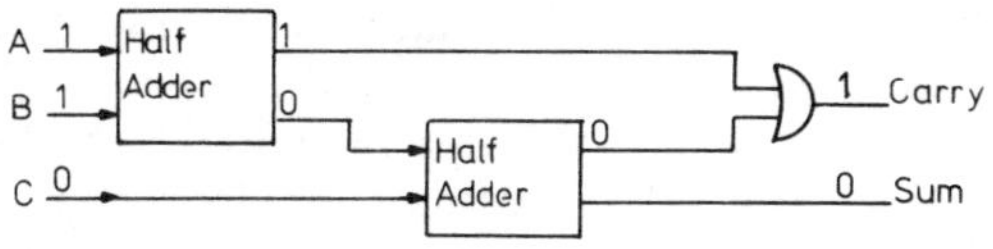

Figure 2.38 Full adder circuit with A = 1, B = 1 and CARRY = 0

If the inputs are changed to produce A = 1, B = 1, CARRY = 1 a sum and a carry of 1 is produced, see *Figure 2.39*.

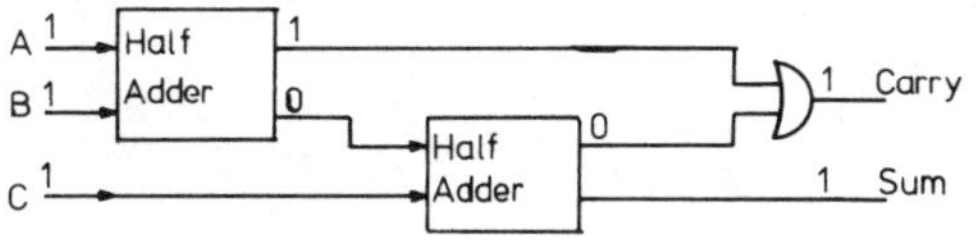

Figure 2.39 Full adder circuit with A = 1, B = 1 and CARRY = 1

If the reader goes through all the possible input combinations of the full adder, the truth table shown in *Figure 2.40* will be produced. So far in this examination of binary addition only two bits of data plus a carry have been considered. In practice more than two bits will be added, for example two four bit nibbles or two eight-bit bytes. In this situation a parallel binary adder is required to solve the problem.

Inputs			Outputs	
A	B	C	S	C
0	0	0	0	0
0	0	1	1	0
0	1	0	1	0
0	1	1	0	1
1	0	0	1	0
1	0	1	0	1
1	1	0	0	1
1	1	1	1	1

Figure 2.40 Truth table with full adder combinations

Parallel binary adder

Consider the addition of two four bit nibbles as shown in *Figure 2.41*. When adding the least significant bits only two binary digits are involved, (A1 and B1). Therefore the bit 1 column of the addition sum will only require a half adder circuit. Every other column in this problem will require the use of a full adder to cope with a possible carry input from the previous column.

	A4	A3	A2	A1
+	B4	B3	B2	B1
S5	S4	S3	S2	S1

Figure 2.41 Addition of two 4-bit nibbles

A block schematic of a four-bit parallel binary adder is shown in *Figure 2.42*. To illustrate the operation of this adder we will consider the addition of the decimal numbers 10 and 7 as shown in *Figure 2.43*. Ten in binary is 1010 which will form the A input to the adder while the B input will be 0111. The A and B inputs to the half adder will produce a sum of 1 and a carry of 0. The A and B inputs to the second full adder will produce a carry of 1 and a sum of 0. The A and B inputs of the third full adder will produce 1 plus the carry from the previous full adder resulting in a carry of 1 and a sum of 0. The A and B inputs to the final adder will produce 1 plus the carry from the previous full adder resulting in a sum of 0 and a carry of 1. Note that the carry output of the most significant full adder forms the MSB of the answer. The binary answer is therefore 10001 which is the number 17 in decimal.

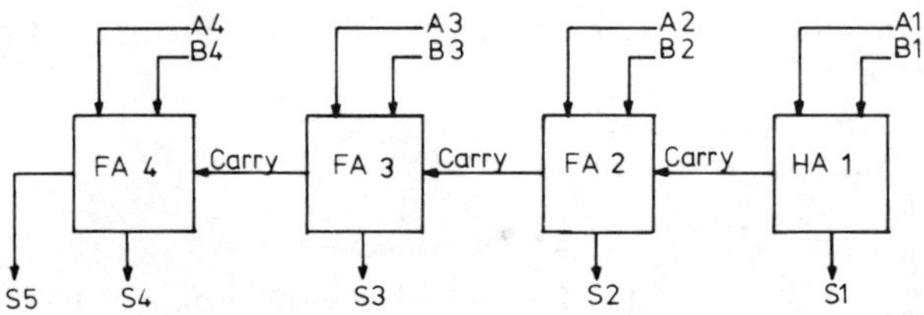

Figure 2.42 Schematic of four-bit parallel binary adder

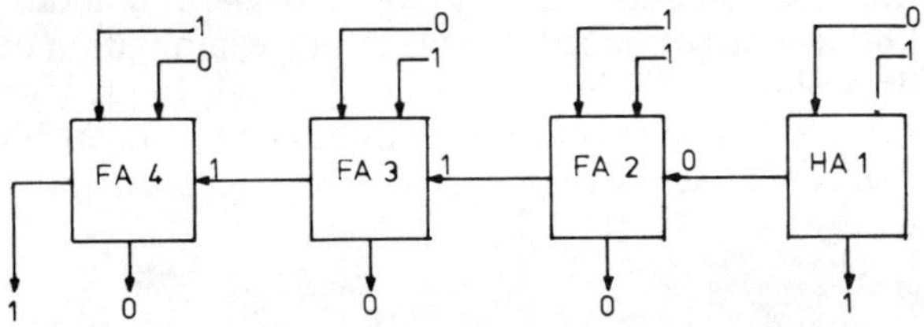

Figure 2.43 Circuit of Figure 2.42 with numbers 10 and 7

The addition of four bits has been used here so that the example is kept as simple as possible. It is of course possible to add binary numbers of any magnitude simply be adding the required number of full adder circuits. For example if two eight-bit bytes were to be added then seven full adders and one half adder would be required.

A/D and D/A conversion

The basic idea of ADCs and DACs was explained in chapter 1, and here we will examine both of these concepts in greater detail. These circuits form a very important part of digital processing because there are many situations where a digital system has to interface with analogue signals. A typical example is the joystick control in a TV games unit. Out of the two processes digital to analogue conversion is the easiest to understand and will therefore be described first.

2.10 Digital to analogue conversion

The fundamental problem of digital to analogue conversion is to convert each bit of the digital input into a correctly weighted analogue output voltage.

2^2	2^1	2^0	Weighting
0	0	0	0/7
0	0	1	1/7
0	1	0	2/7
0	1	1	3/7
1	0	0	4/7
1	0	1	5/7
1	1	0	6/7
1	1	1	7/7

Figure 2.44 Table for digital to analogue conversion

To clarify this concept examine the table shown in *Figure 2.44*. Here three bits are used to produce eight different numbers from 0 to 7. Suppose we want to change the eight possible digital signals shown in this diagram into equivalent analogue voltages. If the maximum analogue output voltage is 7 V then each increment in the binary number from 000 to 111 will represent an increase of $\frac{1}{7}$th of the output voltage. Consequently 000 = $\frac{0}{7}$ths, 001 = $\frac{1}{7}$th, 010 = $\frac{2}{7}$ths etc. From this table we can represent in fractional form the weighting of each bit of the digital signal as shown in *Figure 2.45*.

Bit	Weight
2^0	1/7
2^1	2/7
2^2	4/7

Figure 2.45 Table for digital to analogue conversion

Resistive ladder

From *Figure 2.45* a resistive ladder can be constructed which has three inputs (one for each bit), and one output (the analogue voltage). This resistance network must do four things in order to change the digital input into an equivalent analogue output voltage.

1. The 2^0 bit must produce an output $\frac{1}{7}$th of 7 V;
2. The 2^1 bit must produce an output of $\frac{2}{7}$ths of 7 V;
3. The 2^2 bit must produce an output of $\frac{4}{7}$ths of 7 V.
4. These three voltages representing the digital bits must be summed together to form the analogue output voltage.

From this specification we can produce a truth table for the resistor network and this is shown in *Figure 2.46*, and a circuit which will satisfy this table is illustrated in *Figure 2.47*. Resistors R0 to R2 form the potential divider network with RL providing the output load resistance. The value of RL must always be significantly

Digital Input			Analogue Output
0	0	0	0 V
0	0	1	+1 V
0	1	0	+2 V
0	1	1	+3 V
1	0	0	+4 V
1	0	1	+5 V
1	1	0	+6 V
1	1	1	+7 V

Figure 2.46 Truth table for resistor network

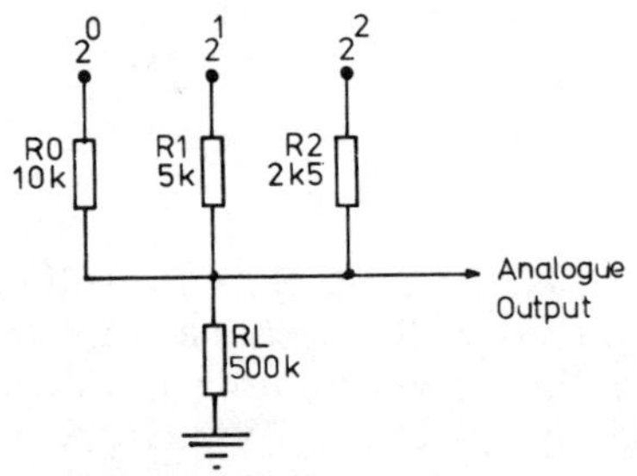

Figure 2.47 Circuit for Figure 2.46

greater than R0 so that the potential divider is not heavily loaded. The values of the resistors used in the potential divider follow their binary weightings, for example:

R0 = 10K, R1 = 10K/2, R2 = 10K/4.

Assume that a digital input signal of 001 (MSB written first), is applied to the three inputs. As a logic '0' = 0 V the 2^1 and the 2^2 inputs are connected to 0 V producing the equivalent circuit shown in *Figure 2.48*. R2 and R1 are now connected in parallel producing an equivalent total resistance of 1K66 ohms. The circuit can now be further simplified to a two resistor potential divider as shown in *Figure 2.49*. The output from this circuit is taken across the lower resistor and this voltage can be calculated as follows:

I = 7 V / (10K Ω + 1K6 Ω) = 0.603 mA.
O/P = 0.603 mA * 1K6 Ω = 0.965 V = 1 V.

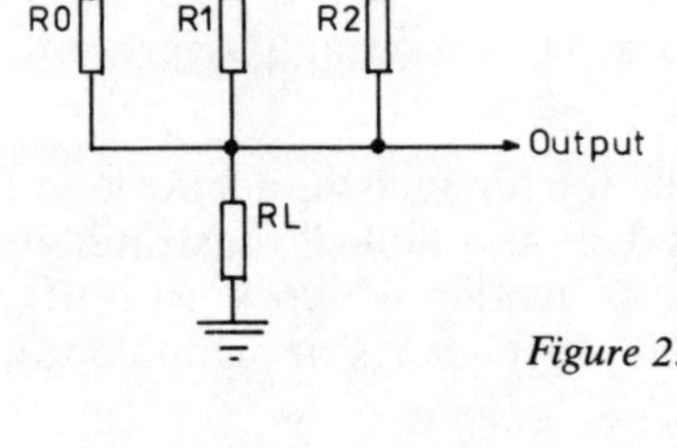

Figure 2.48

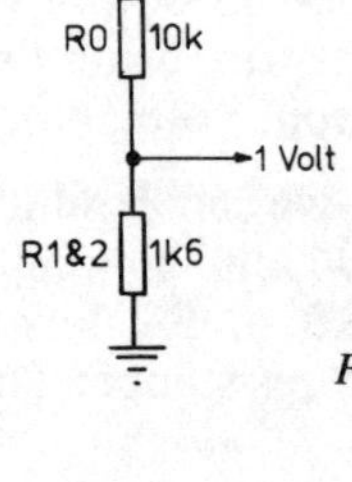

Figure 2.49

It can be seen from these calculations that an output voltage of 1 V is produced with an input of 001. This satisfies the 2nd line of the truth table in *Figure 2.46*. This process can be repeated for any combination of digital inputs and it will be found that the correctly weighted analogue output voltage will be produced.

This circuit has several disadvantages. These are:

1. The values of the resistors used must be very accurately determined, and because each resistor in the network is a different value the complete network has to be constructed using (precision) close tolerance resistors.
2. The resistor used for the MSB will pass a much greater current for the resistor used for the LSB. For example, in a ten-bit system the current flowing through the MSB, resistor will be approximately 500 times larger than the current through the LSB resistor. Consequently component values have to be chosen very carefully.

Both of these problems can be overcome by using a resistance network of a different configuration, known as the binary ladder.

The binary ladder

A binary ladder is a resistive network that has an output voltage which is the correctly weighted sum of the digital inputs. An example of a four-bit binary ladder is illustrated in *Figure 2.50*.

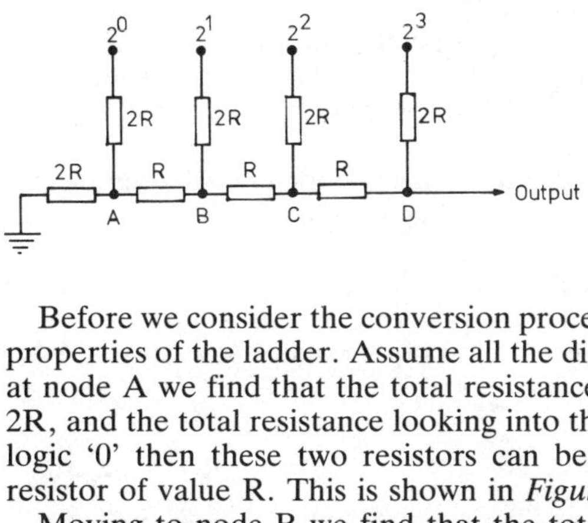

Figure 2.50 A four-bit binary ladder

Before we consider the conversion process it is necessary to examine the resistive properties of the ladder. Assume all the digital inputs are at logic '0' (0 V). Starting at node A we find that the total resistance looking into the terminating resistor is 2R, and the total resistance looking into the 2^0 input is also 2R. If the 2^0 input is at logic '0' then these two resistors can be combined in parallel to form a single resistor of value R. This is shown in *Figure 2.51 (a)*.

Moving to node B we find that the total resistance looking towards node A is again 2R, as is the resistance when looking towards the 2^1 input. Because the 2^1 input is at logic '0' we can again simplify the circuit in exactly the same way as before to produce the circuit shown in *Figure 2.51 (b)*.

Continuing to point C the total resistance looking back to point B is again 2R and the resistance looking towards the 2^2 input is 2R. As before these two resistances are in parallel producing a total resistance of R Ω as shown in *Figure 2.51 (c)*.

Moving to the final node D, which is also the output from the ladder, the same equivalent resistances exist. Looking back towards node C a resistance of 2R is present, and looking towards input 2^3 a resistance of 2R is present. Input 2^3 is at logic '0', placing these two resistances in parallel producing a final output

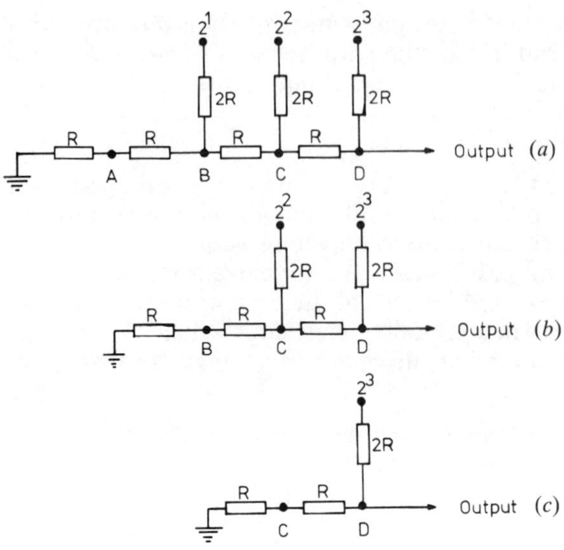

Figure 2.51 Resistive properties of the ladder

impedance of R Ω. From the above we can conclude that the total resistance looking from any node back towards the terminating resistor or out towards the digital input is 2R.

When each digital input is taken to a logic '1' instead of a logic '0' the above statement is also true, because we can assume that the internal resistance of an ideal power supply is 0 Ω, and that the digital inputs are ideal voltage sources. In practice this is not strictly true, but the very low resistance of the power supply makes little difference to the circuits operation.

To summarize we can say that the impedance from any node in the ladder back to the terminating resistor or out to the input voltage source will always be 2R regardless of the logic state of the inputs.

Now let us examine the operation of this binary ladder with an input of 1000, (MSB first). The circuit of the ladder with these input conditions is shown in *Figure 2.52*. With this combination of inputs there are no input voltage sources to the left

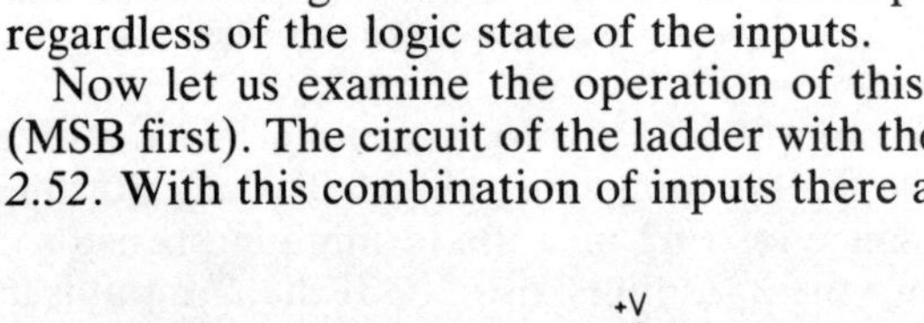

Figure 2.52 Binary ladder with an input of 1000

of node D, and consequently the entire circuit to the left of this node can be replaced by a single resistance of 2R producing the equivalent circuit shown in *Figure 2.53*. From this equivalent circuit we can deduce that the output voltage will be half the supply voltage. Thus a logic '1' in the MSB position will produce an output voltage of V/2.

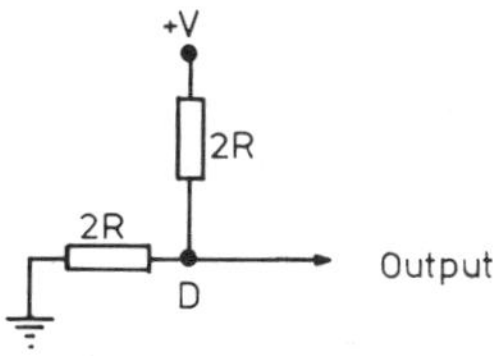

Figure 2.53 Equivalent circuit of Figure 2.52

If the input is now changed to place a logic '1' on the 2^2 input the circuit now appears as shown in *Figure 2.54*. Using the theory already established we can say that the resistance looking back along the ladder from point C is $2R$, and the resistance looking towards +V is also $2R$. Thevenin's theorem allows us to replace this part of the circuit with a single resistor with a value of R and a voltage source of $V/2$. When this is added to the remainder of the circuit the situation in *Figure 2.54 (b)* exists. As with the previous example when this situation is further simplified a two resistor potential divider can be produced with the output analogue voltage taken at the junction of the two resistors. However, in this example the supply voltage to the potential divider is now $V/2$, which will mean that the output voltage will be one half of $V/2$ or $V/4$.

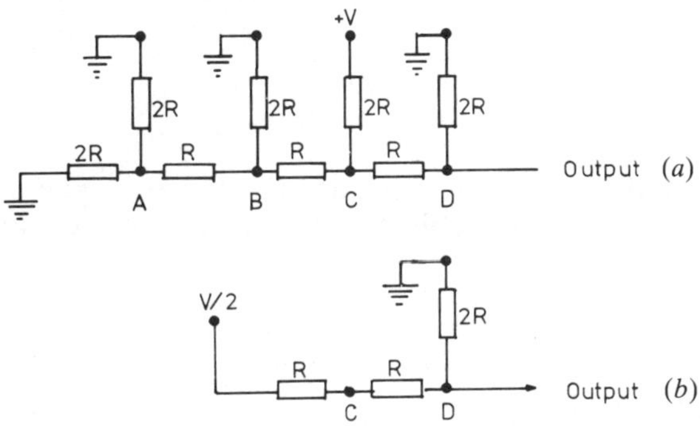

Figure 2.54

Bit Position	Binary Weight	Output Voltage
MSB	1/2	V/2
2nd MSB	1/4	V/4
3rd MSB	1/8	V/8
4th MSB	1/16	V/16
5th MSB	1/32	V/32
|	|	|
|	|	|
|	|	|
Nth MSB	$1/2^N$	$V/2^N$

Figure 2.55 Binary ladder output voltages

This process can be continued to show that the third MSB will produce an output voltage of $V/8$. The fourth MSB produces an output voltage of $V/16$, and so on.

The output voltages for any binary ladder are summarized in *Figure 2.55*, showing that each digital input is translated into a correctly weighted output voltage.

Practical D/A converters

Either of these two resistor ladders can form the heart of a practical D/A converter, but additional electronics will be required before the theory of the ladder can be put into practice.

In chapter 1 we saw that a simplified schematic of an A/D converter could be represented as shown in *Figure 2.56*. The eight-bit digital input is converted into an analogue output during the presence of a positive pulse from a sampling clock oscillator. When the sampling pulse falls back to a logic '0' the analogue output voltage is maintained while the digital input changes between sample pulses. Once the digital input is stable again a pulse from the sampling clock oscillator is applied to the converter, causing the converter to produce a new equivalent analogue output.

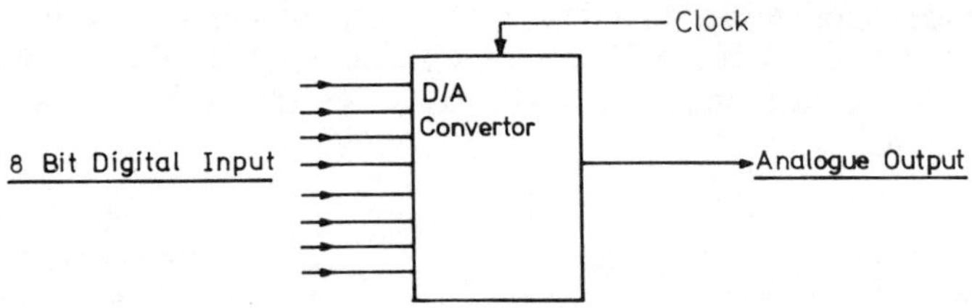

Figure 2.56 Conversion of eight-bit digital input

A practical circuit for such a digital to analogue converter is shown in *Figure 2.57*, note in this case that the circuit has been simplified and uses only a four-bit digital input.

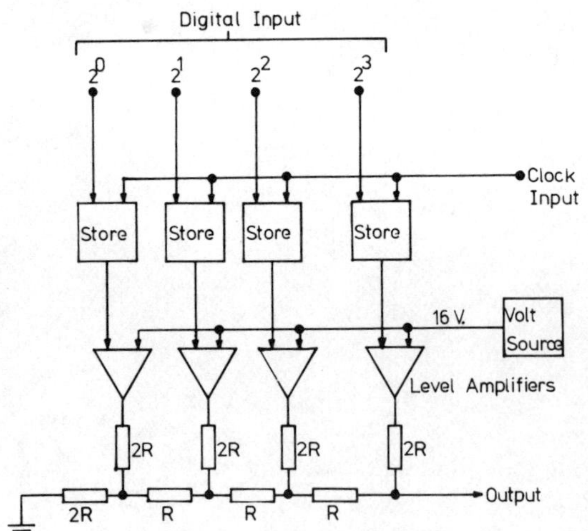

Figure 2.57 Circuit for digital to analogue converter

The binary ladder can be easily recognized at the base of the circuit diagram, and each input to the ladder is produced by an accurately controlled level amplifier. These amplifiers have two inputs. One comes from a precision voltage source which represents the amplitude of a logic '1' input to the binary ladder. This voltage can be any value, and the exact voltage required is determined by the maximum peak to peak amplitude of the output analogue voltage. In this example it is +16.

The second input to each level amplifier comes from a digital storage circuit. In practice this block could be any type of suitably modified flip flop (e.g. D type). When the digital input to each store is stable a clock pulse from the oscillator causes each store to transfer the logic level present at its input to its output. When the clock pulse falls back to a logic '0' the input of each store will be held. This action will provide each level amplifier with the correct static logic level while the digital input to the converter changes to its next state. When the input data is again stable the clock pulse causes the stores to operate again and store the new pattern of input data. A logic '0' at the input of a level amplifier will produce 0 V at its output, while a logic '1' at an input will produce +16 V at its output.

Let us now examine the operation of the D/A converter with the circuit handling a practical waveform. The example chosen here is a simple one where the input to the converter is a four-bit nibble which increments from 0000 to 1111 once every millisecond, and each change in input data is synchronized to the clock oscillator. The four-bit digital input will produce a total of sixteen different codes, each code will be changed by the converter into a separate output voltage. If the precision voltage source produces a reference of +16 V for the level amplifiers then the analogue output from the converter will be increased by 1 V every time the digital input is incremented by one bit. The resulting analogue output voltage will be a sixteen increment step waveform rising from 0 to 16 V as shown in *Figure 2.58*.

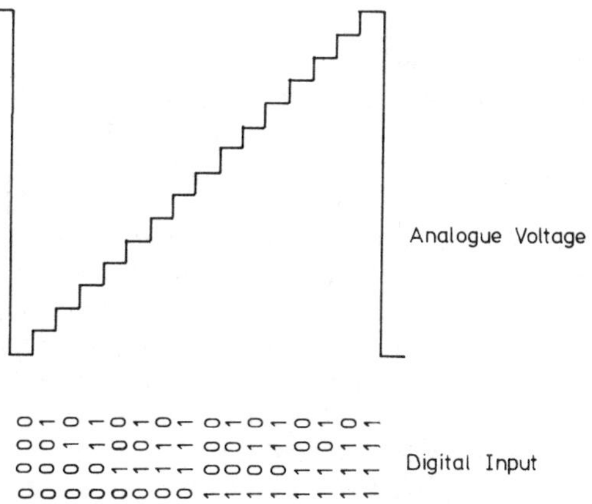

Figure 2.58 Sixteen-step increment waveform

Once the digital input to the converter reaches 1111 it will return to 0000, causing the analogue output voltage to fall from +16 V to 0 V. The sequence then repeats itself. Using four bits in this way will produce an analogue signal with a minimum voltage differential of 1 V and, of course, in practice much greater accuracy is required. If the D/A converter was designed to accept an eight-bit input and the voltage reference source was kept at +16 V then the output voltage from the converter would have a minimum incremental change of 1/256 of 16 V = 62 mV. If the system complexity was further increased to handle a twelve-bit input then an accuracy of 4 mV will be obtained.

2.11 Analogue to digital converters

The process of converting an analogue signal to a digital signal takes place in a circuit known as an analogue to digital converter. This process is more complex than digital to analogue conversion and a number of techniques have been developed. The simplest of these is known as simultaneous analogue to digital conversion.

Simultaneous A/D conversion

This method of A/D conversion is based on the use of a number of comparator circuits, each of which will indicate when the analogue input voltage exceeds the amplitude of a predetermined reference voltage. A simplified simultaneous converter is shown in *Figure 2.59*.

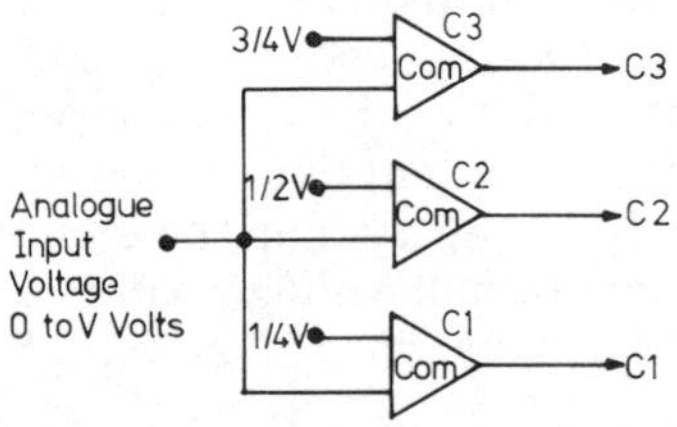

Figure 2.59 Simultaneous converter

Assume the analogue input voltage will vary between 0 and +V volts, and the reference voltages to each comparator are as shown in *Figure 2.59*. When the analogue input is at 0 V it will be below the threshold point of each comparator, consequently each comparator will give 0 V output. As the input voltage rises it will approach the reference voltage of comparator C1, and as soon as this voltage is reached or exceeded comparator C1 will produce an output indicating that the input is at or in excess of *V*/4. As the input voltage continues to rise comparator C1 will maintain its output voltage, and eventually a point will be reached where the input voltage will be at or in excess of the reference voltage to comparator C2. When the input voltage is at or exceeds *V*/2 comparator C2 will produce an output. With further increase in input voltage the reference voltage to C3 will be exceeded producing an output at C3. The truth table of this circuit is shown in *Figure 2.60* indicating that only four voltage ranges can be detected by the converter. This resolution is very poor and by increasing the number of comparators the accuracy will be improved. For example eight comparators will enable the detection of nine input levels, sixteen comparators will allow the detection of seventeen levels etc.

	Comparator Output		
Input Voltage	C1	C2	C3
0 to 1/4 V	0	0	0
1/4 to 1/2 V	1	0	0
1/2 to 3/4 V	1	1	0
3/4 to 1 V	1	1	1

Figure 2.60 Truth table

The outputs of the comparators cannot be used as a direct digital equivalent of the input because they are not coded in binary. All they are indicating is that the input has reached or exceeded a particular level. Some form of coding circuit will therefore be required to take the three outputs from the comparators and code them into a binary signal. In this particular example the coding circuit will accept the comparator outputs and convert them into two digital signals. These two output lines will be able to represent in binary code the four separate levels detected, e.g. 00, 01, 10, 11. Once a correctly binary coded version of the input has been produced each bit of the binary signal has to be held in a temporary store to produce a latched digital output every time the circuit is clocked. A simplified diagram showing this arrangement is shown in *Figure 2.61*.

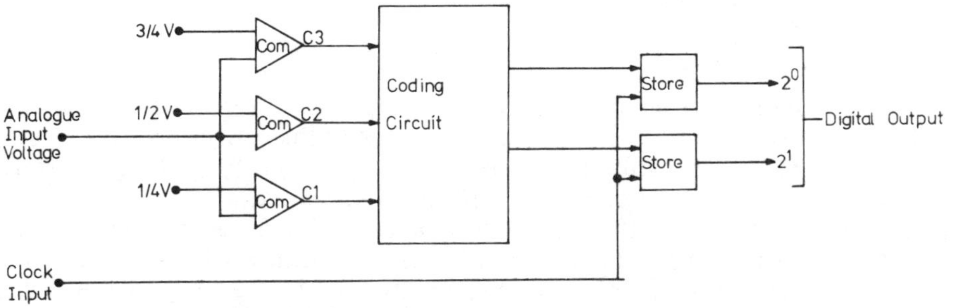

Figure 2.61 Simplified simultaneous A/D converter

As the resolution of the digital output is improved the circuit complexity increases enormously. In general it can be said that $(2^n)-1$ comparators are required to produce a digital signal having n bits. For example if an eight-bit output was required then $(2^8)-1$ comparators are required or 255. A converter of this complexity, although very fast in operation, is very heavy on components and tends to be unreliable. For these two reasons a viable alternative is required.

A/D conversion—counter method

A high resolution A/D converter can be constructed using only one comparator, provided that a variable reference for this comparator is available. This reference voltage can then be applied to the comparator, and when the reference voltage is equal to the analogue voltage the comparator will produce an output. The basic arrangement for such a converter is shown in *Figure 2.62*.

To the right of the diagram is a group of circuits which form a digital to analogue converter. The digital input to this stage comes from a binary counter which will produce a binary coded output of n bits which increments by 1 every time the counter receives a clock pulse from the clock oscillator. The operation of this converter will be similar in principle to the example explained in *Figure 2.57* with an output from the binary ladder taking the form of a step waveform rising in equally timed steps. This step waveform provides the reference voltage for the comparator.

When the amplitude of the stepped reference voltage equals the amplitude of the analogue input voltage the comparator produces an output which operates a gate

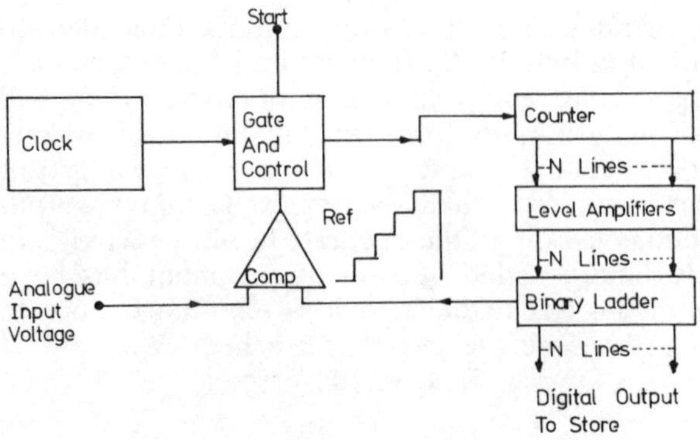

Figure 2.62 Arrangement of A/D converter

and stops the counter. The binary coded output from the counter at this point will be the digital equivalent of the analogue input voltage to the converter. This digital signal is then stored temporarily, to produce the final output from the converter. The counter is then reset to 0 before the process is repeated.

One major disadvantage of this type of system is the time it takes to sample the input voltage. Consider the following situation. If the clock oscillator produces a frequency of 1 MHz, each pulse will be generated at 1 microsecond intervals. Consequently the binary counter will be incremented by 1 every microsecond. If the counter produces an eight-bit output, and the analogue input is at its maximum voltage, it will take 255 microseconds to complete the count from 00000000 to 11111111 and sample the input. This is, of course, the worst possible case where the reference voltage to the comparator will have to reach its maximum value.

The time it takes for the reference to reach the same voltage as the analogue input will be directly proportional to the amplitude of the input voltage. In many situations this type of converter, with its long sampling time is unacceptable, and the design of the converter has to be improved so that the time required between samples is reduced.

Many advanced analogue to digital conversion techniques are available, but one possible improvement to the circuit just described is to eliminate the need to reset the counter each time a conversion has been made. When this is done the counter would not start at zero after each sample but would begin at the value of the last converted voltage. This means that the circuit must be capable of predicting whether the analogue input voltage is rising or falling, and then increment or decrement the binary counter accordingly. Additional logic circuitry will be required of course to decide whether to count up or down. This method is known as continuous conversion and such a converter is therefore called a continuous type A/D converter.

The television industry has been quick to make use of the more advanced A/D converters, together with computer techniques, to produce extremely versatile visual effects. The heart of these effects units is a very large memory which is capable of storing a digital equivalent of a complete TV picture, (known as a frame store). An extremely accurate A/D converter receives the composite colour signal and samples the video amplitude at very high speed. A digital equivalent is therefore produced which is then stored.

Once a complete frame has been stored in the memory, computer technology takes over. The stored digital information is used to reconstitute pictures which can be compressed in size, revolved on their axis etc. It is obvious that as ADC and DAC techniques improve greater use will be made of this sort of facility, and the application of D/A and A/D converters in this field indicates the importance that these circuits play, and will play in future television systems. It is quite possible that in the not too distant future TV transmissions will take place using digital techniques only, and the receiver will use either a digital display or use a DAC to reconstitute the video picture before it is fed to a conventional CRT.

Chapter 3

Sequential logic circuits

All sequential logic circuits produce outputs in a predetermined sequence. Each circuit must contain some form of memory device which enables the circuits to retain the required information on the sequence of input signals.

Sequential logic circuits can be subdivided into two major categories known as *synchronous* and *asynchronous*. A synchronous system is one in which all the changes in the circuit take place simultaneously. To achieve this, some form of synchronizing signal is required which will time all the elements within the system. This synchronizing signal is commonly known as a clock because it determines the precise time at which any changes will take place.

In an asynchronous system a change that takes place in one section will cause further changes in other sections. All the changes that take place in an asynchronous logic system propagate through the system in a manner which is determined by the speed at which each section operates.

Nearly all the sequential logic systems are based on a simple circuit known as a *bistable*, because the circuit has two stable states which can be switched from one to the other by one or more inputs to the circuit. Another term which is widely used is flip-flop, because the circuit will flip or flop between its two stable states. The term flip-flop is the most widely used and for this reason it has been chosen for this text.

3.1 The set reset (SR) flip-flop

The SR flip-flop is the simpliest of the sequential logic circuits. There are several ways in which this circuit can be constructed, and the example shown in *Figure 3.1 (a)* uses two NAND and two NOT gates. Its two inputs have been labelled S (set)

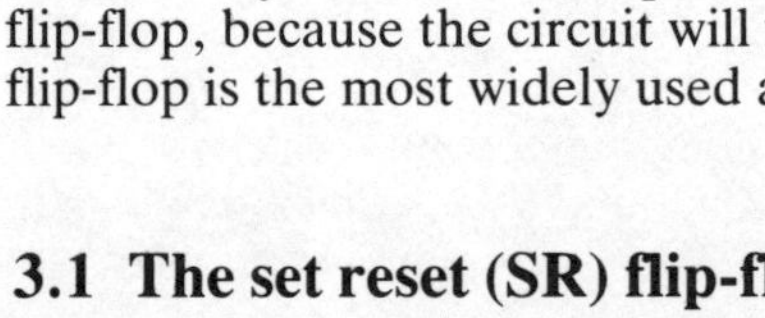

S	R	Q	$\overline{Q}$
0	0	NC	
0	1	0	1
1	0	1	0
1	1	?	?

NC=No Change

? = Indeterminate

Figure 3.1 SR flip-flop circuit diagram

and R reset and the two outputs Q and $\overline{Q}$. The $\overline{Q}$ output is so named because it is a NOT function of the Q output. Hence if Q = 1 the $\overline{Q}$ = 0 and if Q = 0 the $\overline{Q}$ = 1. Note that the outputs of the two NAND gates are cross coupled. This causes the outputs to latch at either a logic '0' or '1'. The operation of this circuit will be considered in three stages.

S = 1 R = 0 (set)

Assume that S = 1 and R = 0. These two inputs are inverted by their respective NOT gates and applied to the two NAND gates, thus the signal at A will be a logic '0' and at B there will be a logic '1'. If A is at logic '0' then the Q output from gate 3 must be at a logic '1'. This logic '1' is cross coupled to an input of gate 4 producing two logic '1s' at its input (Q and B). The $\overline{Q}$ output from this gate will therefore be at logic '0'. This logic '0' is cross coupled to one of the inputs to gate 3, keeping Q at a logic '1',

S = 0 R = 1 (reset)

If the conditions on the S and R inputs are reversed to produce S = 0 and R = 1 the flip-flop outputs will change state resulting in Q = 0 and $\overline{Q}$ = 1. This can be confirmed by examining the input conditions on the NAND gate 4. If R = 1 then B will be a logic '0', and this is applied to one of the inputs to gate 4. If this NAND gate has a logic '0' at any of its inputs its output will be a logic '1' causing $\overline{Q}$ to rise to a logic '1'. The inputs to gate 3 will now be two logic '1s' causing Q and also the second input of gate 4 to fall to a logic '0'. The circuit will therefore be in the opposite state to the first example.

S = 0 R = 0 (latch)

If S and R are both at logic '0' then the outputs from the two NOT gates will both be at logic '1', producing a logic '1' at A and B. Under these conditions the Q and $\overline{Q}$ outputs will remain unchanged from their previous logic state. To clarify this assume that S = 1 and R = 0 thus Q = 1 and $\overline{Q}$ = 0, if S and R both change to a logic '0' points A and B both rise to a logic '1'. Under these conditions gate 3 will have two inputs of (A = 1) and ($\overline{Q}$ = 0), and gate 4 will have two inputs of (Q = 1) and (B = 1). The output conditions of the two NAND gates will therefore remain static.

The three separate conditions listed above can be summarized in the truth table shown in *Figure 3.1 (b)*. The input condition S = 1 and R = 1 is invalid because this combination of inputs implies that both the Q and $\overline{Q}$ outputs are being forced to a logic '1'. Under this condition the circuit will cause both outputs to rise to a logic '1', but problems will arise if S and R change simultaneously from 1 to 0.

Examination of the circuit will show that if S and R = 0 the circuit should produce an output of Q = NOT $\overline{Q}$. We have seen from the truth tables that if S and R = 0 the output conditions depend on the previous state. In the situation being examined this is of course impossible and what will happen in practice is that the output conditions set up on Q and $\overline{Q}$ will be unpredictable and will be dependant on the switching speed of the gates. Because of this unpredictability the input condition S = 1 and R = 1 is never used.

Clocked SR flip-flop

An SR flip-flop can also be clocked by an external timing waveform which will determine the exact point at which the flip-flop will be latched. This facility enables the circuit to be used in synchronous situations and the circuit diagram of a clocked SR flip-flop is shown in *Figure 3.2*. Note that the two NOT gates used in *Figure 3.1* have been replaced with two input NAND gates and the clock input which determines the precise time at which the circuit operates has been applied the inputs of these two gates. The clock line in most circuit diagrams is abbreviated to CK.

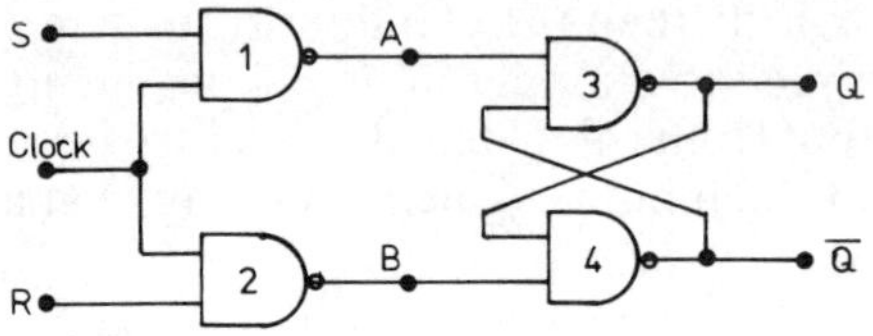

Figure 3.2 Clocked SR flip-flop circuit diagram

If the clock input is at a logic '1' then the two NAND gates operate in exactly the same way as the NOT gates from the previous example, inverting the S and R inputs. Operation of the clocked SR flip-flop in this mode, with the clock input at a logic '1', is identical in every respect to the description of the simple SR flip-flop.

When the clock is at logic '0', points A and B will both be at logic '1' regardless of the conditions at the S and R inputs. In this mode of operation the two cross coupled NAND gates will remain latched in their previous condition. This mode of operation is identical to S = 0 and R = 0 in the simple SR flip-flop.

To summarize, S and R are only inverted and applied to points A and B respectively when the clock input is at logic '1'. Under these conditions the flip-flop can be either set or reset depending on the S and R input states. With the clock at logic '0', A and B both rise to a logic '1' leaving the flip-flop latched in the previous condition regardless of the input conditions present at S and R. As with the simple SR flip-flop the input condition S = 1 and R = 1 with CLOCK = 1 is illegal. The truth table for the clocked SR flip-flop is shown in *Figure 3.3*. Because the flip-flop can be clocked, the truth table indicates the logic states before and after a clock pulse. Qn and $\overline{\text{Qn}}$ indicate the output before the arrival of the clock pulse and

tn				tn+1		
S	R	Qn	$\overline{\text{Qn}}$	Qn+1	$\overline{\text{Qn+1}}$	
0	0	0	1	0	1	No Change (Latched)
0	0	1	0	1	0	
1	0	0	1	1	0	Set
1	0	1	0	1	0	
0	1	0	1	0	1	Reset
0	1	1	0	0	1	
1	1	0	1	?	?	Operation Indeterminate
1	1	1	0	?	?	

Figure 3.3 Clocked SR flip-flop truth table

Qn+1 and $\overline{Qn}$+1 indicate the output after the arrival of the clock pulse. Consequently each input condition at S and R has to be listed twice, to indicate the action of the circuit with both possible output conditions at Q and $\overline{Q}$.

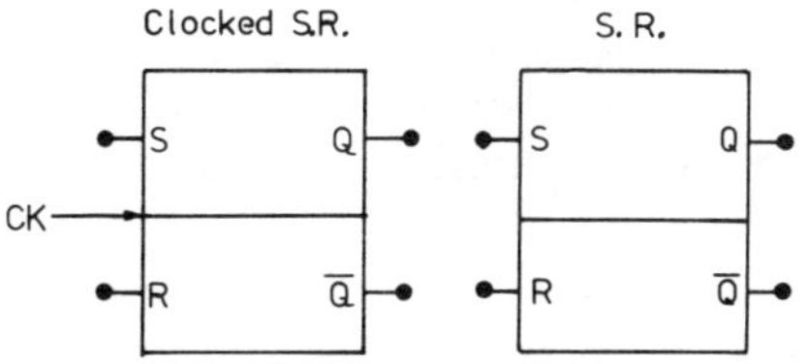

Figure 3.4 Clocked SR flip-flop logic diagram

It is not common practice to draw the circuit diagram of the SR or the clocked SR flip-flop in detail when a large number of these elements are used in circuit diagrams. The circuit diagram of the SR flip-flop may be condensed into a simplified form to make reading the circuit diagram much easier. The schematics of the SR and clocked SR flip-flops are shown in *Figure 3.4*.

3.2 The D-type flip-flop

SR and clocked SR flip-flops can only have their outputs set or reset depending on the conditions present on the SR inputs. In this form the circuit has limited use and possibly one of the simplest ways in which the circuit can be made more versatile is to add a single NOT gate to produce the D-type flip-flop shown in *Figure 3.5*. A NOT gate is connected so that the R input to the clocked SR flip-flop is always an inverted S signal. The circuit now has a single input known as the data input hence the name D-type flip-flop. Because R is always $\overline{S}$ it is impossible for the illegal condition S = 1 and R = 1 to arise.

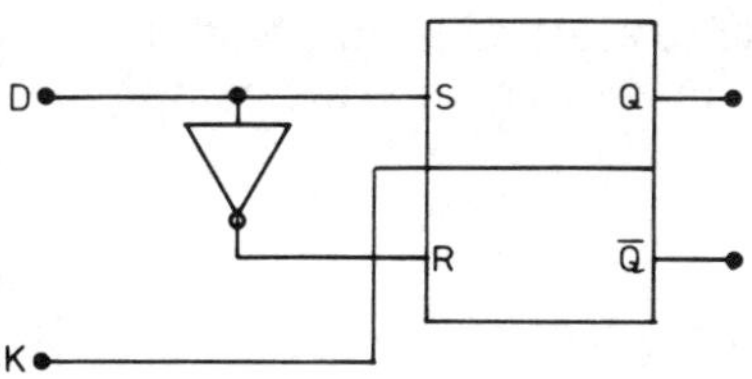

Figure 3.5 D-type flip-flop logic diagram

Consider the operation of the circuit with the D input = logic '1'. As long as the clock input is at a logic '0' the inputs at S and R will not effect the output conditions at Q and $\overline{Q}$. As soon as the clock waveform becomes positive, the logic '1' at the D input will cause the Q output to be set to 1. The NOT gate will invert the D input to produce a logic '0' at the R input resetting the $\overline{Q}$ output to a logic '0'. When the clock falls to a logic '0' the SR flip-flop will remain latched with Q = 1 and $\overline{Q}$ = 0.

If D is at logic '0' and the flip-flop is again clocked by a positive pulse, Q will be cleared to a logic '0' and $\overline{Q}$ will be set to a logic '1'. Again when the clock falls to a logic '0' this output condition will remain latched until the D input changes and the flip-flop is clocked by another positive pulse.

From the above description it can be seen that the D-type flip-flop acts as a simple store because the logic state present on the D input will be transferred to the Q output during the presence of a positive clock pulse. When the clock pulse returns to a logic '0', the Q output maintains its logic state regardless of the input condition at D.

A single D-type flip-flop is therefore capable of storing one bit of data and if it is required to store a complete nibble of four bits, then four such circuits would be required driven from a single clock signal. One possible arrangement is shown in

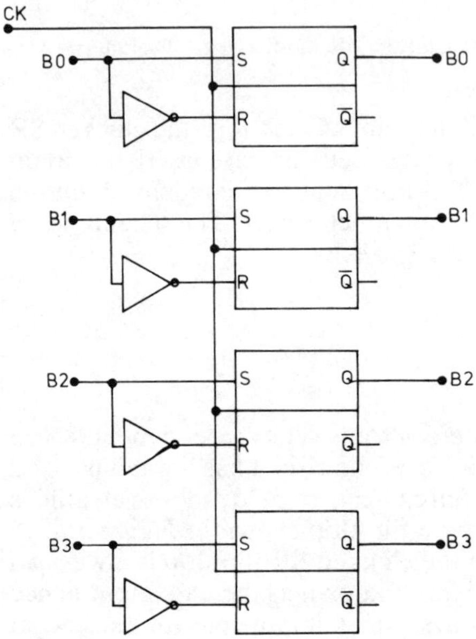

Figure 3.6 4-bit D-type latch

Figure 3.6. During the presence of a positive clock pulse each D-type flip-flop will store the bit present at its input. When the clock returns to a logic '0' the stored data will remain at the Q outputs until the circuits are clocked again. Note that the $\overline{Q}$ outputs are not normally used in this application unless of course the complement of the stored data is required.

3.3 Positive and negative edge triggering

All the clocked flip-flops discussed so far have been triggered on the positive edge of the clock waveform, and are therefore known as positive edge triggered circuits.

In most commercial applications of synchronous sequential logic all the circuits are negative edge triggered, causing each circuit element to change state on the occurrence of the negative edge of the clock pulse. The reason for this choice will become apparent when the master slave concept is discussed later in this chapter, but its introduction has been left until now so that the descriptions of the flip-flops covered so far could be kept as simple as possible.

To convert any of the flip-flops that have been examined so far from positive to negative edge triggered, a single NOT gate is added at the clock input, causing the circuit to be clocked when the waveform falls to a logic '0'.

3.4 The JK flip-flop

This is the most versatile of the flip-flops because a single JK flip-flop can be used in any of the modes previously described. With this circuit the J input replaces S and the K input replaces R. The circuit of a simple positive edge triggered JK flip-flop is

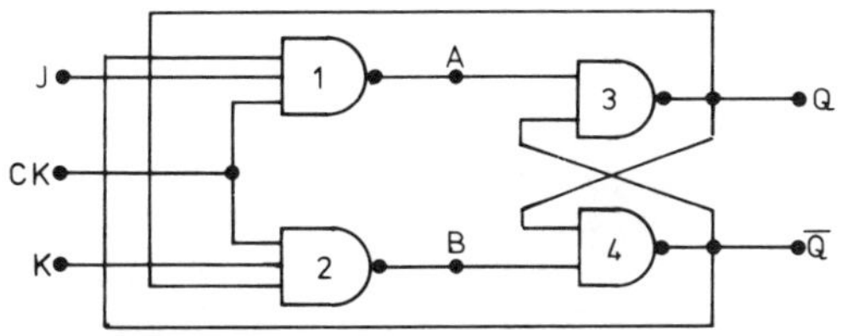

Figure 3.7 JK flip-flop circuit diagram

t_n				t_{n+1}		
J	K	Q_n	$\overline{Q_n}$	Q_{n+1}	$\overline{Q_{n+1}}$	
0	0	0	1	0	1	No Change (Latched)
0	0	1	0	1	0	
0	1	0	1	0	1	Reset
0	1	1	0	0	1	
1	0	0	1	1	0	Set
1	0	1	0	1	0	
1	1	0	1	1	0	Toggle
1	1	1	0	0	1	

Figure 3.8 JK flip-flop truth table

shown in *Figure 3.7*. A full truth table showing its operation is illustrated in *Figure 3.8*. The operation of the JK flip-flop can be summarized using the following four cases:

1. J = 0 and K = 0

Under this condition the J and K inputs force the outputs of gates 1 and 2 to a logic '1' irrespective of the logic state of the clock pulse or the feedback from the Q and $\overline{Q}$ outputs. This ensures that the Q and $\overline{Q}$ outputs do not change and the flip-flop remains in the latched state.

2. J = 0 and K = 1

NAND gate 1 is disabled forcing point A to a logic '1'. NAND gate 2 is enabled forcing point B to a logic '0' only if Q and clock are at a logic '1'. Because of this the next positive going edge of the clock pulse will reset the flip-flop to Q = 0 and $\overline{Q}$ = 1 if it is not already in this state.

3. J = 1 and K = 0

NAND gate 2 is now disabled forcing point B to a logic '1'. NAND gate 1 is enabled only if $\overline{Q}$ and clock are at a logic '1'. On the occurrence of the next positive going edge of the clock pulse the flip-flop will be set to Q = 1 and $\overline{Q}$ = 0 if it is not already in this state.

4. J = 1 and K = 1

If Q is at logic '0' NAND gate 2 is disabled and NAND gate 1 is enabled. The next positive going edge of the clock pulse will then set the flip-flop. On the other hand if Q is a logic '1' and $\overline{Q}$ = logic '0', gate 2 is enabled and NAND gate 1 disabled. The next positive going edge of the clock forces the flip-flop to reset. Hence if both J and K are at logic '1' clock pulses cause the JK flip-flop to toggle. In this mode it is imperative that the duration of the positive clock pulses are shorter than the propagation delay of the circuit, otherwise the circuit will oscillate.

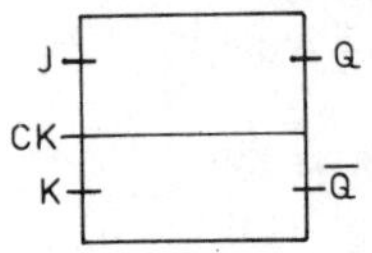

Figure 3.9 JK flip-flop logic symbol

A simplified logic symbol for the JK flip-flop is shown in *Figure 3.9*. Remember that if the input to any logic circuit is left disconnected it will naturally assume a logic '1'. For convenience, the J and the K inputs are often left unconnected and it should be noted that under these conditions J and K will be at logic '1' putting the flip-flop in the toggle mode.

Race conditions

A race condition will exist in a digital system whenever the inputs to a logic gate or circuit attempt to change state simultaneously. During this period of time the output from the gate or circuit in question will not be predictable.

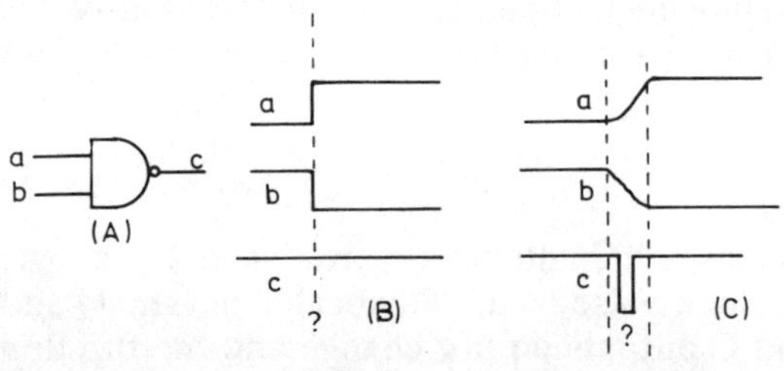

Figure 3.10 Race condition waveforms

Consider the situation shown in *Figure 3.10*. When input (a) is at a logic '0' and (b) is at a logic '1' then (c) will obviously be at logic '1'. Similarly if (a) is at logic '1' and (b) at logic '0' again the output (c) will be at logic '1'. A problem arises however during the transition period of the two waveforms because they both seem to change state at the same time. If this transition period is expanded as shown in *Figure 3.10 (c)* it can be seen that the two waveforms do not change instantaneously, causing the output (c) to be indeterminate.

Using the JK flip-flop in *Figure 3.7* as an example waveform (a) in *Figure 3.10* could be the feedback signal from $\overline{Q}$ to gate 1 and waveform (b) could be the clock waveform. In this were the case then the flip-flop would behave unreliably because the two waveforms change state at the same instant in time.

One major disadvantage with the simple JK flip-flop in the toggle mode, is that the width of the clock pulse must be less than the propagation delay of the flip-flop, otherwise a race condition will exist causing the circuit to become unstable. One of two things must be done therefore to make this circuit of any practical use.

1. Ensure that the clock pulses are shorter than the propagation delay of the circuit in use.
2. Increase the propagation delay of the circuit so that it becomes longer than the clock pulse being used.

Both these solutions have disadvantages. In the first case, extra circuitry may be required to shorten the clock pulse in order to keep its width less than the propagation delay of the flip-flop. Even if this were a practical proposition, many flip-flops have propagation delays of only a few nanoseconds and a very narrow pulse would be required to clock the circuit.

In the second case, increasing the propagation delay of the flip-flop would reduce its operating speed, and in many situations a high operating speed is required. The clock frequency in a Teletext decoder is 6.9 MHz!

To prevent these situations from arising, and to remove the possibility of a race condition a further concept may be explored. This is to prevent the output from the flip-flop from changing until the clock pulse has returned back to a logic '0'. If this simple rule is obeyed then all the problems outlined so far will be overcome. *Figure 3.11* shows diagrammatically the basics of this idea. The delay between the negative edge of the clock pulse and the flip-flop output will be fixed regardless of the clock pulse width and frequency.

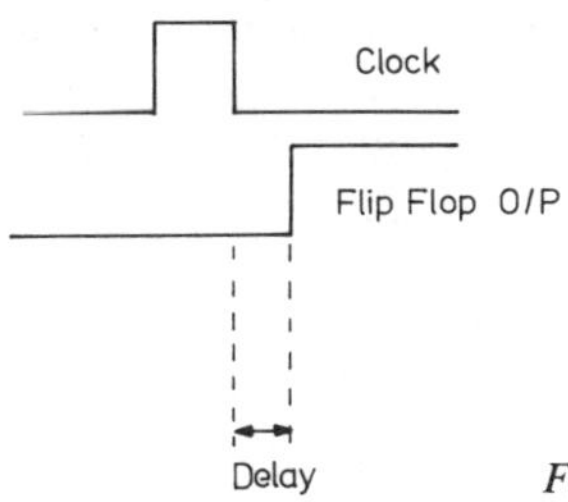

Figure 3.11 Master slave delay waveforms

A flip-flop which overcomes all of these problems is known as a master slave flip-flop and in most practical applications of sequential logic circuits master slave flip-flops are used.

3.5 The master slave SR flip-flop

The master slave SR flip-flop is in fact two flip-flops connected in cascade. Such an arrangement is shown in *Figure 3.12*.

The first stage of this flip-flop is known as the master and consists of four NAND gates coupled to form a positive clocked SR flip-flop. Assume that S = 1 and R = 0. When the clock pulse rises to a logic '1' the conditions at S and R are coupled to points A and B causing the Qm output to rise to logic '1' and the $\overline{Qm}$ to fall to logic '0'. These two outputs form the S and R inputs to a second flip-flop known as the slave. This flip-flop is negative edge triggered, and while the clock input to the master is at a logic '1' the clock input to the slave will be at logic '0' due to the inverting action of the NOT gate 9. Under these conditions the slave is inhibited and its Qs and $\overline{Qs}$ outputs will remain unaffected.

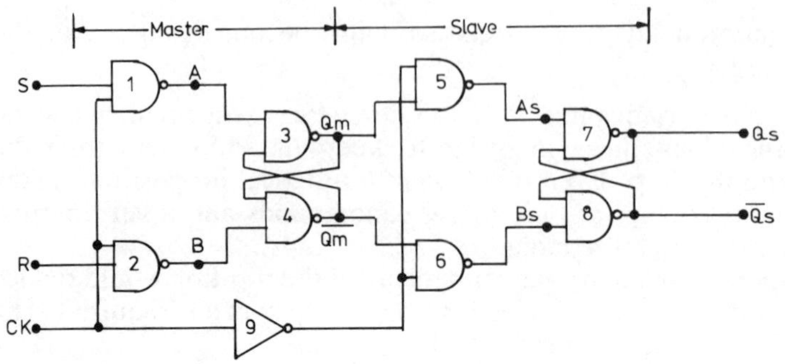

Figure 3.12 Master slave SR flip-flop

When the clock input to the master falls to a logic '0' the outputs from the master flip-flop will remain latched at Qm = 1 $\overline{Q}$m = 0. The clock to the slave now becomes positive causing its S and R inputs to be complemented and coupled to points A and B respectively, causing the Qs and $\overline{Qs}$ outputs to change state.

Examination of the waveform in *Figure 3.13* shows that the S and R inputs are clocked through to the Qs and $\overline{Qs}$ outputs in two stages. When the clock is positive S and R are clocked into the master flip-flop, and when the clock is negative S and R are clocked into the slave flip-flop. It is impossible for the Q and $\overline{Q}$ outputs of a master slave flip-flop to change state before the clock pulse has fallen to a logic '0'.

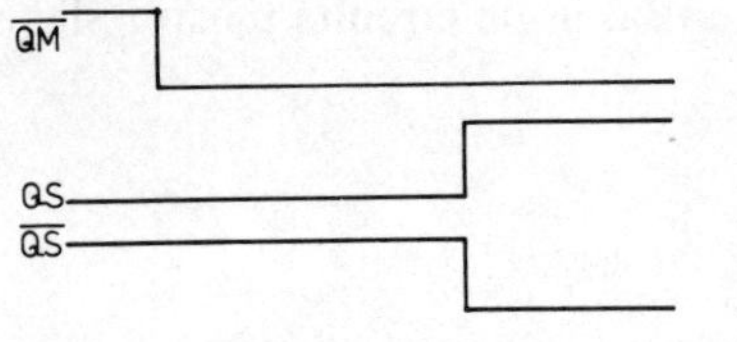

Figure 3.13 Master slave SR flip-flop waveforms

This circuit configuration has produced a flip-flop which has a propagation delay which is always slightly longer than the duration of the positive clock pulse irrespective of the clock pulse width. It effectively has a variable propagation delay which equals the width of the clock pulse plus the propagation delay of the slave flip-flop.

The master slave concept is used in practically every application of a sequential logic system because it overcomes nearly all the problems which are encountered with race conditions.

3.6 The master slave JK flip-flop

The circuit of a master slave JK flip-flop is shown in *Figure 3.14*. As with the simple JK flip-flop already described, the Q and $\overline{Q}$ outputs are cross-coupled and fed back to the input NAND gates 1 and 2 together with the J, K and clock signals. The truth table of the circuit is identical to that previously illustrated in *Figure 3.8* but by using the master slave concept it is possible to use a clock input with any mark to space ratio without encountering a race condition.

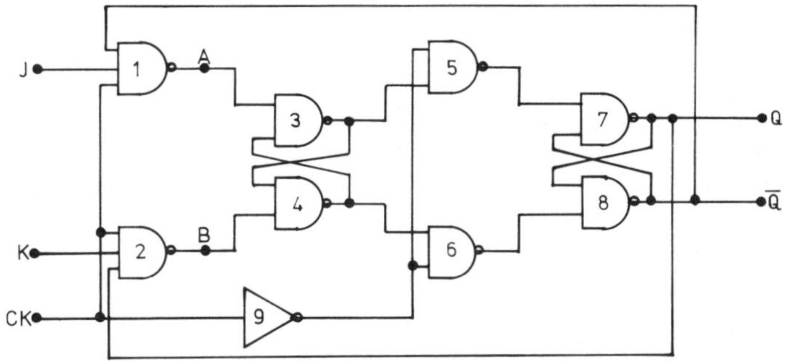

Figure 3.14 Master slave JK flip-flop circuit diagram

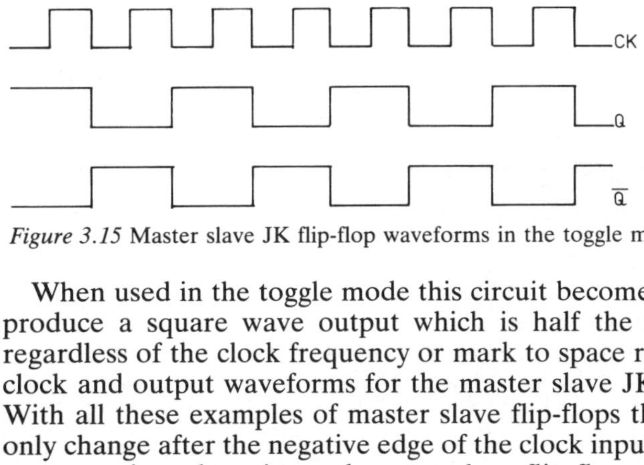

Figure 3.15 Master slave JK flip-flop waveforms in the toggle mode

When used in the toggle mode this circuit becomes very versatile because it will produce a square wave output which is half the frequency of the clock input regardless of the clock frequency or mark to space ratio. *Figure 3.15* shows typical clock and output waveforms for the master slave JK flip-flop in the toggle mode. With all these examples of master slave flip-flops the Q output of the circuit will only change after the negative edge of the clock input, hence the circuits are known as a negative edge triggered master slave flip-flops.

3.7 Clock oscillators

We have seen that a large number of sequential logic devices require clock pulses to synchronize all the elements used in the circuit. One or more clock oscillators may be used, and these can be synchronized or free running. Several types of oscillator circuit can be used to produce the required clock pulses and these can be broadly divided into two main categories:

(a) Oscillators which produce a direct digital output.
(b) Oscillators which produce an analogue signal which is then converted into a digital output.

The first of these two types is the easiest to implement because its output will interface directly with the logic circuits in the system. The second category offers much wider scope to the designer of a system because any well tried oscillator circuit may be used e.g. Hartley, Colpits etc. The analogue output waveform can then be converted into a digital signal by using a Schmitt trigger or other circuit which will 'square up' the analogue waveform. Analogue and digital oscillators that can be free running or synchronized will be discussed in the descriptions below.

3.8 The astable multivibrator

An astable multivibrator has two unstable states, causing it to switch back and forth between these two states and produce a square wave output. If the supply potential to the multivibrator is at the same potential as a logic '1' then the oscillator will produce alternating logic '1s' and logic '0s' naturally at the output without the need for further electronics. *Figure 3.16* shows a transistor astable multivibrator. The

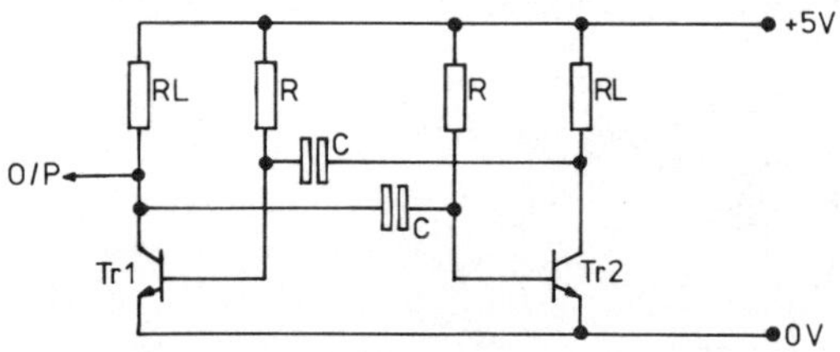

Figure 3.16 Astable MV circuit diagram

transistors alternately saturate and cut off at a frequency which is determined by the values of R and C. This frequency can be calculated by using the formula:

f approx $= 0.7/CR$ (assuming the values of R and C are equal on both sides)

Where f = frequency in Hertz;
R = value of base resistor in Ω;
C = value of coupling capacitor in Farads.

The periodic time of the waveform can be calculated using the formula:

t approx $= 1.4 \times CR$

Providing the values of the base resistors and the coupling capacitors are equal the circuit will produce a symmetrical square wave output because each transistor stays off for the same amount of time. In some circumstances an asymmetrical output is required and this can be obtained by using either two different base resistors or two different capacitor values.

3.9 Logic oscillator

If the frequency of the clock waveform is not critical, and does not have to be accurately controlled, then a very simple oscillator can be constructed using two NOT gates and a CR network, see *Figure 3.17*.

Assume the capacitor is discharged. The input to NOT gate 1 will be a logic '0' and the output a logic '1'. In this state the top end of the resistor R will be at logic

'1' and C will be able to charge towards the logic '1' potential via R. This causes the voltage on the input of NOT gate 1 to rise at a speed determined by C×R. A point will be reached when this voltage exceeds the threshold point between a logic '1' and a logic '0'. At this point the output from gate 1 will fall to a logic '0', producing a very low impedance across C and R. C will now discharge via R and the low impedance at the output of gate 1 until the voltage at the junction of C and R falls below the threshold point for a logic '1'. The output of gate 1 now changes state

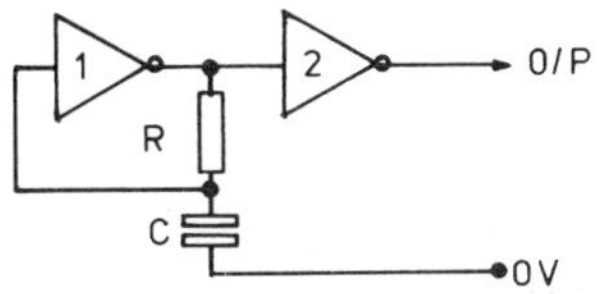

Figure 3.17 Logic oscillator circuit diagram using NOT gates

back to a logic '1'. C now recharges via R and the process will repeat itself. NOT gate 2 is used to isolate the CR network from the output, and also to provide a cleanly switched output waveform. When using this form of circuit it is common practice to use gates which incorporate Schmitt triggers, so that clean and fast switching between one state and the other is achieved.

With this simple circuit it is not possible to accurately control either the frequency or the marked space ratio of the output waveform, but it is very useful in situations where a digital square wave of an approximate frequency is required.

Analogue oscillators

Sine wave, sawtooth and parabola oscillators have been in common use for many years and some of these are used in digital as well as analogue systems, especially where a high frequency or extreme accuracy is required. Perhaps the most widely used of analogue oscillators in digital systems is the crystal oscillator, which may be voltage controlled.

3.10 Crystal oscillator

One possible circuit configuration of the crystal oscillator is shown in *Figure 3.18*. The parallel tuned circuit which determines the frequency of the oscillator is made up by the components C1 and XTL, placing the resonant tuned circuit in the base circuit of this common emitter amplifier.

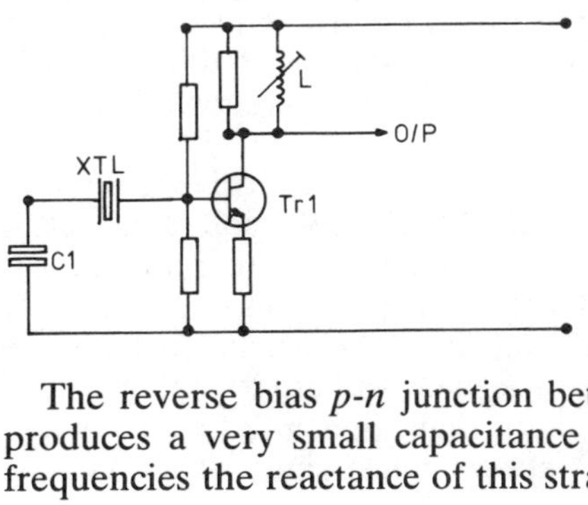

Figure 3.18 Crystal oscscillator circuit diagram

The reverse bias *p-n* junction between the base and collector in the transistor produces a very small capacitance which is typically a few picofarads. At high frequencies the reactance of this stray capacitance falls producing a feedback path between collector and base of the transistor. When a purely resistive collector load is used this capacitance provides negative feedback because the signal voltage on the collector is 180 degrees out of phase with the signal voltage on the base.

If a variable inductance is added to the collector load circuit the phase of the collector signal can be changed by the inductance until the feedback between the collector and base become positive. At this point the circuit will oscillate. L is therefore adjusted to provide maximum output from the collector. It does not determine the frequency of oscillation.

Synchronizing an oscillator to data

In many cases the clock waveform has to be synchronized to a h.f. digital signal. It would be possible to synchronize a multivibrator to provide the synchronism required, but the multivibrator is not a suitable circuit for high frequency operation. When situations like this arise it is normal to use a digital signal to synchronize an analogue oscillator first, and then convert the resulting sine wave into a digital clock waveform.

This sort of problem presents itself in a Teletext decoder where the Teletext data which is transmitted in serial form has to synchronize a 6.9 MHz clock oscillator before the data can be decoded. One possible method of providing the synchronized waveform is illustrated in *Figure 3.19*.

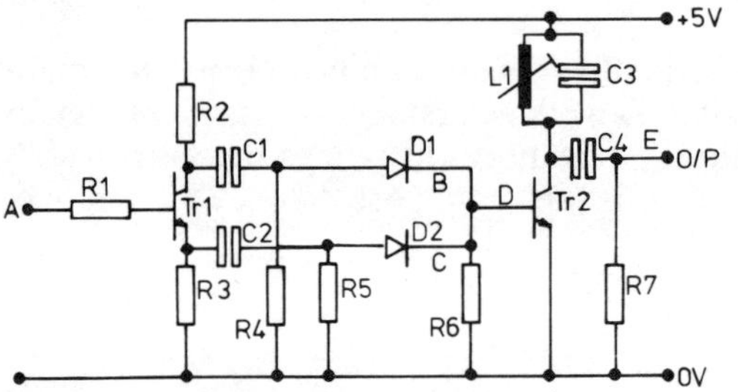

Figure 3.19 Data locked LC oscillator circuit diagram

Data is d.c. coupled to the base of TR1 which will saturate during a logic '1' and cut off during a logic '0', TR1 is a simple phase splitter providing a signal that is in phase with the input data on its emitter, and a 180 degrees out of phase with the input data on the collector. These two waveforms are differentiated by C2 and R5 and C1 and R4 respectively. D1 and D2 clip the differentiated waveforms producing positive trigger edges at points B and C respectively, as shown in *Figure 3.20*.

These two waveforms combine at point D and momentarily drive TR2 into conduction. The resulting pulses of current which flow in TR2 collector circuit cause the high Q parallel tuned circuit L1 C3 to oscillate, and providing data is present on the base of TR1 the tuned circuit will maintain oscillation. Consequently on the collector of TR2 there will be a sine wave which will be synchronized to the bit rate of the incoming data. L1 will adjust the phase and amplitude of this signal.

It is impossible to describe every type of oscillator that can be used with a digital system, and those selected for inclusion here represent just some of the circuits

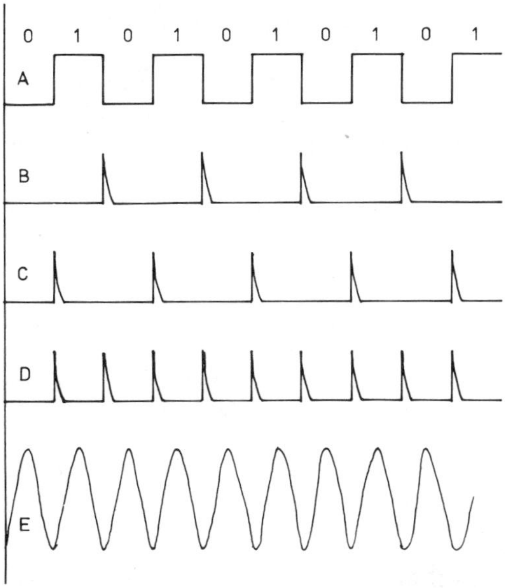

Figure 3.20 Data locked LC oscillator waveforms

which are likely to be encountered. One problem that does exist with an analogue oscillator, regardless of the design, is the interface it requires between its analogue output and the digital system.

The sine wave output from any analogue oscillator cannot be used directly with a digital system, it has to be converted into a square wave before it can be used as a clock signal. There are several ways of 'squaring up' an analogue signal and an example is included here to illustrate one of the many techniques used.

3.11 Schmitt trigger

A Schmitt trigger has two stable states similar to the astable multivibrator previously described, with positive feedback from one stage to the other applied using a common emitter resistor. The circuit requires an input which will cause it to switch from one stable state and another at a very high speed.

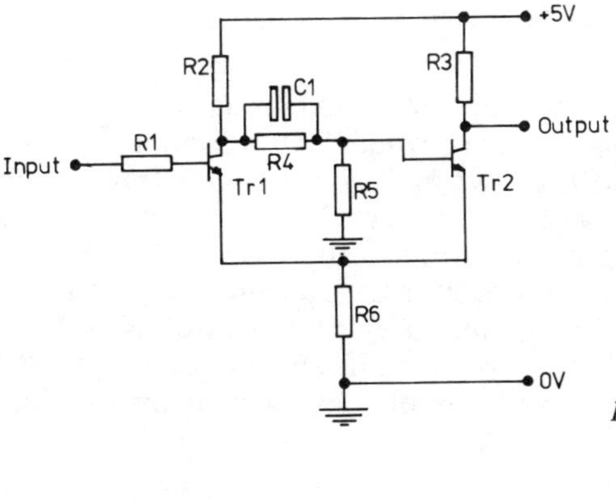

Figure 3.21 Schmitt trigger circuit diagram

Any Schmitt trigger will have a threshold voltage at which point it will change from one stable state and the next. It is therefore possible to apply a sine wave, or any other analogue waveform to the input and produce a square wave output. These properties make the Schmitt trigger an ideal choice for the interface required between the analogue oscillator and the digital system. *Figure 3.21* shows one of the Schmitt trigger circuits which are available.

Assume the input voltage is 0 V. TR1 will be switched off and its collector voltage will be at a positive potential. A proportion of this voltage is fed by the potential divider R4 and R5 to the base of TR2. This base voltage is sufficient to drive TR2 into conduction causing a current to flow via the emitter resistor R6, the low impedance of TR2, R3 to the supply potential. A positive voltage is now present across R6, (approx 2 V), which in turn will also be present on the emitter of TR1. TR1 has now been further reversed biased because the *p*-type base is at 0 V and the *n*-type emitter has risen to +2 V, ensuring the transistor remains cut off.

If the input voltage to the base of TR1 slowly rises, the Schmitt trigger will remain stable until the base voltage of TR1 rises above its emitter voltage by approximately 0.6 V. When this point is reached TR1 will start to conduct and its collector voltage will fall. TR2 is now being driven towards cut off by two factors:

1. If TR1 conducts an increased current will flow through the emitter resistor R6. This will increase the emitter voltage of TR2 thus lowering TR2 base/emitter voltage.
2. The fall in voltage on the collector of TR1 will reduce the base bias to TR2 (a process that is speeded up by the presence of C1).

TR2 will therefore rapidly switch off, reducing the current it draws through R6 as it does so. VR6 therefore falls increasing the forward bias on TR1, ensuring that TR1 saturates rapidly. TR1 will now be conducting and TR2 will be cut off causing the output voltage to rise from its previous low state to a high state.

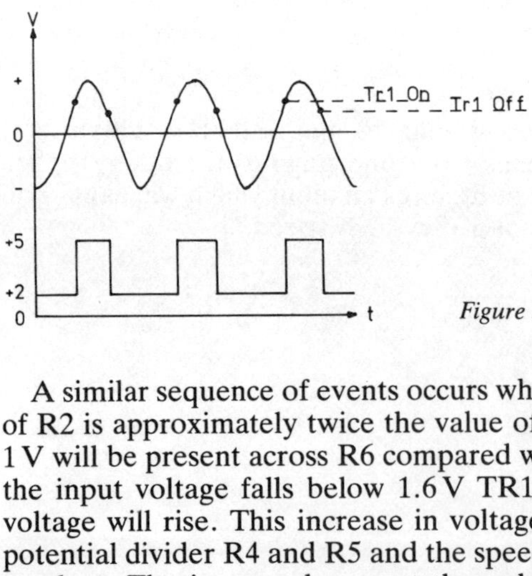

Figure 3.22 Schmitt trigger waveforms

A similar sequence of events occurs when the input voltage falls again. The value of R2 is approximately twice the value of R3, hence when TR1 is conducting only 1 V will be present across R6 compared with 2 V when TR2 was conducting. When the input voltage falls below 1.6 V TR1 will start to switch off and its collector voltage will rise. This increase in voltage is passed on to the base of TR2 by the potential divider R4 and R5 and the speed up capacitor C1, causing TR2 to start to conduct. The increased current through the common emitter resistor R6 causes TR1 emitter voltage to rise driving TR1 further into cut off, and consequently TR2 further into saturation. This situation escalates rapidly resulting in the original condition of TR1 cut off and TR2 conducting. The output voltage will therefore drop rapidly back to a low state.

Typical input and output waveforms for this type of circuit are shown in *Figure 3.22* and it can be seen that the Schmitt trigger does produce a true digital signal (i.e. it only has two states) and in this example a logic '0' = 2 V and a logic '1' = +5 V.

Some additional minor processing would be required if the digital system used the levels logic '0' = 0 V and logic '1' = +5 V, and this processing would have to take place in two stages:

1. The signal can be coupled to a d.c. restorer which would set the logic '0' reference to 0 V.
2. With the logic '0' reference accurately set it is a simple matter to add a diode clipper to remove any voltage in excess of +5 V. A simplified schematic of this process is shown in *Figure 3.23*.

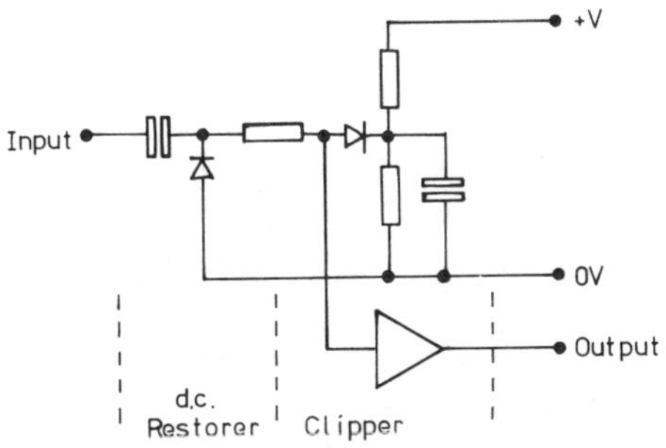

Figure 3.23 Schmitt trigger to digital system interface circuit

3.12 Practical applications of sequential logic circuits

So far in this chapter we have examined individual sequential logic circuits for both synchronous and asynchronous applications. We will now examine some practical applications of these circuits and pay particular attention to those which are used in digital television applications.

Any digitally orientated television system receives data in serial form whether it be Teletext, Viewdata or remote control signals. Sequential logic circuits must therefore play a very important role in any prospective digital system for use in TV because they are universally used to deserialize data (convert data from serial to parallel form) so that it may be processed by the combinational logic elements of the system. This important application of sequential logic will therefore be examined first of all.

3.13 Serial to parallel conversion

Converting a serial data input to a parallel data output or *vice versa* is a process that takes place in a group of circuits known as shift register. In practically all TV orientated digital systems deserialization of data is carried out synchronously, hence the circuits require a clock waveform which is accurately synchronized to the bit rate of the incoming serial data. Shift registers can be constructed using a variety of logic circuits but the two most ideally suited are the D type flip-flop or the JK flip-flop. Of these two, the master slave JK flip-flop is preferred because it eliminates the possibility of a race condition arising. We will next examine the application of both of these circuits starting with the D type flip-flop.

3.14 Shift register using D-type flip-flops

Earlier in this chapter we saw how a D-type flip-flop can be used as a store which will transfer the logic state present at its D input to the Q output during the presence of a positive clock pulse. The circuit will then maintain this logic level at the Q output until another clock pulse is received.

A shift register uses this property to great advantage by connecting a number of D-type flip-flops in cascade as shown in *Figure 3.24*. In this example four flip-flops are used but in practical applications the size of the parallel output will determine the number of flip-flops required.

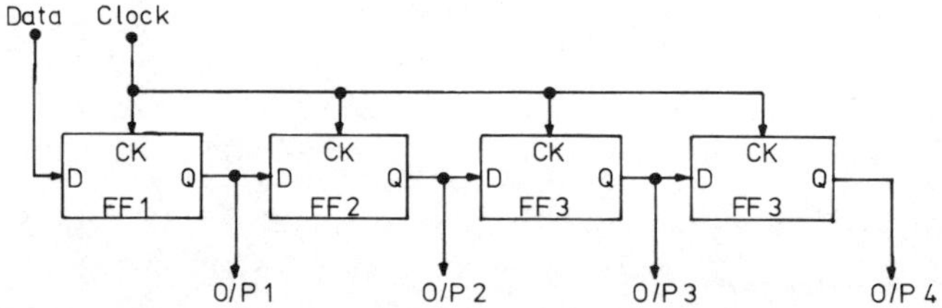

Figure 3.24 Logic diagram of a shift receiver using D-type flip-flops

Assume the serial data pattern illustrated in *Figure 3.25 (b)* is present on the data input to the shift register and a synchronized clock waveform as shown in *Figure 3.25 (a)* is applied to the clock input on each flip-flop. Before clock pulses are applied to this circuit, the state present at the Q output of each flip-flop will be indeterminate because it depends in which state the flip-flop latched when power was applied to the circuit. In practice this random pattern of data present at the output makes no difference to the operation of the circuit but to simplify the explanation we will assume that all four outputs are at logic '0'.

When the clock pulse *t1* is applied to the shift register the data present at the D input of each flip-flop will be transferred to its respective Q output. The logic '1' present on the data input will therefore appear at output 1, the 0 that was present at the Q output of flip-flop 1 will be transferred to output 2, the 0 that was present on the Q output of flip-flop 2 will be transferred to output 3 and so on.

It can be seen that the data that was present at the D input of each flip-flop at the instant the clock pulse arrived was shifted one place to the right, hence the name 'shift register'. As these flip-flops are not of the master slave type, the clock pulse duration will have to be shorter than the propagation delay of each flip-flop. If it is not there is a very great danger that the logic '1' present on the data input will ripple through the series of flip-flops while the clock pulse is at a logic '1'. This problem is common to all forms of synchronous shift register and consequently master slave flip-flops are always used when there is a danger of this situation arising.

Between periods *t1 and t2* when the clock pulse is at logic '0' and the circuit is inactive, the output from the shift register will be 1000 as shown in *Figure 3.25 (c)*, and it will remain in this state until the next clock pulse.

At *t*2 the data input to the shift register has fallen to a logic '0' and during the clock pulse period this logic level will be transferred from the D input to the Q output of flip-flop 1, while the D inputs of all the other flip-flops will be transferred to their respective Q outputs. Examination of the waveform in *Figure 3.25 (d)* shows that after the second clock pulse *t*2 the logic conditions set up on the output of the register will be 0100. Again the data that was present at the D input to each flip-flop has shifted one place to the right.

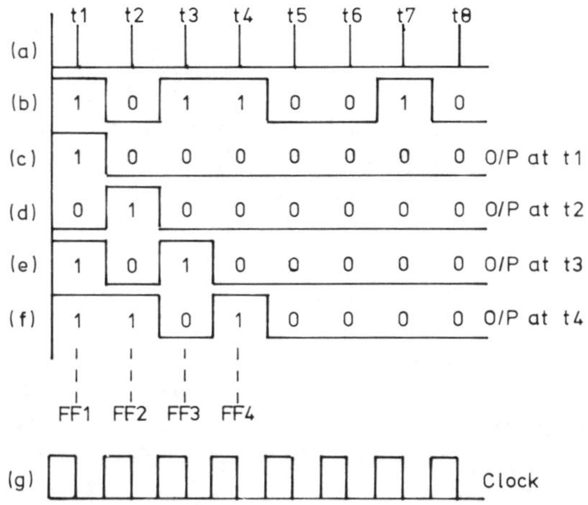

Figure 3.25 Waveforms for shift register using D-type flip-flops

This process continues until the fourth clock pulse, at which point the shift register is full, containing the first four bits of the received data as shown in *Figure 3.25 (f)*. Because data shifts through the register from left to right, output 4 will contain the first bit received, and output 1 will contain the last bit. The first four bits of the data which were received in serial form are now set up in parallel form at the outputs of the shift register. They will remain static until the next clock pulse *t*5 arrives, at which time all the data will shift one place to the right.

Some means of storing the output from the shift register must take place between *t*4 and *t*5 otherwise the data will be lost. This can easily be arranged by using a four bit latch which is described in detail later in this chapter.

3.15 Shift register using master slave JK flip-flops

The basic principles of the shift register just outlined still applies to the register shown in *Figure 3.26*, but there are a few changes in the circuit configuration.

The first flip-flop in the chain has a NOT gate on its K input converting it from the conventional master slave JK flip-flop into a master slave D-type flip-flop. The Q and $\bar{Q}$ outputs of each flip-flop are connected to the J and K inputs of the next. Hence the first flip-flop in the series is a master slave D-type flip-flop, and the remainder are master slave JK flip-flops.

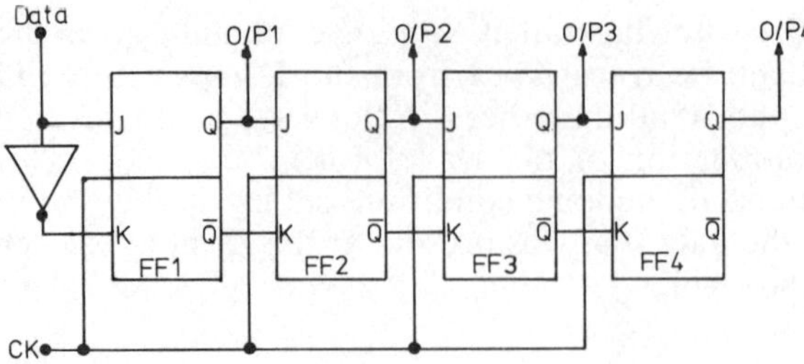

Figure 3.26 Circuit diagram of shift register using D-type flip-flops

Remember that with the master slave concept, data is transferred into the master flip-flop of each stage when the clock pulse rises to a logic '1' and then into the slave flip-flop when the clock falls back to a logic '0'. Consequently the output of each element in the register will only change after the negative edge of the clock pulse. This means that the clock pulse can be any width providing it does not exceed the time occupied by one bit of the serial input data. The output waveforms shown in *Figure 3.25* still hold true for this type of shift register, and a typical clock waveform is included in *Figure 3.25 (g)*.

The complete diagram of a shift register can be greatly simplified by removing all the interconnections between the stages, producing a simple single block as shown in *Figure 3.27*. It is easy to forget the complexity of the electronics contained in this simplified diagram. Each master slave flip-flop will contain the nine gates of *Figure*

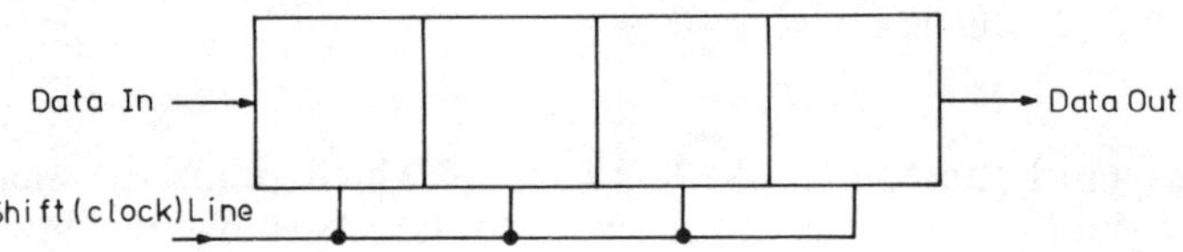

Figure 3.27 Logic diagram of shift register

3.12 giving the shift register a total of 37 gates including the first input NOT gate. Each gate uses a minimum of four transistors giving a minimum total of 148 transistors. If the shift register was for an eight-bit byte this semiconductor count would have to be doubled to 296 devices in a single IC. When this exercise is carried out for a typical digital system it is easy to see why such systems only began to make an impact once the large scale manufacture of integrated circuits was possible.

3.16 Parallel to serial conversion

We have seen from past examples that there are many applications where data has to be relayed from one system to another as a serial signal, and consequently circuits have to be devised to convert data from parallel to serial form. It again can be done with a shift register providing extra circuitry is used, but before explaining how such a circuit is produced it is necessary to add a little extra theory to the established operation of the master slave JK flip-flop.

Preset/clear

The master slave JK flip-flop is an excellent workhorse for nearly all sequential logic systems, but it versatility can be increased by providing two extra inputs known as Preset and Clear. With these two additional inputs at logic '1' (or open circuit) the flip-flop behaves in exactly the same way as previously described.

If the preset input is taken to a logic '0' two things will happen:

1. The operation of the flip-flop is inhibited regardless of the state of the JK or clock inputs.
2. The Q output will be preset to a logic '1' and the $\bar{Q}$ to a logic '0'.

These conditions will remain fixed until the preset input returns to a logic '1' at which point the Q output will remain set to a logic '1' until the circuit is clocked. Normal operation of the flip-flop is then resumed.

When the clear input is taken to a logic '0', again operation of the flip-flop is suspended but on this occasion the Q output is cleared to a logic '0' and the $\bar{Q}$ output assumes a logic '1'. As with the previous condition when the clear input returns to a logic '1' normal operation resumes after the next clock pulse.

These two added facilities make a very useful contribution to this circuit. For example, with the shift register using JK flip-flops illustrated in *Figure 3.26* we can clear the outputs of the complete register to a logic '0' by the application of a negative clear pulse to each flip-flop. This will clear all the data contained in the register ready for the receipt of the serial data input.

The preset and clear facilities come into their own when we require a shift register to accept data in parallel form and then shift it through the register to produce a serial output. In the example that follows master slave JK flip-flops with preset and clear facilities are used to produce a four bit parallel to serial conversion of data.

3.17 Parallel to serial conversion using JK flip-flops

A full circuit diagram of a parallel to serial register is shown in *Figure 3.28*. The circuit has been devised using logic elements already described but it is possible to produce a complete register in a single integrated circuit. Although the circuit looks complex at a first glance, (as do most unfamiliar logic diagrams), its operation is relatively simple.

The heart of the register is the four master slave JK flip-flops FF1 to FF4. These are connected in a similar way to those used in the serial to parallel converter of *Figure 3.26* apart from the J and K inputs to FF1. These have been connected to J = 0 and K = 1 which means that every time this flip-flop is clocked its Q output will remain at logic '0' and $\bar{Q}$ will be at logic '1'. This will ensure that when the register is clocked, 0s will be loaded in from the left-hand side and propagate through the register to the right-hand side. This is not strictly necessary but it is nice and tidy.

To explain the concept of this circuit we will assume for the moment that the Q output of FF1 has been preset to a logic '1' FF2 to logic '0' FF3 to a logic '1' and FF4 to a logic '0'. When the first clock pulse is received the data present on the J input of each flip-flop is shifted one place to the right in exactly the same way as in the previous description. If the process is repeated, the data will move steadily to the serial output and a series of logic '0s' will follow the data in from the left-hand side,

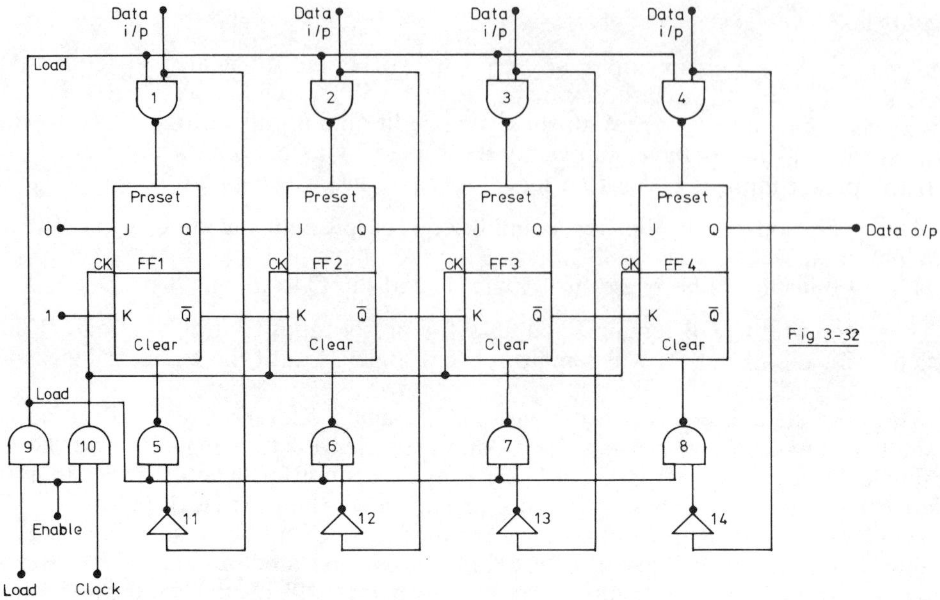

Figure 3.28 Circuit diagram of parallel to serial shift register

as shown in *Figure 3.29*. The serial data from the register is taken from the Q output of FF4, and if this is examined it can be seen that the parallel data input of 1010 appears in serial form one bit at a time at a speed determined by the frequency of the clock pulse.

CK	QFF1	QFF2	QFF3	QFF4
L	1	0	1	0
1	0	1	0	1
2	0	0	1	0
3	0	0	0	1

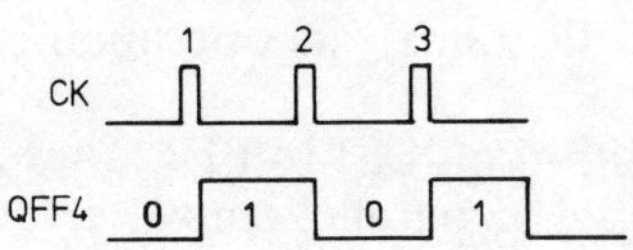

Figure 3.29 Action table for parallel to serial shift register

So much for shifting the data out in serial form, but how do we load the register with a parallel byte of data before it is clocked through? This is achieved by using the preset and clear inputs to each flip-flop. Using the byte of data (1010) that was described above we can see that when the register is loaded with a parallel byte of data FF1 has to be preset, FF2 cleared, FF3 preset, FF4 cleared to produce the pattern 1010 in the register.

This process is carried out by the NAND gates numbered 1 through to 8. When the load input to gates 1–8 is at logic '0' the output from each gate will always be a logic '1' resulting in a logic '1' on both the preset and clear inputs of each JK flip-flop. Under these conditions each flip-flop will operate in the shift mode. When the load pulse is at a logic '1' each NAND gate will pass the complement of its second input. For example, NAND gate 1 will have two logic '1s' at its input (load pulse and logic '1') producing a logic '0' at its output. This will preset FF1 to Q = 1 and $\bar{Q}$ = 0. FF2 will have a logic '1' on the load input and a data input of logic '0', therefore its output will remain at a logic '1' keeping the preset input of FF2 at a logic '1'. However the data input to this stage is also inverted by the NOT gate 12, complemented by NAND gate 6 and fed to the clear input of FF2 as a logic '0', clearing the Q output of FF2 to a logic '0'. In a similar manner FF3 is preset to a logic '1' and FF4 cleared to a logic '0'.

Remember that when the flip-flops are being either preset or cleared during the presence of a load pulse the clock input has no effect on the circuit. The data will only be shifted through the register when the load pulse returns to a logic '0'.

Both the load and the clock inputs to the register are fed via AND gates 9 and 10 together with an enable signal. When the enable is at a logic '1' load and clock pulses can be applied to the circuit but if the enable falls to a logic '0' the outputs of gates 9 and 10 will also stay at logic '0' disabling the circuit in both the load and shift modes of operation.

3.18 Latches

There are many situations where a byte of data has to be held in a store so that it can be interrogated, decoded or complemented, etc, Circuits which provide temporary storage of data are known as latches. We have already met them in the form of the D-type flip-flop earlier in this chapter, and we have seen that a JK flip-flop can also be used as a store by adding a NOT gate between the J and K inputs.

Latches may be used individually to store a single bit of data or, as is more common, to store a complete byte. We will now examine how a four-bit latch may be used to store a four-bit nibble of data and then apply the circuit together with the

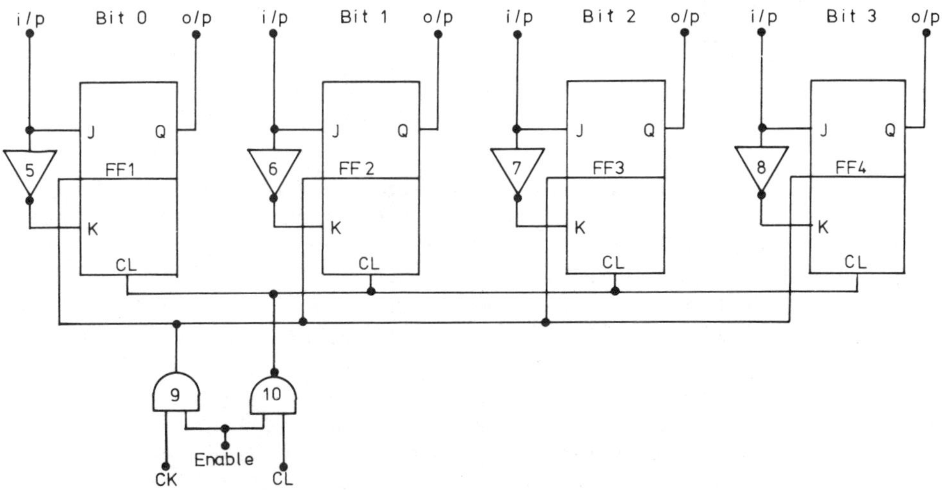

Figure 3.30 Circuit diagram of 4-bit latch

serial/parallel shift register discussed earlier to form a complete circuit of a serial to parallel converter.

The circuit shown in *Figure 3.30* shows one arrangement of a four-bit latch, in this case using master slave JK flip-flops. Each input bit is applied to the J input of the flip-flop, complemented by a NOT gate and applied to the K input. On the application of a clock pulse each flip-flop will be either set or reset depending on the logic level of the input bit. As master slave flip-flops are being used, the Q output of each section of the latch will assume the logic state of the J input immediately after the clock pulse has returned to a logic '0'; this output will be held until the next clock pulse arrives. The input circuits to the register are arranged in such a way that the data remains stable during the presence of a clock pulse and changes as necessary between clock pulses.

By using master slave JK flip-flops, additional inputs can easily be provided, and in this example we have used an enable and clear. Normally the enable input will be at a logic '1' allowing clock pulses to pass through AND gate 9 and allowing a logic '1' which is complemented by NAND gate 10 to clear the register. In *Figure 3.31* the four-bit latch of *Figure 3.30* is shown in simplified form as it may appear in a circuit diagram.

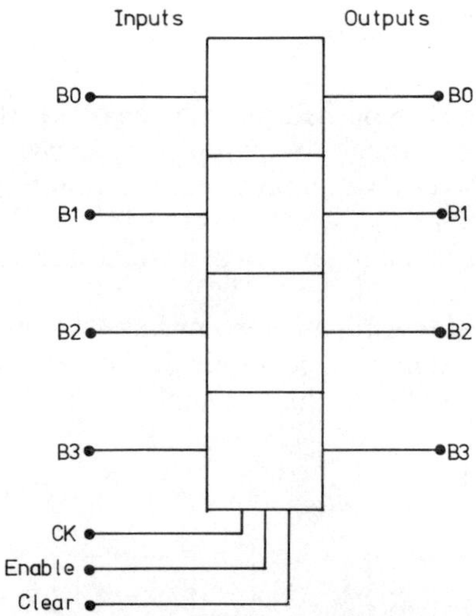

Figure 3.31 Logic diagram of 4-bit latch

We are now in a position to put the serial shift register and the four-bit latch together to produce a complete serial to parallel converter. One possible arrangement is shown in *Figure 3.32*. The serial input data is applied to the input of the shift register and shifted through the register by the synchronized bit clock signal. Each output from the shift register is coupled to an input of the four-bit latch. Data must only be stored in this latch when the shift register is full, i.e. after four clock pulses. Consequently the four-bit latch must only be clocked when the input shift register has a complete nibble of data stable at its output.

To achieve this, the synchronized bit clock signal is divided by four so that data is transferred to the latch after every fourth clock pulse. This synchronism between the two circuits is imperative if accurate deserialization of data is to be achieved, and great care has to be taken to produce clock waveforms which have the correct

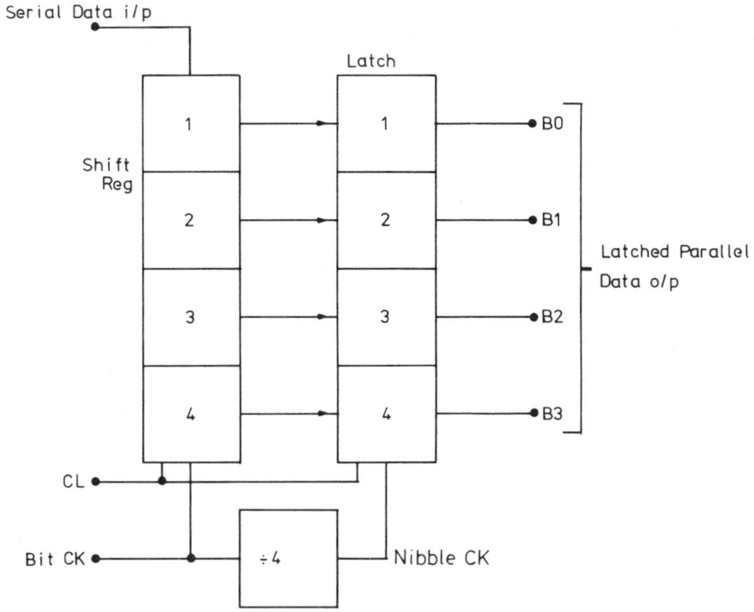

Figure 3.32 Logic diagram serial/parallel converter

timing. If necessary both the serial register and the four-bit latch may have their inputs cleared to a logic '0'.

The only part of this circuit that we have not met is the divide by 4 stage which receives the clock frequency at data bit rate and divides it by four to produce an output clock at data nibble rate. Counters of this type are very common in sequential logic circuits and there are many types available, all of which have their own advantages or disadvantages. Counters are availabe in synchronous or asynchronous form and the one used in this application is of course asynchronous because it does not have an external clock signal, only a single input and output. The final part of this chapter will deal with a limited range of the counters that can be produced.

3.19 Counters

Most TV applications of counters use asynchronous techniques because normally only a single input is required (i.e. the frequency to be divided) to produce the desired frequency at the output. Consequently the counters to be described here are of the asynchronous type and will not require an external synchronizing source such as that used with the shift register.

Earlier theory has showed us that the master slave JK flip-flop will operate in the toggle mode when both the J and the K inputs are at a logic '1'. If J and K are left open circuit, and a square wave frequency applied to the clock input, a signal will be produced at the Q output which has a frequency that is half the clock input. If a series of JK flip-flops are connected in cascade as shown in *Figure 3.33*, each element in the chain will divide its input clock frequency by two. The output at (a) will be the input frequency divided by two, at (b) the input frequency divided by four and (c) the input frequency divided by eight, etc. If the waveforms in *Figure 3.33* are examined this point can be verified and also the effect of negative edge triggering can be seen.

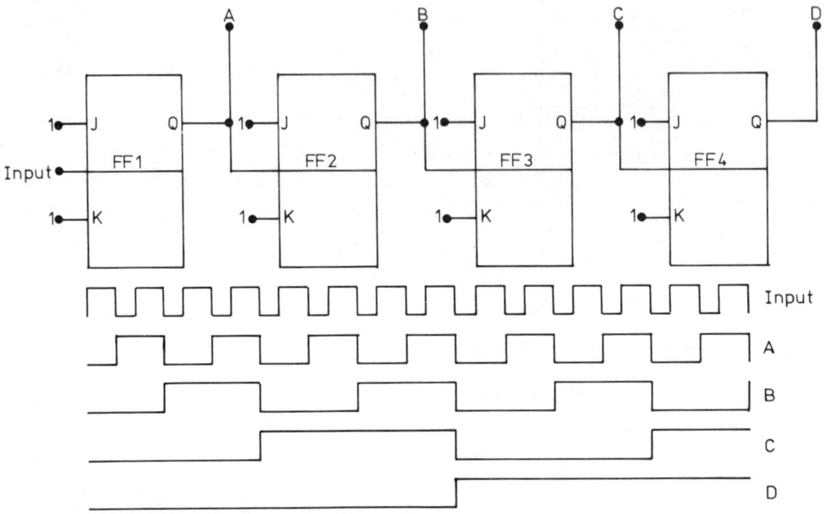

Figure 3.33 Circuit diagram binary counter

Comparison of the input and output waveform of any stage in the counter will show that the output only changes state on the negative edge of the input waveform. This form of counter will operate over a wide range of frequencies but its high frequency performance is limited by the total propagation delay of each flip-flop. In television orientated digital circuits clock frequencies of up to 10 MHz are used and this form of counter will operate as a divider quite happily in excess of this frequency.

Adding additional JK flip-flops to the circuit will increase the value of the divisor in binary notation (e.g. 2, 4, 8, 16, 32, 64, etc.). Providing one wishes to divide by these numbers this form of counter could be used. Two stages would be ideal for the divide by four counter in *Figure 3.32* but if the designer wishes to divide by say 5, 7, 11, etc. additional circuitry must be added to modify the operation of the counter.

Before describing how this is achieved, we will examine the waveforms associated with the counter in a little more detail. *Figure 3.34* illustrates these waveforms, but this time they are turned on their end with the input waveform on the right-hand side and waveform (d) on the left-hand side. Individual logic levels have also been included with the waveforms to show that a binary count progresses through the circuit on the occurrence of each negative edge of the input clock pulse.

Assuming that all the flip-flops were cleared before the start of the count it will start off with all the signals, including the input, at logic '0'. On the first clock pulse a binary count of one is present. On the next clock pulse the count increases to two and so on until finally the (a), (b), (c) and (d) waveforms are all at logic '1' giving a count of fifteen.

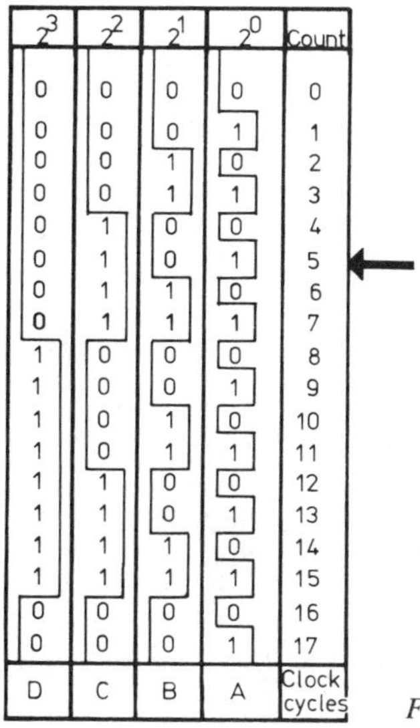

Figure 3.34 Waveforms of pure binary counter and logic states

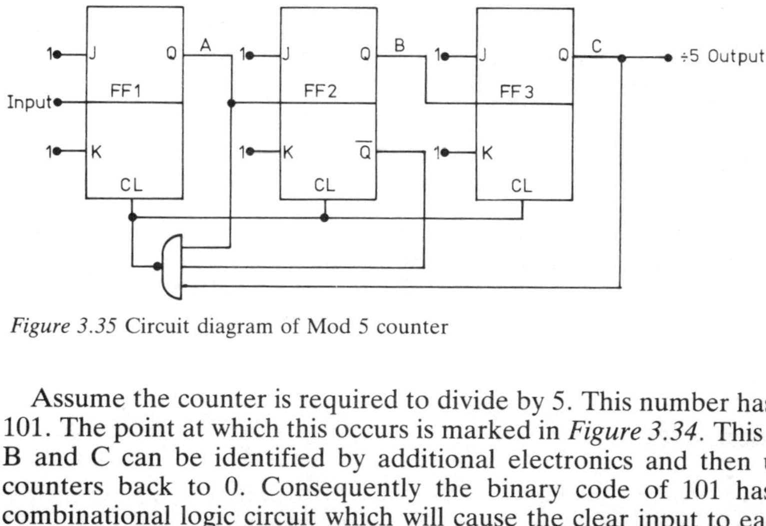

Figure 3.35 Circuit diagram of Mod 5 counter

Assume the counter is required to divide by 5. This number has a binary code of 101. The point at which this occurs is marked in *Figure 3.34*. This code at points A, B and C can be identified by additional electronics and then used to reset the counters back to 0. Consequently the binary code of 101 has to be fed to a combinational logic circuit which will cause the clear input to each JK flip-flop to fall to a logic '0'. The technique used to achieve this was outlined in chapter two (refer to *Figure 3.35*). The logic '1s' at A and B are fed to two inputs of a three input NAND gate. The logic '0' present at B is complemented at the $\overline{Q}$ output of FF2 and this is also fed to the remaining input of the NAND gate. The output of this NAND gate will now fall to a logic '0' when all three of its inputs are at logic '1'. It will therefore clear all the flip-flops to a logic '0' when C, $\overline{B}$ and A are at logic '1', and on the occurrence of the next input pulse counting will resume from 0. The output of the counter is taken from the Q output of flip-flop 3.

Chapter 4

Storage devices

Storage of data takes place in a group of circuits known as the memory. The application, and therefore the design of a particular memory, varies considerably depending on the function of the system in which the memory is being used. Many different types of storage device exist, and in this chapter we will examine those most commonly used in the field of digital electronics, paying particular attention to the range of electronic memories which are found in Teletext decoding and microprocessor based systems.

4.1 Semiconductor memories

Semiconductor memories are capable of storing and retrieving data. Storing data is referred to as *writing*; retrieving data is known as *reading*. The types capable of read and write operations and are commonly known as Random Access Memories (RAM). Random access memories fall into two major types, these being *dynamic* and *static*.

4.2 Dynamic memories

In its simplest form a dynamic memory cell consists of two switches and a capacitor connected as shown in *Figure 4.1*. Data is written to the memory cell by closing S1

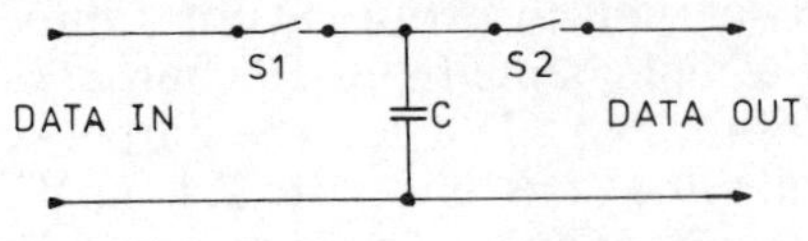

Figure 4.1 2-switch dynamic RAM

and allowing the capacitor to assume the logic state of the input data signal. Assume the memory cell is being used in a system where a logic '1' is represented by +5 V and a logic '0' by 0 V. To write information to the memory cell, S1 is closed and the capacitor will follow the voltage at the data input, i.e. it will charge to +5 V for a logic '1', or discharge via the low input impedance if the input to the cell is a logic '0'. Once the data has been written to the memory cell S1 opens, and if the capacitor is perfect it will hold the charge indefinitely. During a memory read operation S2 is closed and the voltage stored across the capacitor is coupled to the output of the memory cell.

Unfortunately the capacitor used in the dynamic memory is not perfect, and will lose its charge over a period of time, (perhaps a few hundred milliseconds); thus the contents of the cell must be restored or 'refreshed' regularly. Refreshing a dynamic memory is carried out by reading the contents of the cell into a temporary store, (possibly a JK flip-flop), and then writing this information back in to the same memory cell.

In practice the two switches and the capacitor are replaced by MOS transistors as shown in *Figure 4.2*. If an enabling voltage is applied to the write input then TR1 will conduct. The logic state present on the write data input will therefore be coupled via TR1 to the gate of TR2, and the capacitor storage element of the cell. When TR1 conducts the capacitor will charge or discharge until it assumes the same state as the input data. When the write select input to the cell is removed TR1 switches off disconnecting the input data line from the capacitor.

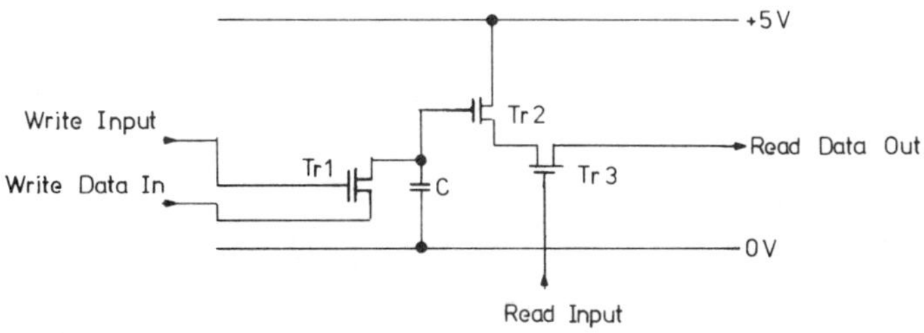

Figure 4.2 Dynamic RAM

TR2 acts as a source follower to the input stage and therefore presents a high impedance to the capacitor preventing the stored charge from leaking away. At the same time it is able to offer a low impedance output voltage corresponding to the charge on the capacitor.

During a read operation TR3 conducts, and connects the data output of the cell, via the low impedance of TR3, to the drain of TR2. Hence the output data line will assume a logic state which corresponds to the charge on the capacitor. With this circuit arrangement the read operation will not effect the charge stored on the capacitor, making it possible to read the data from the memory cell many times between refresh cycles.

In practice a large number of memory cells and their associated circuits are contained in a single integrated circuit, a practical example being the Intel 1103 which contains an array of 1024 memory cells in a 32 × 32 matrix, together with the necessary address and refresh circuitry.

4.3 Static Memory

Static memory cells do not use capacitors as the storage element, but use a flip-flop which can be set or reset as long as power is applied to the circuit. This additional electronics makes each static memory cell much bigger than a dynamic memory cell, hence there are fewer storage elements per chip and it is more expensive. It does however have one advantage over the dynamic memory cell in that it does not require refresh cycles or circuitry.

A typical *n*-channel MOS memory cell is shown in *Figure 4.3*. This example of the static memory cell does not have separate read and write inputs, but has in their place a separate bit or cell select input. When this input assumes a logic '1' both TR5 and TR6 will conduct enabling data to be written into the cell or read from the cell. Consider the operation of this circuit during a memory write sequence. The gate of TR5 will be held positive by the cell select input, causing the data at the input to be coupled by the low impedance of TR5 to the gate of TR2. If it is assumed that the input data is at a logic '1' then TR2 will be driven into conduction causing its drain voltage to fall to approximately the source voltage. This low potential is cross coupled to the gate of TR1 which in turn switches off causing its drain voltage to rise to the supply potential (the same potential as the incoming data).

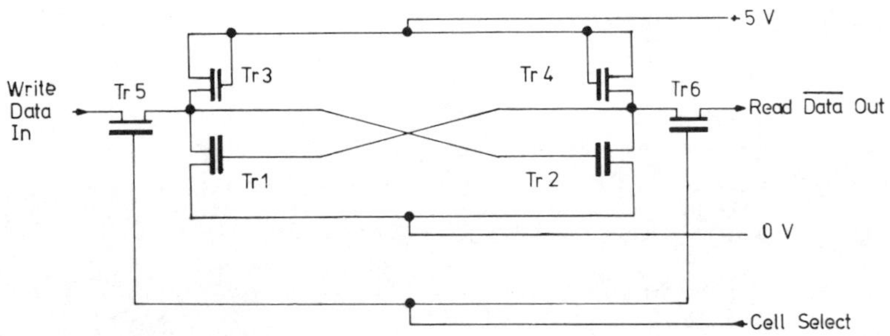

Figure 4.3 Static memory cell

Transistors TR3 and TR4 form active (pull up) drain loads for TR1 and TR2 respectively. These load transistors are permanently forward biased because the gate and drain of each transistor is connected together. The resulting low resistance between source and drain forms the load resistance for TR1 and TR2. It should also be noted that the data output from the cell, which is coupled via TR6, will be the inverse of the stored data, i.e. the cell output is a NOT function of the cell input. A simple inverting process using a NOT gate added to the output of the cell is used to remedy this if necessary.

During a memory read sequence incoming data is disconnected from the cell and the cell select input is made positive. Both TR5 and TR6 conduct, but as there is no input to the cell the stored data in the cell is unaltered. TR6 couples the inverted stored information to the output circuitry, leaving the data stored in the cell unaffected by the reading process.

Both dynamic and static memory cells are volatile, that is if power is removed the contents of the cell will be destroyed unless, of course, some back up power supply is arranged. This type of memory is commonly used for storing temporary data such as pages of Teletext or the results of calculations, etc. carried out by a microprocessor. They are not generally suitable for storing data which is required on a permanent basis, e.g. character generation in a Teletext decoder or for permanent programme storage in microprocessor based systems.

4.4 Read Only Memories (ROM)

As their name implies read only memories or ROMs contain pre-stored data which can only be read from the memory cell. The range of ROMs is non-volatile, therefore the contents of the memory will not be lost if power to the circuit is removed.

Several types of ROM exist, the main types being the Mask Programmable Read Only Memories (ROM), Programmable Read Only Memories (PROM), and the Erasable Programmable Read Only Memory (EPROM).

4.5 Mask Programmable Read Only Memory

The data stored in a Mask Programmable Read Only Memory is specified by the IC purchaser (normally the manufacturer of the equipment in which the memory is to be used), and introduced to the Read Only Memory during the manufacturing process of the integrated circuit. As the name of the memory implies the contents of each memory cell is determined during the masking process of the semiconductor device. It is a relatively simple process to permanently wire each individual storage element to produce the desired output of either a logic '0' or a logic '1'.

This type of memory array is widely used in Teletext decoders for the generation of the alphanumeric characters. Data representing the shape of each character is stored in the memory using the mask programmable technique. When a particular group of memory cells is read, the logic content of the cells is used to generate the dot pattern of the required video signal for the character. A widely used application of this type of memory is in microprocessor systems, where the manufacturers programme (known as firmware), is required to permanently reside in the system to provide control instructions for the microprocessor.

Mask Programmable Read Only Memories are therefore used for large volume production of a manufacturer's programme or data. Providing large volume production is maintained this type of memory is inexpensive, but it is totally unsuitable for low volume production or prototype work. If low volume production of a Read Only Memory is required then one of the following types is normally used.

4.6 Programmable Read Only Memory

There are several types of Programmable Read Only Memory, some using MOS devices and others bipolar devices. In the example shown in *Figure 4.4* a semiconductor diode which is connected in series with a fusible link is used as the storage element. When the memory array is purchased in a single integrated circuit it is wired with all the fuse links intact, giving a logic '1' at the output of each cell when the appropriate address line is at a logic '1'.

The data or programme is written into the memory using a special PROM programmer, which applies a high voltage pulse, normally in the order of 12–26 V, for a specified period of time to the selected cell. This action 'blows' the fusible link between the address line and the anode of the diode. If the cell is now read by applying a logic '1' to the appropriate address line the voltage can no longer pass to the output data line. This results in a logic '0' being read from the cell.

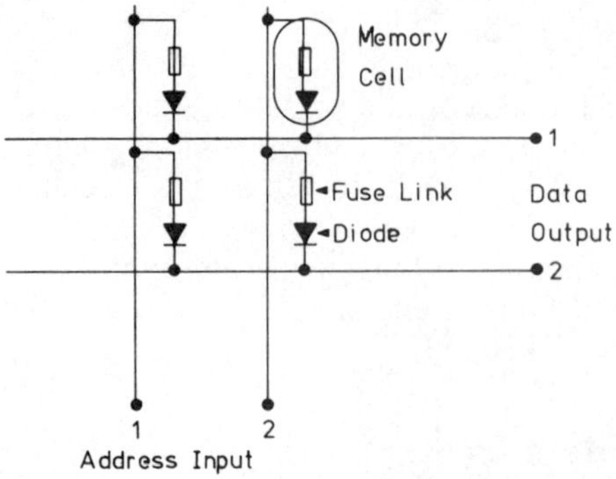

Figure 4.4 PROM diode

By blowing the fusible links in a set pattern, determined by the data or programme to be stored in the memory array, the manufacturer can 'programme' the memory to his own requirements. Normally this operation takes place with the use of a small microprocessor based system operating a PROM programming printed circuit card. Once a fusible link has been destroyed it is impossible to replace it, and hence the only time information can be written to the PROM is during the initial programming sequence. When the PROM is placed in a digital operating system, the pulses used to read the contents of the PROM do not affect the fusible links because they are of a much lower voltage than the pulses used in the PROM programmer.

Erasable Programmable Read Only Memories

This type of memory has the ability to be reprogrammed, but the integrated circuit must be removed from the operating system to enable reprogramming to take place. The erasing process is carried out by exposing the chip to ultra-violet light for a specified period of time. A transparent window on the top of the package to allows the light to pass through.

The result of the erasing process is to set the output of each memory cell in the integrated circuit to a logic '1'. The IC is then inserted into a special EPROM programmer which is under the control of a small microprocessor. The logic '0s' of the new programme are then written into the desired locations of the memory array. Once the memory has been reprogrammed the transparent window must be covered to prevent the ultra-violet light in the sun's rays from erasing the memory over a long period of time. It is standard practice to insert details of the data stored in the memory on the cover of the glass window. The integrated circuit can then be replaced in the operating system where it functions as a conventional Read Only Memory.

4.7 Memory addressing

In the previous examples of storage elements, the descriptions have been associated with a single storage cell, i.e. the storage and retrieval of a single bit of data. In practice it is necessary to store many thousands of bits. We would for example require more than 6000 memory cells for the storage of a complete page of Teletext data, and a computer uses many thousands of storage cells. Even a simple system having a 4K byte memory requires 32000 memory cells if it is using an eight-bit data bus.

Individual memory cells must therefore be arranged in a logical order so that the system working with the memory can select a particular memory cell quickly and efficiently, and then be able to write or read data to or from the cell. This process is known as *memory addressing*.

Memory cells are connected in an array of rows and columns. The number of cells in a particular row or column is normally a direct binary multiple, i.e. 2, 4, 8, 16, 32, etc. *Figure 4.5* shows a 4 × 4 matrix containing a total of 16 cells. Assume it

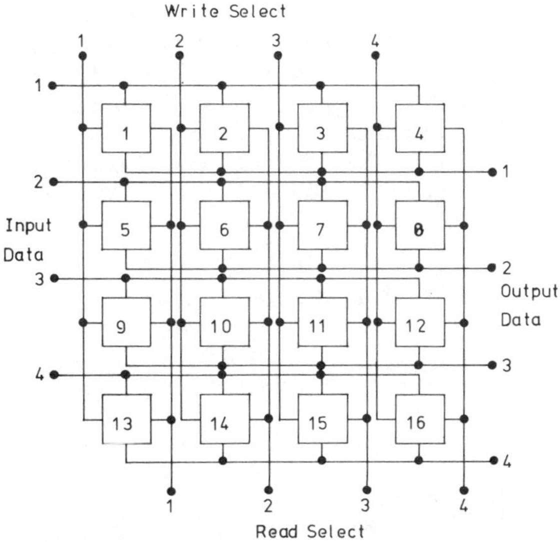

Figure 4.5 Memory array

is required to write some data to cell 11. This cell is contained in horizontal row 3 and the data to be written to the cell would be connected to input data point 3. Once the data is present at the input to cell 11, write select input 3 is made active, and the incoming data is stored. Data is not written into cells 9. 10 and 12 in the same row because the write select lines feeding these cells have not been activated. Using the same technique, data is not written into cells 3, 7 and 15 in the vertical column because the data input lines are all open circuit, so although the cells have been write selected the data in the cells remains unaltered.

Selecting a cell for a memory read function is carried out in a very similar way. Again if it is assumed that the data from cell 11 is to be read then read select input 3 would be activated and the data from cell 11 would appear at output data point 3. Although the outputs of all cells in this row are interconnected only data from cell 11 appears at the output because the other cells in the row have not been read selected.

Arranging memory cells in this manner is the most economical way of organizing the input and output data lines and also the read and write select lines. If the size of the array is increased to a 32 × 32 matrix the memory will contain a total of 1024 cells. A memory of this size would require 32 write select, 32 read select, 32 input data connections and 32 output data connections. A total of 128 connections to the matrix. This is clearly far too many for a single integrated circuit and some means of condensing this number is required if the semiconductor memory is to be viable.

Only one data input and output to the array is required at any instant in time, and with the addition of the necessary switching circuitry inside the IC the number of data input and output connections can be reduced to 2, leaving 32 read and 32 write lines to address the 1024 cells.

4.8 Binary coded addressing

The decimal number 1024 can be represented in binary using a maximum of 10 BITS (i.e. a count from 0000000000 to 1111111111). If the address of the cell is coded in binary form, then a maximum of 10 address lines will be required to select any single cell in the 32×32 memory array. Five binary inputs will be required for the row address of the matrix and a further five for the column address. This will reduce the number of connections to and from the integrated circuit to a manageable level, but will complicate the internal circuitry of the IC because the binary coded address will have to be decoded by the integrated circuit before the requested cell can be selected. *Figure 4.6* shows the block diagram of a 16×16 memory array together with the necessary address and control circuits.

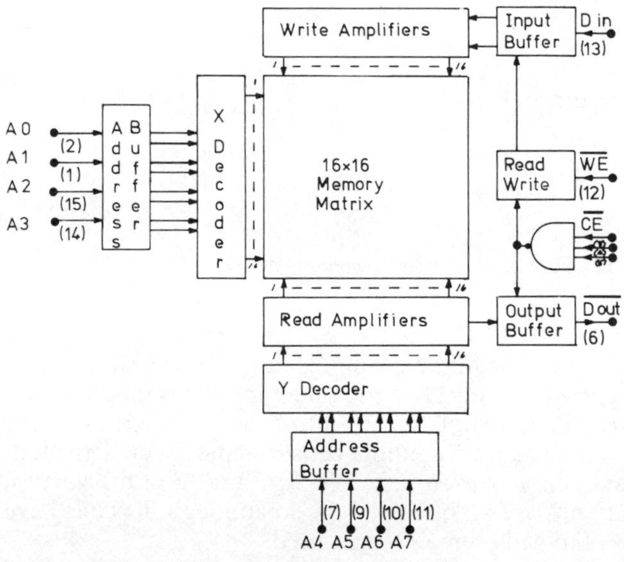

Figure 4.6 Block diagram of 256 bit memory chip

Address inputs A0 to A3 contain the binary information which determines the row address of the memory cell to be selected. Similarly address inputs A4 to A7 contain the binary information of the column address. Each address input is coupled to a buffer which provides a buffered address output and also a buffered NOT address output. These signals are then coupled to address decoders which convert the binary inputs to decimal outputs which in turn select the required row and column in which the selected cell resides.

Once the address signals are stable and the desired cell selected, a write enable signal is applied to the $\overline{WE}$ input. If this signal is set at a logic '0' then the internal logic of the IC will enable the memory cell to be written to, and if the $\overline{WE}$ input is at a logic '1' then the internal logic selects a memory read function. A memory read or memory write will only take place if the integrated circuit is enabled. This is achieved by making the chip enable inputs on pins 3, 4 and 5 assume a logic '0'. If any of these three inputs rise to a logic '1' then the write or read function will be inhibited. This facility gives the designer of a memory system the ability to connect a large number of memory circuits to common address lines, and select which integrated circuits will respond to the address and write enable signals by using separate chip select signals.

The memory addressing techniques that have been described so far have been concerned with storing a single bit of data in an array of memory cells of a particular size. In practice it is required to store a large number of bits simultaneously, for example a character of Teletext is represented by 7 bits of data and it would require 7 separate memory cells, all activated at the same time, to store the complete character. To achieve this a system known as parallel addressing is used.

4.9 Parallel addressing

Parallel addressing is a technique used to enable the storage of more than one bit of data at a single memory address. Using *Figure 4.7* as an example we will examine the storage of a 4 bits.

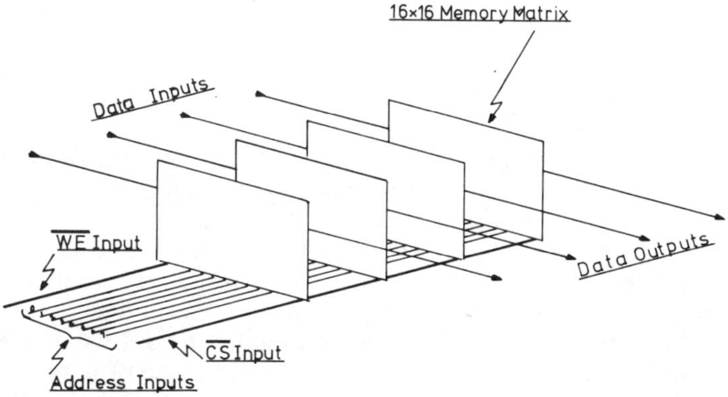

Figure 4.7 Parallel addressing

Each memory IC in this example has a 16 × 16 memory array exactly the same as the integrated circuit of *Figure 4.6*. All the address, write enable and chip select inputs are connected in parallel so that a single signal on any of the input lines is coupled to all four integrated circuits. The data input and output connections remain separate, hence the complete 4 × 16 × 16 bit store will have the following connections.

4 data input lines.
4 data output lines.
1 $\overline{\text{write enable}}$ input.
1 $\overline{\text{chip select}}$ input.
8 memory address input lines.

This makes a total of 18 input/output connections to a group of memory circuits capable of storing 1024 bits of data (excluding power supply requirements).

If it is required to store a four-bit nibble of data at location 1 in the store the following sequence of events would take place.

1. The four bits of data would be applied to the four data inputs.
2. Memory address (1) would be selected by placing 00000001 on the address lines. (MSB first).
3. The $\overline{\text{write enable}}$ input would be set to a logic '0'.
4. The $\overline{\text{chip select}}$ input would be set to a logic '0'.
5. The input data would now be written to cell No. 1 in each IC.
6. The $\overline{\text{chip select}}$ input is set to a logic '1' to inhibit the ICs.

The memory cells considered in this chapter so far have been electronically operated, and in every case the contents of the memory will reside in the equipment, forming an integral part of the operating system's electronics. In many situations it is desirable to store data using a more transportable, non-volatile medium. This will enable the long term storage of data and also enable data stored by one system to be read by another. This method of data storage is used by microprocessors and large computers but, at the time of writing, it is not used by the digital circuitry associated with television receivers.

It is inevitable, as technology progresses, that pages of Teletext, TV games and other television based data will be stored using a magnetic medium. For this reason tape and disc storage have been included in this chapter on storage devices.

4.10 Magnetic tape storage

A very compact, cheap and convenient storage medium for large amounts of data or programme material is undoubtedly the cassette tape. The cassette format has been developed and tested over a long period of time in the audio market, along with the transport system and the electronics with which it is used. With such a convenient and well tried storage medium available, the designers of microprocessor systems have seen little point in developing a low cost alternative, and in practice it has proved to be sufficiently reliable. Unfortunately it does have one major disadvantage and this is the access time to any individual piece of data.

With an electronic memory it is quite possible to search, find and read a single piece of data in a few microseconds, but this is not possible with the encumbrance of the mechanical tape transport system used with the cassette recorder. Information stored on cassette tape normally consists of large blocks of data which can be 'played back' and subsequently written into an electronic memory. Once the data has been transfered from the tape into an electronic memory it can then be manipulated at the high speeds at which the electronic system operates.

If the data is to be stored on a cassette tape two major electronic processes have to take place.

1. The data must be converted from a digital to an analogue signal because the cassette recorder is designed to process an analogue audio signal and not the two state digital signal.

2. The speed at which the data is transferred from the electronic memory to the cassette tape must be very slow if it is to fit into the limited audio bandwidth of the cassette recorder's electronics and recording medium, (normally no more than 10 KHz for a low cost machine).

Several other problems also have to be taken into consideration when using cassette tape. These are as follows:

1. The tape recording system has poor signal to noise ratio.
2. If poor quality tape is used there may be 'dropouts' (sections of the tape where the magnetic particles have been removed because of tape wear).
3. Varying head to tape pressure caused by the limitations of the tape transport will produce variations in signal level during both the record and playback modes.
4. The tape speed may vary considerably during both record and playback, and also between one tape recorder and another. This will result in a change of data frequency during the playback mode.

Any interface circuitry must be able to compensate for all these inherent disadvantages and still be able to produce error free recording and playback.

It is quite possible that the audio cassette will be used for the long term storage of Teletext data. The cost of LSI components is falling rapidly, and with the aid of a microprocessor, a small ROM, and a minimum of interface circuitry it is possible to read the contents of a Teletext decoder memory and transfer the data to tape at a very low cost.

4.11 Magnetic disc storage

Magnetic or floppy discs have been in use for several years with microprocessor systems and they are currently one of the most popular methods of storing large quantities of data or programmes for micro-computer systems. They are however considerably more expensive than the cassette tape system and their use would be restricted to business users of Teletext or by the home user who would require storage of text orientated computer data. Floppy discs have a number of highly desirable features which include:

1. Transfer of data between the system and storage medium is faster than cassette tape.
2. The access time to any given piece of data is much faster than cassette tape.
3. The discs are easy to handle and reasonably cheap.
4. Each disc is capable of storing between 80 K and 1 M byte of information depending on its size and packing density.

A floppy disc is shown in diagrammatic form in *Figure 4.8*. It consists of a circular flexible disc which has been coated on one or both sides with a magnetic recording medium similar to that used in cassette or reel to reel tape. This circular disc is permanently housed in a square envelope made of a flexible plastic based material. A central hole is cut in either side of this envelope to allow the disc drive mechanism to grip the centre of the disc so that it can rotate inside the envelope. Also cut into each side of the plastic envelope is a horizontal slot which allows a magnetic record/replay head to gain access to the recording medium.

The head can record or replay a number of concentric tracks on the disc, each track being separated from its neighbour by a small guard band. Using a 5.25 inch disc as an example there will be a total of 35 tracks, each track containing a maximum of 2.304 K bytes making a total of 80.6 K bytes maximum per disc using single sided single density recording techniques. The rotational speed of the disc is 300 rev/min and this enables the head to read or write data at a speed of 125 K bits per second. If this size of disc was used for the storage of Teletext pages the disc would be capable of writing or reading the data at a speed of 17 pages per second. A single disc of this size would be able to store 79 pages of text.

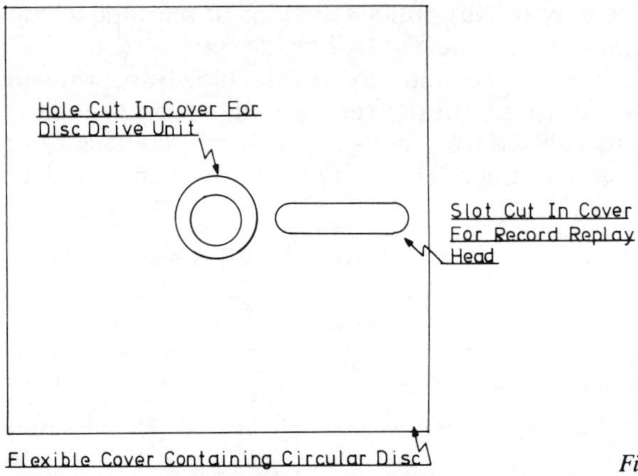

Figure 4.8 Floppy disc

A floppy disc storage unit requires a microprocessor based operating system to provide the necessary control signals which enable the transfer of data to and from the disc. The mechanics of the disc unit, although relatively simple, is quite expensive (from approximately £200). To store Teletext data on disc the user would require the heart of a complete microprocessor based computer, plus the Teletext decoder. The only part of this system which would be in the price range of the domestic user is the floppy disc itself which is approximately £2. Until these costs come down dramatically the only users of this type of system are going to be commercial firms which require Teletext data to be stored on a regular basis. Possible examples are financial houses, banks, insurance companies, the larger hotels and possibly educational establishments.

The range of memories described in this chapter form the basis of the storing techniques which are used in the range of digital equipment that is currently available. A brief description of each medium has been given in order to familiarize the reader with the range of storage devices and techniques that are currently in use. Detailed descriptions of individual storage techniques will be made later, as and when each technique is met.

Chapter 5

Display devices

It has been felt necessary to devote a complete chapter of this book to display devices because they already form an important part of the digital circuitry used in TV receivers. As technology progresses, bringing down the price of display devices, they will be increasingly used to relay information to the user. The c.r.t. is, of course, the most important display device in any TV receiver and its primary function is to convert the analogue television waveform into a moving colour picture.

The c.r.t. is equally happy with a digital input and can be used to display any amount of information produced by a digital system such as a computer, Teletext or Viewdata decoder, or a TV games unit. It is not the intention to cover the range of cathode ray tubes in this chapter because it is assumed that the reader is fully conversant with these devices and their techniques. An introduction to the Video Display Unit (VDU) is included however, because it is becoming increasingly used with the Viewdata system, and will form a very important part in the marriage between the home computer and the TV receiver which is rapidly developing as the computer market expands.

The aim of this chapter is to explain the theory and application of several digital display devices which have become available to the manufacturers of TV receivers. Those which are widely used are the Light Emitting Diode (LED), the Liquid Crystal Display (LCD), and the use of these two to form the seven segment display. All of these devices have become very popular because they interface directly with a digital system, and this interfacing can be carried out with the minimum of additional electronics, thus keeping the cost down.

5.1 The light emitting diode (LED)

The LED is a semiconductor device consisting of a single *p-n* junction. The semiconductor chosen is of the direct gap type such as gallium arsenide (Ga As). This type of semiconductor is used because when the free electrons in the *n*-type semiconductor re-combine with the positive holes in the *p*-type semiconductor radiation in the visible and infra-red spectrum is produced. Diodes which emit light energy in the visible spectrum are of course used as display devices, and those which emit infra-red radiation are widely used in TV receiver remote control transmitters.

A cross section through an LED is shown in *Figure 5.1*. Light is only emitted at the *p-n* junction, where the holes and electrons re-combine, consequently the junction must be very close to the surface of the diode so that as much light as possible can be radiated. The material used for the anode connection must not obscure this radiated light, and it must be bonded in such a way to the *p*-type material so that as much light leaves the anode surface of the diode as possible.

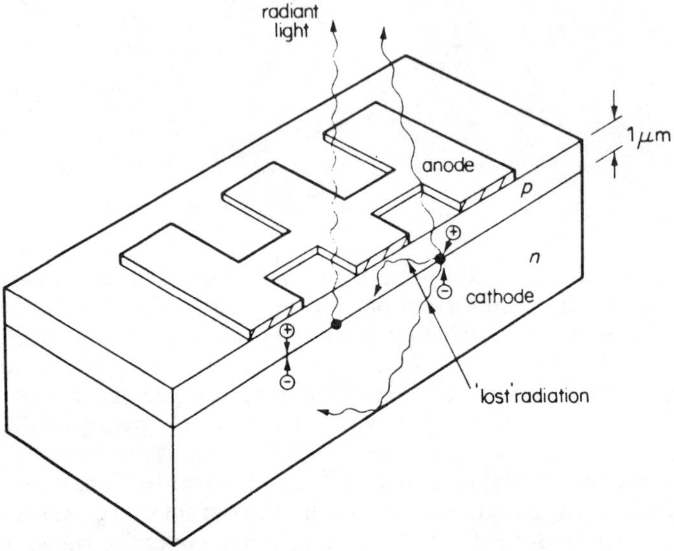

Figure 5.1 Cross-section of LED

One of the greatest problems in manufacturing LEDs is getting all the radiated energy from the electron hole recombination to leave the semiconductor material. The amount of light that leaves an LED is determined by two major factors:

1. The transparency which the semiconductor material has to the emitted wavelength of light.
2. The high refractive index of the semiconductor material.

The first of these two causes the light radiated from the *p-n*junction to be absorbed by the semiconductor before it has chance to leave the surface of the diode. The second causes emitted light to be reflected from the surface of the diode back into the semiconductor. This is shown in *Figure 5.1*. In a material such as gallium arsenide phosphide (Ga As P), any light striking the surface at an angle greater than approximately 17° to the perpendicular is reflected back into the diode. This problem is further aggravated because some of the light that is perpendicular to the surface is also reflected back, leaving a small margin of approximately ±1–17° in which light produced at the junction can actually leave the diode.

The colour of the light emitted depends on the semiconductor used. Popular colours are red, orange, yellow and green, and each diode normally has an inbuilt filter to enhance the wavelength of the light emitted. The light output of an LED is directly proportional to the forward bias current that passes through it, and this current can be quite high (up to 40 mA). In portable equipment this high current can become a considerable problem if battery economy is to be maintained. One solution is to provide current pulses of a high value to the diode. This will produce a high light output, yet keep the mean value of the current at an economical level.

All LEDs require some form of current limiting device included in their drive circuits. If a diode were connected directly across a supply voltage the resulting high forward current would destroy it completely. It is therefore common practice to include a series resistor which limits the current flowing through the diode, thus protecting it from damage. The value of this resistor can be calculated by using the formula:

$$R = Vs - (Vf/If)$$

Where R is the value of the resistor in Ω.
Vs is the value of the supply voltage.
Vf is the value of the forward voltage across the diode.
If is the forward current flowing through the diode.

Values of If and Vf are dependant on the type of diode and are normally determined by the colour of the emitted light. For example red LEDs operate between 1.6 and 2.5 V with currents of 5–25 mA respectively, and orange, green and yellow types operate between 2 and 3.5 V with currents of between 10 to 40 mA respectively. Restricting the forward current of the diode is of course very important, and some manufacturers include a series resistor in the diode encapsulation so that current limiting becomes automatic.

One feature of LEDs which is not commonly realized is that they have a very low reverse breakdown voltage of about 3–11 V PIV. When the diode is used with a d.c. supply this characteristic does not present any problems because the diode is never in a situation where it becomes reversed biased. However, if the diode is being used in conjuction with an a.c. supply added reverse protection is sometimes required. This added protection can come in the form of a reverse parallel connected diode as shown in *Figure 5.2*. An alternative form of protection is

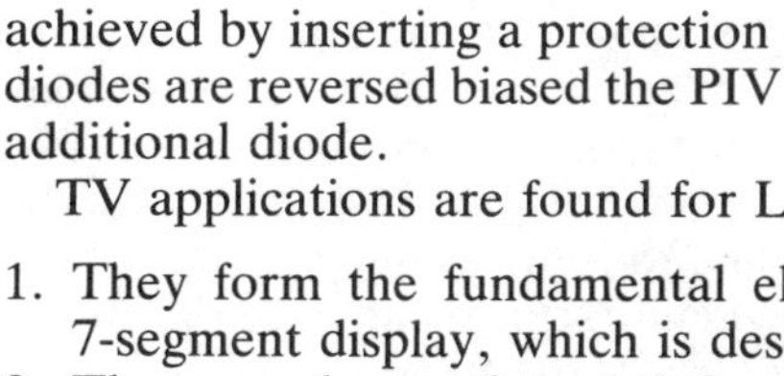

Figure 5.2 LED with series resistor and parallel diode

achieved by inserting a protection diode in series with the LED so that when both diodes are reversed biased the PIV across the LED is limited by the insertion of the additional diode.

TV applications are found for LEDs in two major areas.

1. They form the fundamental element of more complex displays such as the 7-segment display, which is described later in this chapter.
2. They are also used as an infra-red light source in remote control transmitters.

If an LED is used in a TV receiver the disadvantage of heavy current consumption becomes of little consequence because the LT supplies in the receiver will be more than capable of supplying the additional current required for the display. In remote control transmitters the high power consumption presents a few problems in the design of appropriate drive circuitry, but the effect of high power consumption is reduced by using pulse coded modulation for the transmission of the remote control instructions. With a PCM signal fed to the diode the average continuous current drawn by the LED is reduced to tolerable levels. When this is added to the fact that the LED is only operational when a remote control command is transmitted, total power consumption of the battery powered unit is very small indeed.

5.2 The liquid crystal display (LCD)

The Liquid Crystal Display (LCD) is not a semiconductor device at all and unlike the LED it does not radiate light but polarizes it.

A typical field effect LCD consists of a thin film of organic liquid, which is approximately 6–10 microns deep, and is contained in a small cell sandwiched between two plates of glass. See *Figure 5.3*. On the internal face of each sheet of glass is a very thin transparent conductive coating which is etched in the shape of the digit, character or pattern which is to be displayed. Whenever a potential is applied to a pair of electrodes which have liquid crystal between them an electric field is set up across the liquid filled in the cell. When the liquid crystal is subjected to this field it will polarize the light waves passing through the cell.

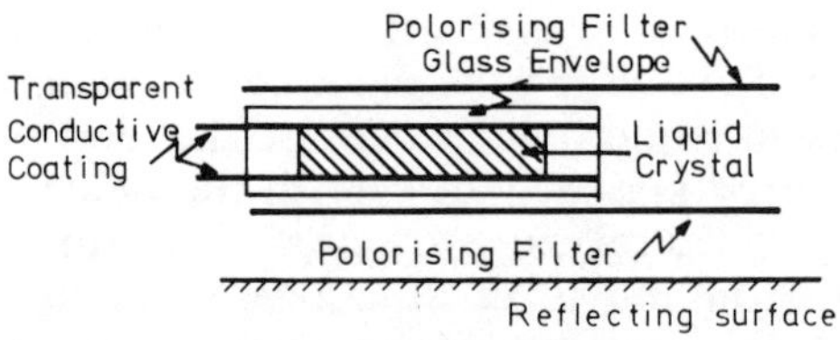

Figure 5.3 Cross-section of LCD

This characteristic of the liquid crystal does not enable it to absorb light by itself, but if the crystal is viewed through a pair of polarizing filters on the front and rear of the cell surfaces then the polarization of the light becomes visible. This situation is normally enhanced by placing a reflective surface behind the display so that light passes through the polarizing filters and the cell once, and is then reflected back through the cell a second time before reaching the viewer.

With the polarizing films above and below the display correctly orientated, a transparent image on an opaque background, or an opaque image on a transparent background can be produced, giving a white on black or black on white display respectively. In nearly all practical applications the polarizing films are arranged to produce a black display on a transparent background. The polarizing filters can be incorporated on the upper and lower plates of glass to form an integral part of the display device, or they can be separate so that the manufacturer can have the choice as to whether the characters or the background absorb the light.

As LCDs are made of glass they should be handled with respect, and should not be unduly stressed or twisted. Sometimes when handling displays which have the polarizing filters incorporated on the glass plates it is possible to produce a blue/black herring-bone pattern by gently squeezing or twisting the display device. This is caused by pressure on the liquid crystal rearranging its alignment, and consequently polarizing the light passing through the display. This action does not damage the display, and the effect will disappear within a few seconds after the pressure has been removed.

As the liquid crystal display is a field effect device its current consumption is very low indeed making it ideal for battery operated equipment. The equivalent circuit of a liquid crystal cell is shown in *Figure 5.4*. R1 and R2 represent the lead in resistance of the transparent coating above and below the cell and these values are typically 100 K Ω. R3 is the series leakage resistance through the liquid crystal fluid and C1 is the self capacitance between the parallel electrodes above and below the LCD cell. R3 is greater than $10^7 \Omega$ and C1 is about 100 pF. It can be seen that the liquid crystal pocket represents a very high impedance device which will pass very little current indeed.

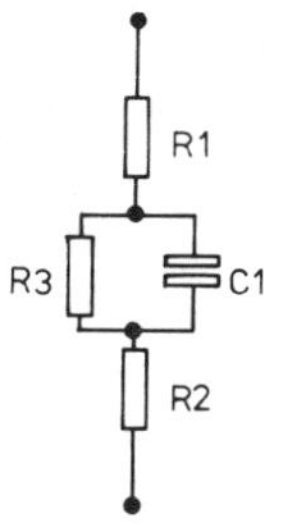

Figure 5.4 Equivalent circuit of LCD cell

Unfortunately the LCD suffers from the disadvantage in that it requires a symmetrical a.c. drive voltage. The cell will work with a fixed d.c. voltage applied across the two electrodes, but over a period of time this will cause straining of the cell walls and will shorten the life of the display. Consequently a square wave a.c. supply is nearly always used for operating LCD displays. This a.c. voltage can vary between 4–17 V r.m.s. and the display will start to operate at between 1.5–2 V r.m.s. A graph showing the absorbtion of an LCD segment *verses* voltage is shown in *Figure 5.5*.

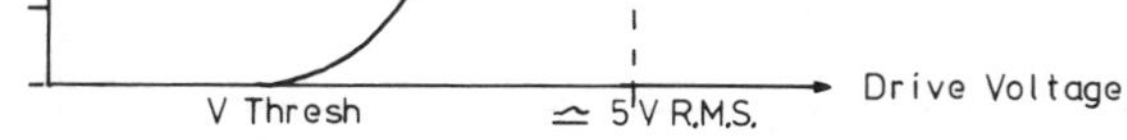

Figure 5.5 Absorption/voltage characteristic of LCD

The major aspect of working with these display devices for the service engineer will be understanding how the drive circuitry to the cell operates. As stated earlier the display requires an a.c. drive of between 4–17 V r.m.s. and at first thought this drive requirement may appear to be difficult to produce if the circuitry is to use digital electronics. In fact it turns out to be a very simple problem to solve.

An oscillator which produces the a.c. drive waveform generates a square wave of between 30 and 120 Hz with an equal mark to space ratio, and this clock frequency is applied to one electrode of the LCD cell and also to one input of a 2 input XOR gate, see *Figure 5.6*. The second input to this XOR gate is the select waveform which determines if the cell is enabled or not. The output of the XOR gate is connected to the other electrode of the LCD cell.

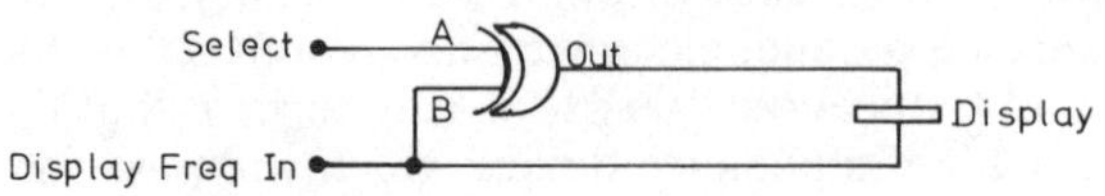

Figure 5.6 LCD drive circuit

When the select input is at a logic '0' the XOR gate will produce at its output a square wave which will be in phase with the clock frequency (refer to the truth table shown in *Figure 5.7*). Consequently the LCD cell has two a.c. waveforms of the same phase and amplitude applied to its upper and lower plates. The electrostatic field developed between the plates will therefore be zero and the cell will be transparent.

A	B	Z
0	0	0
0	1	1
1	0	1
1	1	0

Figure 5.7 Truth table 2 i/p XOR gate

When the select voltage rises to a logic '1' the XOR gate will invert its B input. This produces a square wave at the output of the gate that is 180 degrees out of phase with the clock signal. By examining the waveforms shown in *Figure 5.8* it can be seen that a polarizing voltage is now placed across the segment, and this a.c. voltage will alternate symmetrically above and below 0 V. The cell will therefore polarize the light passing through it at this a.c. frequency and when the cell is viewed through the polarizing filter it will appear dark.

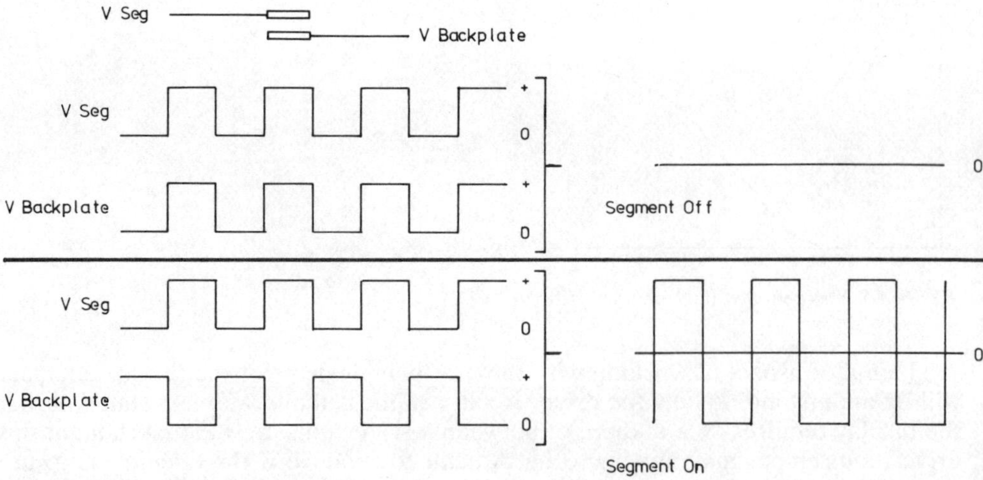

Figure 5.8 LCD waveforms

All that is required then to operate a liquid crystal display is a low frequemcy asymmetrical square wave to provide the drive waveform and an XOR gate to switch the clock frequency in the correct phase to operate the display.

5.3 The 7-segment display

LEDs are widely used either singly or in pairs to replace indicators which until recently had been formed by tungsten and neon lamps. LEDs really come into their own when forming the 7-segment display. As its name implies a seven-segment display consists of seven bar segments arranged to produce the shape of the letter eight as shown in *Figure 5.9*. Each segment of the display contains one or more

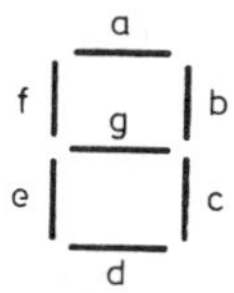

Figure 5.9 A 7-segment display

LEDs which are individually connected to the output pins of the display device. By applying a suitable drive voltage to these pins each segment of the display can be illuminated separately, and when these segments are used together the numeric characters 0 to 9 can be produced, see *Figure 5.10*.

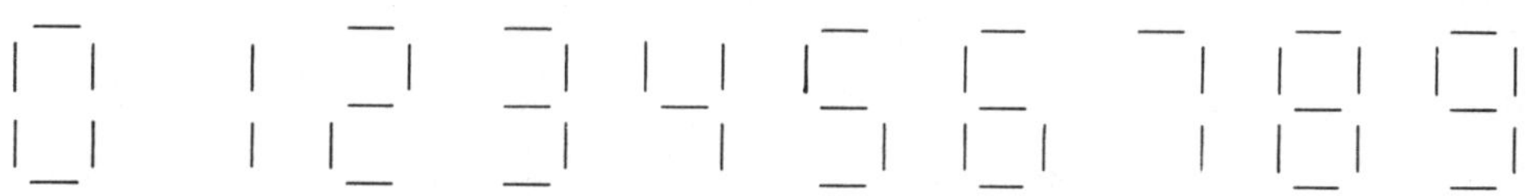

Figure 5.10 0–9 on 7-segment display

LEDs are available in many different formats and therefore the 7-segment display is available in several forms. For example a display may have the equivalent circuit such as the one shown in *Figure 5.11*. Each segment is formed by a single LED having a current limiting resistor in its anode circuit that is connected to a common positive supply. When the cathode of each LED is returned to 0 V the diode emits light illuminating the appropriate display segment. If the cathode is left open circuit, or returned to +5 V, then no current flows through the diode and the segment is switched off. A7-segment display with this circuit diagram is known as a common anode display because the anodes of all seven LEDs are taken to a common positive supply, and their cathodes are connected to the display drive electronics. If a logic gate was used to drive each segment of the display, then a logic '0' at its output would illuminate the segment and a logic '1' would switch it off.

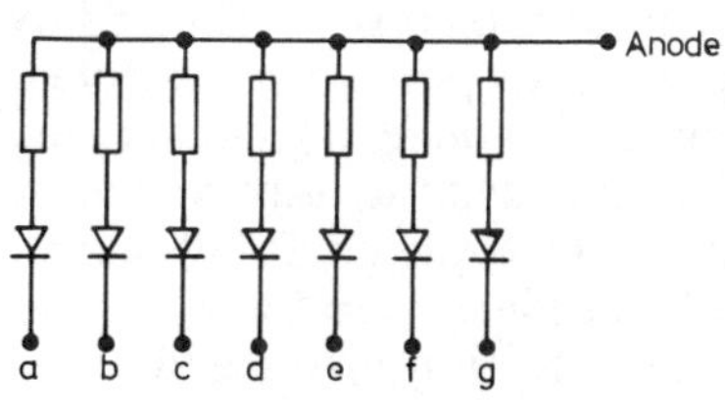

Figure 5.11 Equivalent circuit of 7-segment display

It is quite possible to produce common cathode 7-segment displays where each cathode of the segment LED is returned to 0 V and the drive electronics is connected to the anode of each LED. In this situation a logic '1' from the drive electronics would illuminate its respective segment and a logic '0' would switch off the segment. It is therefore important when replacing 7-segment displays to make sure that the faulty device is replaced with one of the correct type, otherwise the polarity of the signals provided by the display drive electronics will be incorrect.

Common anode or common cathode 7-segment displays may or may not include the series current limiting resistor in the anode of each segment LED. Most equipment manufacturers prefer to use displays that have the limiting resistors because it reduces the size of the PCB and also simplifies the construction of the electronics. In some applications the resistors may be mounted externally, either as a row of discreet components or as a resistor pack. The resistor pack may take the form of a single rectangular encapsulation with 14 pins or a standard 14 pin DIL IC package.

A four bit BCD signal is normally used to represent the digits to be displayed on the 7-segment display, but unfortunately a BCD signal does not represent the pattern for the number to be displayed. *Figure 5.12* shows the four-bit BCD code and its corresponding pattern of '0s' and '1s' required to generate the number for the display. The 7-segment display code shown is for a common anode display, and if a common cathode display were to be used the display code would have to be complemented.

B. C. D.				7 Segment Code						
2^3	2^2	2^1	2^0	a	b	c	d	e	f	g
0	0	0	0	0	0	0	0	0	0	1
0	0	0	1	1	0	0	1	1	1	1
0	0	1	0	0	0	1	0	0	1	0
0	0	1	1	0	0	0	0	1	1	0
0	1	0	0	1	0	0	1	1	0	0
0	1	0	1	0	1	0	0	1	0	0
0	1	1	0	0	1	0	0	0	0	0
0	1	1	1	0	0	0	1	1	1	1
1	0	0	0	0	0	0	0	0	0	0
1	0	0	1	0	0	0	0	1	0	0

Figure 5.12 BCD to 7-segment decoder truth table

Conversion of the 4-bit BCD code to the 7-bit 7-segment display code is carried out in a combinational logic circuit known as a BCD to 7-segment decoder. This circuit is rather complex and a detailed description is beyond the scope of this text. Integrated circuit manufacturers have been able to reduce the size of the BCD to 7-segment decoder to a single IC package which not only includes the decoder but also the driving gates which are required to sink or source the current required by each segment of the display. Consequently it is possible to produce a complete display circuit which will show the decimal equivalent of a 4-bit BCD signal using just two devices, these being the BCD to 7-segment decoder and the display device.

So far we have only considered the drive requirements for a single display device, but there are many situations where several 7-segment displays have to be used to produce for example a three digit number. When this situation arises additional complications appear. Each 7-segment display will have nine input pins (one for each segment, one for a decimal point and one for either the common anode or common cathode connection). Using three such displays, a total of 27 connections would be required between the display and its drive electronics. Furthermore, three BCD to 7-segment decoders will be required, one for each display. As the number of 7-segment displays increases, the problem begins to become a little unmanageable because of the large number of connections between the display and the electronics. For example five digits would require forty-five connections. To reduce the number of conductors between the display electronics and the display devices, and also to reduce the complexity of the drive electronics required, a system known as multiplexing is used.

5.4 Multiplexing 7-segment displays

Examine the display circuit shown in *Figure 5.13*. This shows three 7-segment displays (without a decimal point), connected using a multiplexed address scheme. Each segment is connected to the same segment of its neighbouring display, making a total of 7 segment connections for all three display devices. The common anode connections of this diagram are kept separate making a total of 10 connections to the display panel as opposed to 24 connections if the displays were not multiplexed.

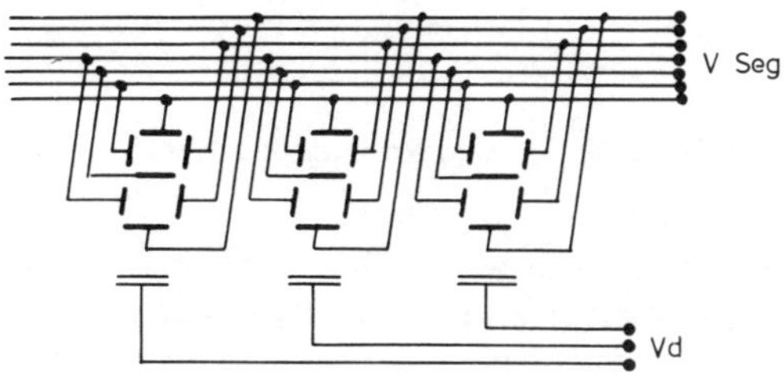

Figure 5.13 Multiplexed display

Assume the pattern for 0 was set up on the V Seg input lines. In this situation segments a, b, c, d, e, f would all receive a logic '0' and segment (g) of each display would receive a logic '1'. When the three common anode connections are taken to logic '1' then the number 000 would be displayed. Under these conditions the drive gates operating the 7 segments would have to be capable of providing a maximum of three times each individual segment current in order to power the three display devices successfully.

It has been mentioned earlier that LED currents are quite high and each segment drive gate would have to sink approximately 60 mA of current for each of the V Seg inputs to the display. Obviously some means of reducing the current requirement of the gate is necessary. This is achieved by scanning each display device in a similar way as a TV line is scanned on a CRT. A logic '1' is first applied to the anode connection of the left-hand display, while logic '0s' are applied to the common anode connections of the remaining two. This means that the left hand display is operating while the remaining two are not. Consequently the V Seg drive gates are only providing current for one 7-segment display

After a short period of time, say 20 mS, the left-hand display is switched off and the centre display is operated by placing the pattern of 010 on the Vd input to the display panel. The centre segment is now operational while the outer two are switched off. Finally the right-hand segment is illuminated by using the code 001 at the Vd input, and the other two are switched off. The process now repeats itself.

Providing the scanning speed on the Vd input to the display panel is above the persistence of vision of the human eye, for example 50 scans per second, then the eye will not be able to detect that only one display device is working at a time, but it will see a continuous display. Unfortunately this system will reduce the brightness because each display is only operating for one third of its maximum possible time.

Scanning the three displays in this manner also enables separate numbers to be shown on each display device. For example assume the numbers 1, 2, 3 are to be shown. First of all the code for the number 1 is set up on the V seg input to the display panel and 100 is applied to the Vd input of the panel, see *Figure 5.14*. The left-hand display will now illuminate showing the figure 1. The drive voltage for the anodes then changes over from 100 to 010 so that the centre display is driven. The code on the V seg input to the display panel also changes to the code for the number 2. Hence the centre display now shows the number 2. Finally the Vd input changes to 001 to operate the right hand display while the V seg input code changes for the number 3. As with the previous example, once a horizontal scan has been completed the process repeats itself.

Vd 1	2	3	V Segment a	b	c	d	e	f	g
1	0	0	1	0	0	1	1	1	1
0	1	0	0	0	1	0	0	1	0
0	0	1	0	0	0	0	1	1	0

Figure 5.14 Sequential codes for 1, 2, 3 and Vd input for display panels

Multiplexing signals for 7-segment displays is quite a complex process but it is necessary to reduce the number of connections between the display panel and the electronics to a reasonable level, and also to reduce the power consumed by the complete circuit. If the designer is not careful the total power required for a display of this type can be quite heavy. Examine the following situation. If each segment of the display consumes 20 mA of current then the maximum current drawn by the whole display will be 140 mA for the number 8. Providing the display panel was multiplexed 140 mA would be the maximum current drawn by the display panel at any time because only one display is operating at any point in time. If multiplexing were not used then it would be possible for the display to consume 140 mA × 3 = 420 mA or approximately 0.5 A of current.

If a number of 7-segment displays are multiplexed then each display will only be on for a small amount of time. This will reduce the overall brightness of the display as each segment will only be passing a fraction of the total possible current. This situation can be overcome by increasing the current drawn by each segment. The increased demands made on the power supply using this technique will still be small when compared with the total current consumed by an equivalent non-multiplexed display.

Multiplexing does not therefore just reduce the number of connections it also makes a significant contribution to the reduction in power consumption for the whole display, and the last of these is of considerable importance of the display is battery powered.

Liquid crystal 7-segment displays can also be multiplexed in the same way but unfortunately the multiplexing of LCDs is subject to the following restrictions.

1. The threshold point of their characteristic (see *Figure 5.5*) is not sharp and will vary with temperature.
2. The transient response for the LCDs is slow and to produce a display of an acceptable appearance the drive pulse must reappear before the effect of the previous pulse has decayed appreciably.
3. The high impedance of each LCD segment, together with the fact that each segment must be a.c. driven prevents the use of simple current controlled multiplexing techniques which have just been described.

The above factors limit the number of digits that can be multiplexed when an LCD display is used. The exact number of displays is subjective and will also depend on the normal viewing angle of the display. In general 4 digits represent a practical limit for reflective type displays. This figure can be increased by using transmissive displays which have a diffusing front surface to scatter transmitted light. This type of display does not have an improved contrast but can be used at a wider range of viewing angles which improves its apparent density.

There are many techniques and circuit configurations which will implement the multiplexing of 7-segment liquid crystal and LED displays. The circuit which is about to be described has been simplified to show the multiplexing of a single bit of input data to a three digit display, see *Figure 5.15*.

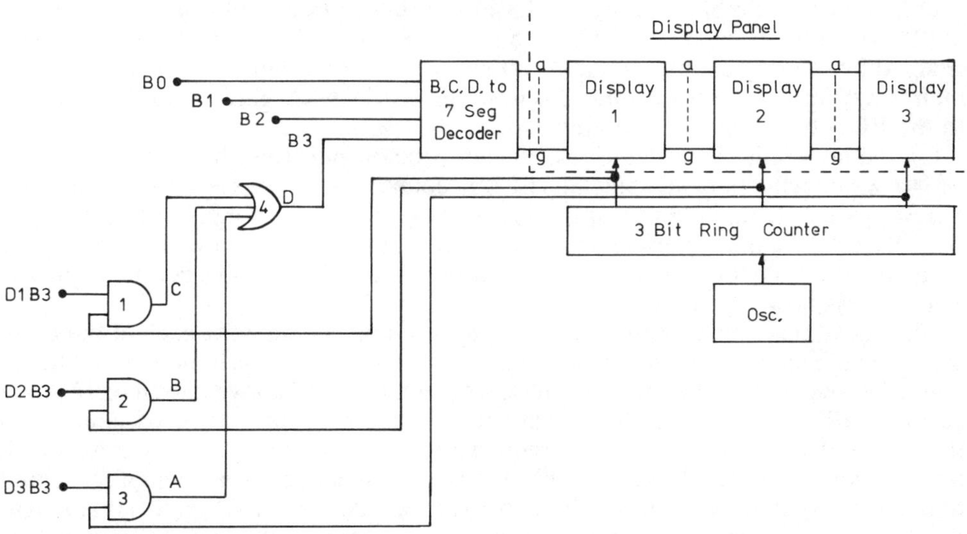

Figure 5.15 Multiplexed LED display

The 7-segment displays are labelled display 1 to display 3 and are of the common anode LED type. The (a) to (g) segments of each display are interconnected with each other as previously described and are driven from the output of a single BCD to 7-segment decoder. The common anodes of each display are connected to their respective outputs of a three stage ring counter which provides the desired scanning sequence for the three displays. The ring counter is clocked by a master oscillator, and its circuitry is so arranged that a logic '1' is shifted one place to the right when the counter receives a clock pulse from the oscillator. Hence after 3 clock pulses the logic '1' will return to the anode connection for the first display. The three BCD inputs for each display enter at the left of the diagram, and are labelled D1 to D3. The electronics for just one bit of each input has been included in the diagram for simplicity's sake.

Assume the ring counter is feeding a logic '1' to the common anode connection of display one, and therefore displays 2 and 3 will be inhibited by a logic '0' on their common anode connections. The logic '1' on the anode of display 1 is also fed to one input of AND gate 1. The other input to this AND gate is bit 3 of the BCD signal which is to operate display D1. The output of AND gate 1 at point C will therefore be the logic level present at the B3 input of this gate, while the other B3 inputs for displays 2 and 3 are inhibited because AND gates 2 and 3 are both receiving a logic '0' from the ring counter. These three signals are fed to a three input OR gate (4), which combines the A, B, C signals. As B and C are both at logic '0' the output from gate 4 at point D will again be the bit 3 input for display 1. This is then fed to the BCD to 7-segment decoder. An identical arrangement to gates 1, 2, 3 and 4 is duplicated for the B0 to B2 inputs of each display, and these will provide the switching for the B0 to B2 inputs of the BCD to 7-segment decoder.

To summarize, when the ring counter operates display 1, B0 to B3 of D1 input will be fed via the combinational input logic gates to the BCD to 7-segment decoder and then in 7-segment code to display 1.

On the next cycle of the clock oscillator the ring counter will produce an output of 010 which will switch off display 1 and switch on display 2. AND gate 2 is now enabled by the ring counter while AND gates 1 and 3 are inhibited by a logic '0' at their lower inputs. B3 for display 2 is now passed via AND gate 2 and the OR gate to the BCD to 7-segment decoder.

On the next pulse from the clock oscillator the output from the ring counter will be 001 which will enable display 3. The ring counter now inhibits gates 1 and 2 and enables gate 3. Data bit 3 of display 3 input can now pass through AND gate 3 and the OR gate to the seven segment decoder. On the next pulse from the clock oscillator the ring counter produces an output of 100 which will operate display 1. The process now repeats itself.

The above description describes the signal path of the MSB for each of the three inputs, and because the circuitry for the remaining three bits is identical the theory applied above will be the same. Using the system outlined above a total of 12 AND gates 4 OR gates 1 BCD to 7-segment decoder a simple clock oscillator and three-stage ring counter would be required to drive the three displays. Although this may appear to be a large amount of logic it is in fact very small when compared to the total number of gates in 3 BCD to 7-segment decoders. When this is added to the power saving, and the ease of construction, it heavily outways the alternative of using a BCD to 7-segment decoder for each display.

5.5 Video display units

Video Display Units (VDU) are becoming increasingly used as the Viewdata system expands and customers require a keyboard interface with the data bank. Secondly, as the home computing market expands, the BBC and the IBA are also expanding the transmission of computer programmes using Teletext as the transmission medium; a technique known as Telesoftware. The VDU will eventually become an integral part of the TV receiver, where the user will simply plug in a keyboard which will enable the receiver to be used as a computer terminal as well as a Viewdata and Teletext system.

A block diagram of a VDU is shown in *Figure 5.16*. This is, in fact, two devices in the one housing. A CRT and its associated video and timebase circuits provide a means of displaying the output from a digital system. The second section of the VDU consists of a keyboard which enables the user to input information to the digital system. This keyboard may be relatively simple with a complexity which is on par with a hand-held remote control transmitter, or it could be a full specification QWERTY keyboard such as those used with computer terminals.

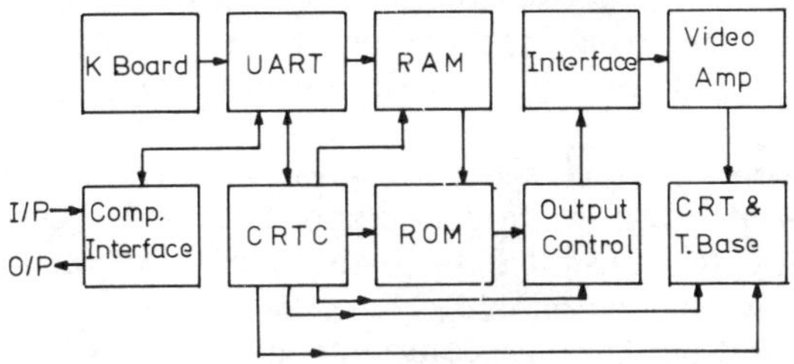

Figure 5.16 Block diagram of VDU

Unfortunately it is not normally possible to subdivide the electronics of such a system into the two clearly defined categories listed above. This is because nearly all the electronics of the VDU receives signals either directly or indirectly from a central source known as the Cathode Ray Tube Controller (CRTC). The CRTC can either be a group of combinational and sequential logic circuits or, as is more often the case, a small microprocessor system. The function of the CRTC is to generate control and timing signals which are used to process the data being handled by the VDU.

Two other groups of circuits are also found in every form of VDU. These are a Random Access Memory which is used to store the data that determines the information to be displayed, and a Read Only Memory which converts the data from the RAM into the character and graphic shapes that are displayed on the screen. The CRTC, RAM and ROM work closely together with each other and form the heart of any VDU.

Before continuing with the description of a VDU block diagram an important difference between the VDU and a conventional TV monitor must be established. It has already been mentioned that the VDU is different in that it contains a keyboard. The other major difference is that the information fed to the VDU is:

(a) A digital signal which can be received in either serial or parallel form.
(b) This signal is not transmitted to the VDU on a continuous basis.

The digital signal fed to a VDU is stored in an internal memory and it is this memory that is scanned or read on a continuous basis to produce the complete field of video information on the CRT. This is a completely different concept to feeding a repetitive video signal to a TV monitor. For example if a TV monitor was to display the written text SERVICING DIGITAL CIRCUITS IN T.V. RECEIVERS the video waveform fed to the TV monitor would not change from one frame to the next. With a VDU the digital signals representing the text SERVICING DIGITAL CIRCUITS IN T.V. RECEIVERS would be transmitted to the VDU once only. These digital signals would be stored in a memory contained in the VDU and this memory would then be 'scanned' to produce the text on the CRT.

There are several reasons for this major difference between the monitor and the VDU. First of all a TV monitor is a dedicated device designed primarily to display continuously moving pictures. A VDU is designed to display information which does not change rapidly, and for large periods of time the information on the CRT will be static. Secondly a VDU always receives its information from some form of digital system. One of the functions of the system will be to supply the VDU with the necessary data to create the desired display, but the system will have many other tasks to perform. For example it could be a computing system in which much of the time is taken up by carrying out calculations or processing data. In this situation the computing system will not want to waste time feeding the same signal to the VDU at a 50 cycle rate. It is far more economical in computing time to feed the necessary data to the VDU once and then update this information when necessary.

Most VDUs receive data in serial form although there are some which will also accept parallel data. In the description that follows it is assumed that the VDU is designed to accept serial data, and also to transmit data to the digital system in serial form, see the block diagram shown in *Figure 5.16*.

Data entering or leaving the VDU will pass via the computer interface. As the VDU may be some distance from the digital system it is common practice to use a digital signal which has a higher amplitude than the signals found within the VDU. For example the RS 232 serial interface system uses a digital signal of 24 V peak-to-peak. When receiving data the computer interface has to reduce and clip the incoming data amplitude down to a level that is compatible with the VDU electronics, and when the VDU is sending data the interface will amplify the data up to the required level for serial transmission.

All data processing in the VDU is carried out in parallel form, making it necessary to convert the received serial data bits into parallel bytes. The parallel bytes of data from the VDU must also be converted into a serial data signal so that it can be sent to the computer. This process of serial-to-parallel and parallel-to-serial conversion takes place in a circuit known as a Universal Asynchronous Receiver Transmitter (UART). A UART is normally a dedicated LSI component which will receive control instructions from the CRTC.

In the block diagram under discussion the UART has one parallel and one serial input and also one parallel and one serial output. When information is being typed on the keyboard the UART will receive the parallel bytes of data produced every time a key is depressed, convert this into a serial data string and feed it to the computer interface so that it can be output from the VDU. In this mode the UART is carrying out a parallel-to-serial conversion. When the VDU is receiving the serial data from the computer the UART converts the serial signal from the computer interface into parallel bytes of data which are then stored in the RAM. In this mode the UART is carrying out a serial-to-parallel conversion.

In most VDU systems the UART is controlled by the CRTC in such a way that it will only write new data to the RAM during the field blanking periods. This enables the RAM to be read during the field scan period and produce the video signal for the CRT. During the field scan period the UART will not be required to update the contents of the RAM so it converts the parallel data from the keyboard into serial form before the keyboard information is fed to the computer interface. The synchronizing of all these events is controlled by the CRTC. The reader will now appreciate why the CRTC normally takes the form of a small microprocessing system so that the complex sequence of events can be carried out with the minimum of components.

The Random Access Memory is used to store the coded data that represents the characters or graphic shapes which are to be displayed on the screen. It also stores control characters which may be used to change the display colour, or produce double height characters etc. The received data from the UART is written into the RAM during the field blanking period, and the RAM is read during the field scan. Control signals which determine if the RAM is to be written to or read from, and addressing signals which identify the memory location being used are all synchronized to line and field pulses which are again generated by the CRTC.

During the field scan period, data is read from the RAM on a character by character basis. Each byte of data represents a single character code, and is fed to the Read Only Memory which in turn converts the data into a video signal for the display.

Assume a complete display consists of 80 characters of text per line and 24 horizontal rows. This will produce a total of $80 \times 24 = 1920$ characters. Each character is represented by a byte of data, and consequently the RAM must have a minimum of 1920 bytes. In some of the more sophisticated VDUs two memories are used. One which displays the normal text on the screen and the other which contains protected characters which cannot be erased or removed from the display unless special control signals from either the keyboard or the computer are received.

The ROM uses the byte of data read from each RAM location as an address signal which will identify a single character location in the ROM. The CRTC also supplies the ROM with a signal which indicates which horizontal line of the character is being scanned. From these two signals the ROM will generate the appropriate digital information which is then used to form the video waveform of the character.

An output control circuit is required to amplify, and clamp the signal from the ROM, and also insert line blanking and field blanking periods. If the VDU has a colour CRT then the one output from the ROM will be split into three signals (RGB), which are then switched under the instruction of special control characters to produce the following colours:

White Red Green Blue Yellow Cyan Magenta Black.

In a monochrome VDU there will, of course, be only one signal feed leaving the output control circuit.

The digital video signal is then fed via an interface to a conventional video amplifier before being applied to the cathode of the CRT. The video amplifier used differs in several respects to the type of video amplifier that is normally found in a domestic TV receiver. First of all there is no need for the 6 MHz intercarrier sound filter. Second, the video signal being generated inside the VDU is not restricted to the bandwidth that is found in domestic TV transmissions.

We have seen from earlier chapters that the ideal digital signal is a square wave and will therefore contain an infinite range of frequencies. It stands to reason therefore that the bandwidth of the digital video waveform generated in a VDU will be significantly higher that the video bandwidth of 5.25 MHz that is found in

TV receivers. In practical terms a display which contains 80 characters per row and 24 rows will require a video bandwidth of approximately 20 MHz, and if the number of characters is reduced to 40 per row with again 24 rows per display the video bandwidth will drop to approximately 10 MHz. These high bandwidth video signals do not normally present any significant problems in the design of video amplifiers as nearly all video amplifier semiconductor devices are capable of handling frequencies much higher that those outlined above. Quality Video Display Units having high resolution displays, require time base and CRT circuits of a comparably high standard for the display to be accurately presented to the user.

A detailed description of the character generation and video processing which is used in Video Display Units is given in chapter 9 on the teletext decoder. The techniques used by the decoders are practically identical to those used in VDUs.

All the synchronizing pulses for the line and field timebases are produced by the CRTC and these will be synchronized to all the other timing waveforms used by the VDU.

The block diagram shown in *Figure 5.16* is designed to introduce the reader to the techniques employed in a typical VDU. The exact techniques used will, of course, differ from one manufacturer to another, and circuit complexity will differ considerably depending on whether the VDU produces colour or monochrome pictures, or is suitable for high resolution graphics, etc.

References

A Low Cost Intelligent VDU Terminal, Mullard Technical Information, No. 75 TP 1698/1, published by Mullard Ltd., Mullard House, Torrington Place, London WCE 7HD (1979)

Multitext, Mullard Technical Information No. 72 TP 1676, published by Mullard Ltd., Mullard House, Torrington Place, London WCE 7HD (1978).

Chapter 6

Data Transmission

Earlier chapters of this book have covered the basic theory of logic circuits, and wherever possible simple examples of their use has been included. We will now consider the techniques employed in transmitting data between one digital system and another before giving detailed examination to the application of logic circuits in television. Basic transmission techniques were very quickly outlined in chapter 1, and this is a good point at which to re-cap on some of those fundamentals before continuing.

Transmission of data can take place in two ways:

1. Over a number of parallel lines one byte at a time.
2. Over a single conductor, or on a carrier, one bit at a time.

Parallel transmission of data is employed where two digital systems are close together, and a multiway cable can be used to inter-connect them. This technique allows very high transfer speeds to be reached because information is passed between systems one byte at a time. For example, parallel data communication is used in microprocessing systems (see chapter 7) and it is common practice to find transfer speeds of 0.5 microseconds per byte or less. This means that data can be manipulated throughout the system at speeds in excess of 2 M bytes/sec.

Communication between circuits using this method of transmission is ideal where the two systems are in close proximity and high transfer speeds are required. If the two systems are some distance apart, or if some form of carrier is to be used to convey the information between a transmitter and receiver, then parallel data transmission is out of the question, and it becomes necessary to transmit the data one bit at a time in serial form.

Serial transmission is used almost exclusively for long distance communication, but unfortunately it is considerably slower than parallel data transmission.

A television set can receive serial data from the following sources:

1. The UHF signal carrying Teletext information.
2. The telephone line carrying Viewdata.
3. The remote control transmitter radiating user instructions from an infra-red or ultrasonic transducer.

The only time a TV receiver processes parallel data is inside a Viewdata or Teletext decoder, and the techniques used in this particular area are described in chapter 9 'Teletext decoders'. Consequently this chapter will examine serial data transmission only.

6.1 Serial data transmission

There are two basic systems in use for serial data transmission. These are known as:

1. Return to Zero (RZ) coding.
2. Non Return to Zero (NRZ) coding.

Figure 6.1 shows the difference between these two coding systems. Return to Zero is shown in *Figure 6.1 (a)* which illustrates a byte of data (01100010) transmitted in serial form. Notice that the waveform returns back to a logic '0' in between the transmission of two successive logic '1s', i.e. the waveform will return to zero every time a bit of information is transmitted.

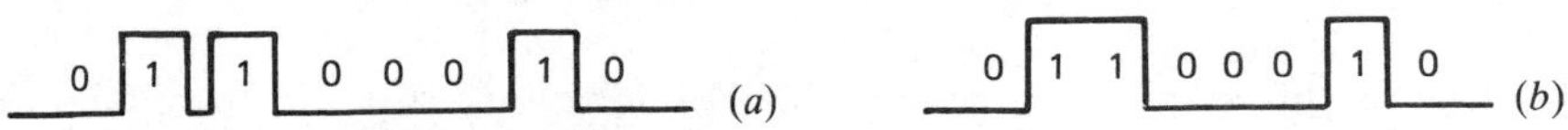

Figure 6.1 RZ and NRZ data coding

Non Return to Zero coding is shown in *Figure 6.1 (b)*, and as its name implies when two or more successive logic '1s' are transmitted the data waveform will not return to zero. In the NRZ system there are fewer transitions from one logic level to another than there are if the same data is transmitted using RZ coding. Consequently, if two identical bytes of data are transmitted using the RZ and NRZ systems, the NRZ signal will require less bandwidth. NRZ coding is used for the Teletext and Viewdata systems because of this bandwidth saving, and it is possible to transmit a digital signal having a transmission speed of 6.9 M bits/sec within a bandwidth of less that 5.5 MHz.

Serial transmission of data can take place using synchronous or asynchronous techniques; both of these are use in digital TV circuits. Asynchronous transmission is used by Viewdata while synchronous transmission is used by both the BBC and the IBA for Teletext.

6.2 Asynchronous transmission

Asynchronous techniques are used where data is to be transmitted in serial form on an intermittent basis. For example where a limited number of bytes are transmitted over a relatively long period of time. Asynchronous coding is shown in *Figure 6.2*.

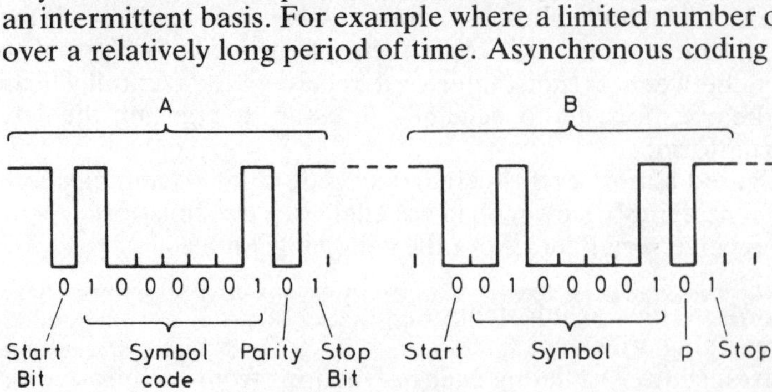

Figure 6.2 A synchronous start/stop coding as used for Viewdata signals

Here two data words are transmitted with a small gap between them. A logic '1' state on the transmission line is normally called a mark while a logic '0' is called a space. Notice that when no data is being transmitted the signal rises to a continuous mark level, and during this time the receiver is idle. Each group of data bits is called a word, and the number of bits in each word will differ between one system and another. Each word contains four distinct groups of data bits. These are:

1. Start bit.
2. Symbol code.
3. Parity bit.
4. Stop bit.

Start bit

At the beginning of every symbol code a space bit is transmitted to indicate the start of the data word to the receiver. Remember that a mark is transmitted between individual words, and if each word starts with a space bit the receiver will be able to identify this change in data level at the beginning of a new word.

Symbol code

After the start bit a sequence of 7 bits is transmitted and these are used to convey the symbol code, in this example 1000001 is the code for the letter 'A' and in the second byte 0100001 is the symbol code for the letter 'B'. The symbol codes shown in *Figure 6.2* follow the Viewdata and Teletext formats. In both of these systems the symbol code contains 7 bits of data followed by an eighth parity bit.

Parity bit

The parity bit is used to enable the receiver to detect a single bit error in the received signal. In the Viewdata system even parity is used, and when the parity bit is added to the 7 bit symbol code the total number of logic '1s', or marks, will add up to an even number.

In the example shown, the symbol codes for 'A' and 'B' both contain an even number of logic '1s' hence both parity bits will be a logic '0'. If the symbol codes contained an odd number of logic '1s' then the parity bit would be a logic '1' making the total number of logic '1s' transmitted an even number.

Stop bit

After the parity bit a stop bit is transmitted. The stop bit will be a logic '1' causing the complete word to end at a mark level.

The electronics designed to receive an asynchronous signal has a clock oscillator that is 'pulled in' to synchronism only when the data word is transmitted. During the mark level present between data words the oscillator will free run. The start bit is used to trigger the receiver oscillator into synchronism so that the following stream of data can be decoded correctly. This form of data transmission is sometimes called Asynchronous Start/Stop, because the receiver is pulled into synchronism with the transmitter at the beginning of each data word.

Once the start bit has been identified a total of 9 data bits (7 for the symbol code, 1 parity bit and 1 stop bit) should be received. If this number of bits is not received the data word has been corrupted, and the receiver will reject the data. If a parity error exists then it is possible for the receiver to identify and possibly correct the corrupted bit of data. The techniques used for parity and Hamming Code correction are fully described later in this chapter.

When data is transmitted manually, for example when the operator of a Viewdata receiver requests a page number, then the gaps in between individual words will vary in length. But when the data base responds to the requested page number it will transmit a continuous stream of data one word after another, without gaps. In this mode the stop bits at the end of each word act as separation bits. After each stop bit the receiver will look for the mark to space transition of the following start bit to detect the beginning of the next data word.

To summarize; asynchronous transmission is ideally suited for situations where data is not transmitted on a regular basis but where words are transmitted in their ones and twos or possibly in short bursts.

6.3 Synchronous transmission

One of the disadvantages with the asynchronous start/stop method is that each data word has additional start and stop bits. For continuous high speed data transmission these extra bits tend to be superfluous and can be removed. This is achieved by transmitting the data words one after another in a continuous stream. *Figure 6.3* shows the letters 'A' and 'B' of the previous example transmitted using the synchronous technique. Each 7-bit symbol code is again followed by its parity bit, and in this example odd parity is used, with each word containing an odd number of logic '1s'. Immediately following the 8-bit code for 'A' is the code for 'B'.

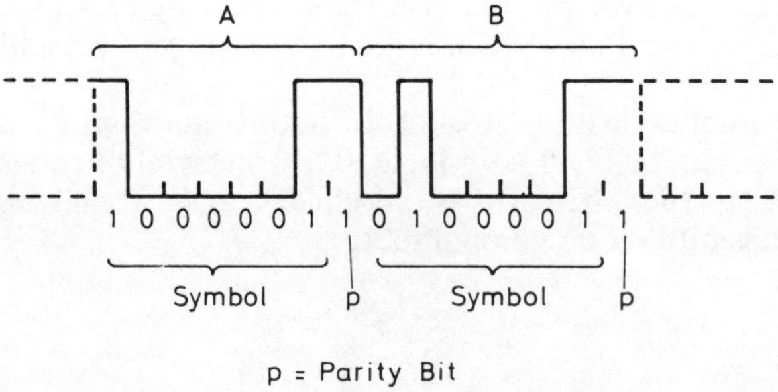

Figure 6.3 Synchronous transmission as used in Teletext

It is still necessary to synchronize the receiver clock oscillator to the transmitted data, and as start and stop bits have been omitted an alternative method of synchronizing is required. This is achieved by transmitting a burst of synchronization signals at the beginning of the data stream, and in the Teletext system this synchronization code is two bytes of alternating logic '0s' and '1s'. At the end of the second byte the receiver clock oscillator will be synchronized to the incoming data.

The receiver is then told when the first valid byte of data ia about to be received by an 8-bit framing code containing a carefully chosen pattern of '0s' and '1s'. In the Teletext system this code is 11100100.

Once the receiver clock oscillator has been synchronized by the first two bytes of data, the framing code can be detected, and used to reset the receiver circuits in preparation for the first valid word. Once reset, the receiver can identify the exact location of every other word in the block by counting the number of bit periods in the groups of 8, starting from the timing reference point produced by the Framing Code.

The synchronization signals occupy three bytes at the beginning of the data stream. Once they have been transmitted the receiver can keep in synchronism with the transmitted data for a limited period of time. It is therefore necessary to transmit the three bytes of synchronizing data at regular intervals.

6.4 Baud rate

An analogue waveform has its timing measured in terms of frequency which indicates the number of cycles generated per second. A digital signal which has been transmitted in serial form also has an equivalent to the analogue frequency, and this is known as the baud rate. The baud rate of a serial digital signal indicates the number of bits transmitted per second. Baud = bits per second. There is a range of standard baud rates used by computing systems from 50 to 19200 baud, and these are listed below.

50	75	110	120	150	200	300	600
1200	1800	2400	3600	4800	7200	9600	19200

If the receiver is to make any sense of the transmitted data it is imperative that the baud rates of both the transmitter and receiver are the same. The Viewdata system uses baud rates of 1200 baud for the transmission of data from the data base to the user, and 75 baud for the return signals from the user's keyboard to the data base. This is very slow indeed when compared with the Teletext baud rate of 6.9375 megabaud.

6.5 The multiplexer

Some of the possible techniques which can be used for producing a serial NRZ signal were quickly outlined as examples in chapter 3 'Sequential logic circuits'. These examples used a shift register to load the data in parallel form and then clock this data out of the register in serial form. Refer to *Figure 3.28* and the associated description. Although this system is quite acceptable for many applications, other techniques are available. The most commonly used, and the one most favoured, is the multiplexer.

A multiplexer can be represented in simplified form as a single-pole multi-position switch as shown in *Figure 6.4*. The data word to be transmitted is set up on the 10 inputs, and the single pole of the switch sequentially scans each input in turn starting with 0 and finishing with 9. On completion of the scan the data is changed and the input scanned again. The speed at which this process takes place is governed by a clock oscillator. Each cycle of the oscillator causes the switch pole to

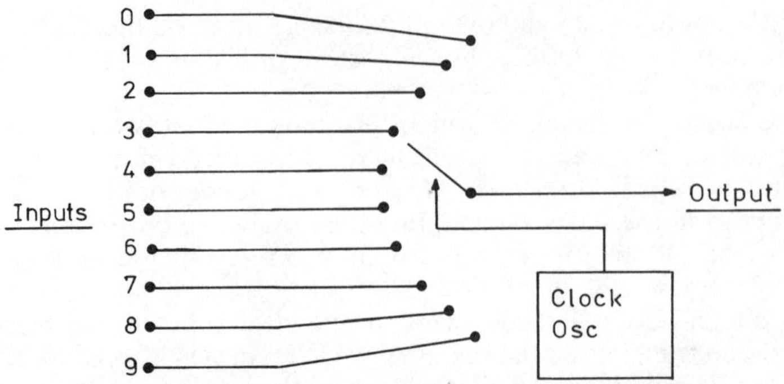

Figure 6.4 Single-pole multiplexer switch

move to the next input position, and when it has reached the 9th input position it will return to the start. The combinational logic equivalent of this simplified idea is shown in *Figure 6.5*.

A parallel data word is applied to the multiplexer inputs X0 to X9. These will be gated in rising numerical sequence to the output X. A 4-bit switch input S0 to S3 is shown at the top of the diagram, and the binary code set up on this input will determine which of the X inputs will be switched through to the multiplexer output. For example if S0,S1,S2,S3 = 0,0,0,0 the top AND gate is enabled, causing the logic level present on the X0 input to be passed by this AND gate and then the OR gate to the output. Examination of the circuit will show that every other AND gate in the circuit will be disabled because at least one of its inputs will be a logic '0'.

If the binary coded switch input is increment to produce S0=1,S1=0,S2=0,S3=0 then input X1 will pass through the circuit to the output, and all other X inputs will be inhibited. It can be seen that as this sequence continues the circuit produces an output level corresponding to the X input whose binary code is present on inputs S0 to S3.

If the circuit shown in *Figure 6.5* is used to produce a serial asynchronous word for the Viewdata system then the X inputs would be set up as follows:

X0	Start Bit	Logic '0'
X1–X7	Symbol Code	
X8	Parity Bit	Set for Even Parity
X9	Stop Bit	Logic '1'

This byte of data will remain fixed at the input to the multiplexer for the duration of its transmission. The S0 to S3 inputs would come from a binary counter, clocked by an oscillator running at the desired baud rate. Every time the clock pulse completes one cycle the output of the binary counter would be incremented, switching the next X input through to the multiplexer output. This process would continue for 10 clock cycles until each of the inputs had been switched through to the output in the correct sequence. At the end of the 10th clock pulse the binary counter supplying the S inputs to the multiplexer would reset to 0000, and at the same time the X inputs would change to the code for the next word to be transmitted. In this example only the inputs X1 to X8 would change because the start and stop bits are fixed logic levels.

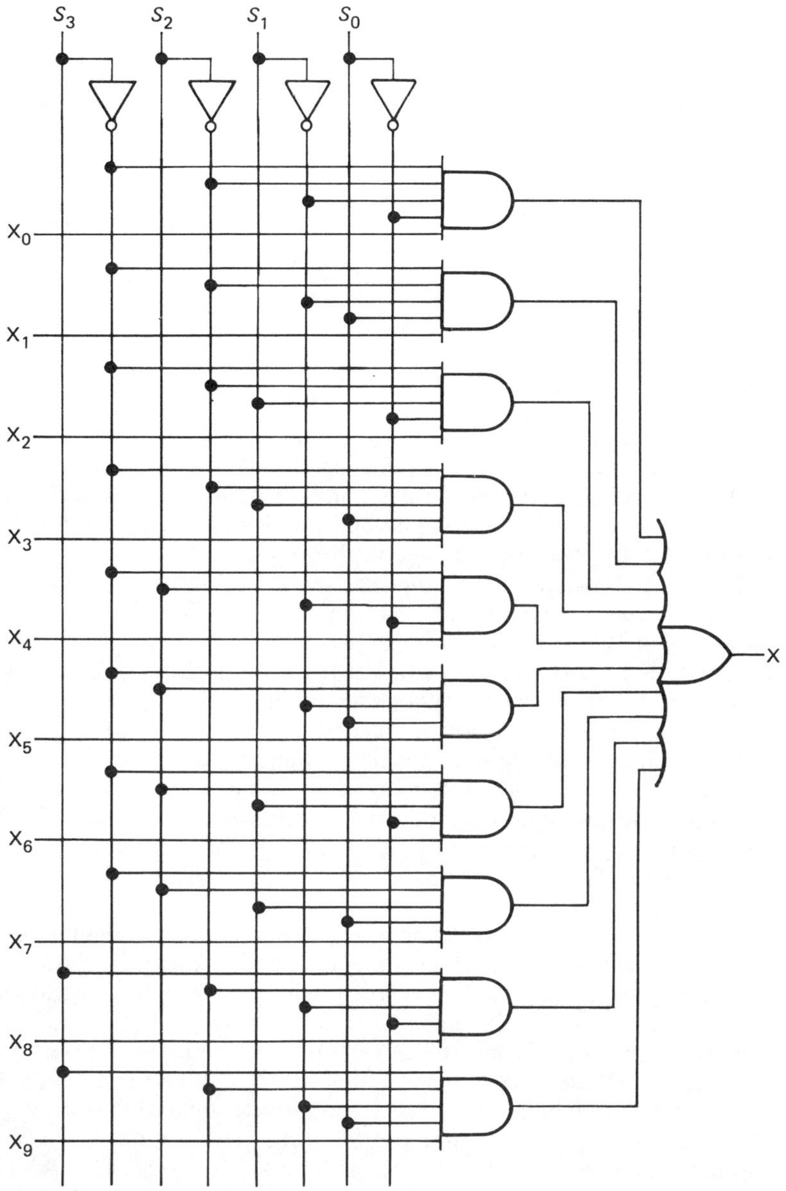

Figure 6.5 Multiplexer circuit diagram

6.6 The demultiplexer

As its name implies, the process of demultiplexing is the reverse of multiplexing and a demultiplexer is a circuit which receives serial data and converts it into parallel data. The circuit used to achieve this can literally be the opposite of the multiplexer. Refer to *Figure 6.6*.

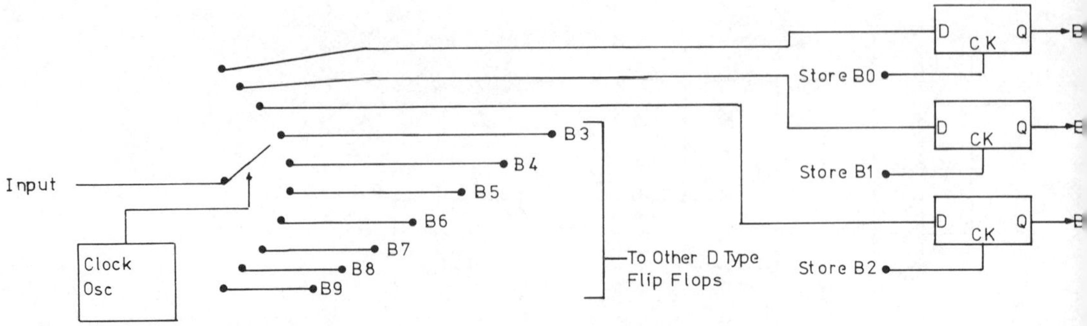

Figure 6.6 Single-pole demultiplexer switch

The incoming serial data is fed to the pole connection of the rotary switch and each output of the switch is fed to a simple D-type flip-flop. As with the previous example the pole of the switch changes position in synchronism with a clock oscillator. The first bit of data received is switched by the demultiplexer to output B0 where it is subsequently clocked into D-type flip-flop and stored.

On the next clock pulse the switch pole changes to the next output position and the second bit of the serial data is stored in the second D type flip-flop. This process continues until all 10 bits of the incoming data are stored.

At this stage there will be a 10-bit parallel word at the output of the circuit, and this can then be processed by the receiver. When the oscillator clocks the circuit a further time, after the 10th clock pulse, the switch pole returns to position B0, and the D-type flip-flops are all cleared ready to receive the next data word.

It can now be appreciated why accurate synchronism between the receiver and transmitter electronics is required for error free reception. If the switching sequences that take place in the receiver do not exactly match those of the transmitter the byte of data produced at the output of the demultiplexer will be invalid, and of no use to the receiver at all.

The circuits of clock oscillators, counters and D-type flip-flops have been dealt with in previous chapters and it is therefore only necessary to give a detailed description of the demultiplexer itself. One possible circuit configuration is shown in *Figure 6.7*.

The serial data stream from the transmitter enters the demultiplexer at point X, and is applied to one input of the 5 input AND gates X0 to X9. The remaining 4 inputs of each AND gate are derived from a binary counter identical to that used in the example for the multiplexer. This counter produces the BCD inputs S0 to S3.

Assume the counter produces an output of S0,S1,S2,S3 = 0,0,0,0. The AND gate producing output X0 will have four logic '1s' on four of its inputs. Its 5th input will be the logic level at point X. This logic level will therefore be present at the output of the top AND gate, while every other AND gate in the circuit will produce a logic '0' out.

During this period the logic level at X0 is latched into its respective D-type flip-flop. On the next clock pulse the next bit of input data will be received, but the S inputs to the demultiplexer will now have the code S0=1,S1,=0,S2=0,S3=0. The AND gate producing X1 output will now be enabled by the BCD signal, causing the logic level generated by bit 1 of the input signal to be present at output X1. It is then latched into its respective D-type flip-flop.

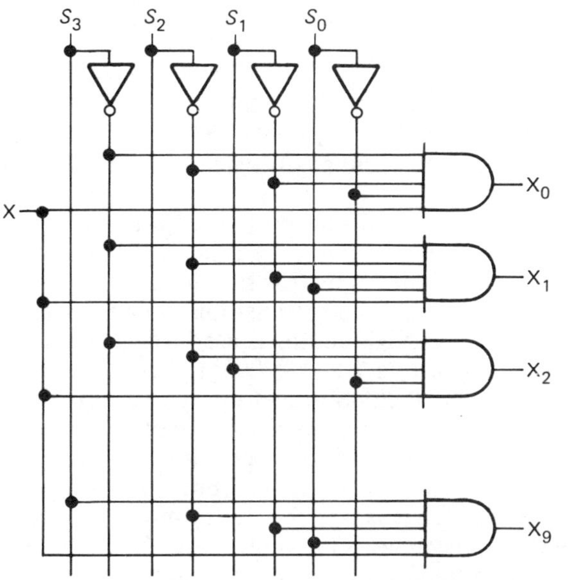

Figure 6.7 Typical demultiplexer configuration circuit

The process continues until all 10 inputs bits have been latched into their flip-flops, producing a 10-bit parallel word at the demultiplexer output. this word will only be present at the demultiplexer output for a small period of time during the 10th clock pulse. At the end of the 10th clock pulse the circuit resets ready to receive the next serial word of data. It is therefore necessary to transfer this 10-bit output to a temporary store so that the data can be processed in parallel form while the demultiplexer receives the next 10 bits from the input.

Note that the speed at which data is either multiplexed or demultiplexed is determined by the baud rate, and the multiplexer and demultiplexer must keep in step. Other circuits in the receiver and transmitter need not necessarily work at the same speed and therefore need not be timed to the baud rate itself. Consequently it is common practice to find that the associated electronics to the multiplexer and the demultiplexer will be carrying out other tasks while data is being received or transmitted.

6.7 Parity checking

We have seen that in the Teletext and the Viewdata systems parity bits are inserted at the end of the symbol code to enable errors to be detected at the receiver. Errors introduced during the transmission of data can be disastrous if not corrected or detected. For example the symbol code for the character 'A' is 1000001, (LSB transmitted first), and if bit 7 is changed say because of a noise pulse the character could become 1000000, causing all the following text to be printed in RED. Errors should be detected and if at all possible corrected.

The addition of the parity bit allows the receiver to identify a single bit error in the symbol code. Most Viewdata and Teletext decoders will ignore a corrupted word which contains a parity error, and will display a blank space rather than print an incorrect character on the screen.

Teletext uses odd parity and hence the total number of logic '1s' received in any symbol code should add up to an odd number. If the receiver counts an even number of logic '1s' then the byte of data will be rejected. Viewdata uses even parity causing the receiver to look for an even number of logic '1s' in the symbol code.

In Teletext, the parity bit also serves another useful purpose. Teletext data is transmitted synchronously and the receiver oscillator is synchronized at the beginning of each line of text by two bytes of data known as the clock run in. Once this oscillator has been synchronised it relies on the transitions in the data to maintain synchronism throughout the line. As NRZ coding is used for Teletext it is possible to transmit a symbol code consisting of 0000000 (reserved control character) or 1111111 (graphic block) causing the data waveform to stay at logic '0' or logic '1' for the duration of the word. If even parity were used then it is possible to transmit 11111111 or 00000000 as single words. If this were the case the clock oscillator would not have any information on which to synchronize during the transmission of these words. Using odd parity ensures that there must be at least one transition in signal level during the transmission of a single word. Consequently at least one synchronizing signal can be generated from the incoming data stream for the clock oscillator.

Odd parity is not required for the Viewdata system because it uses asynchronous transmission containing start and stop bits, the start bit being used to synchronize the clock oscillator at the beginning of every data word.

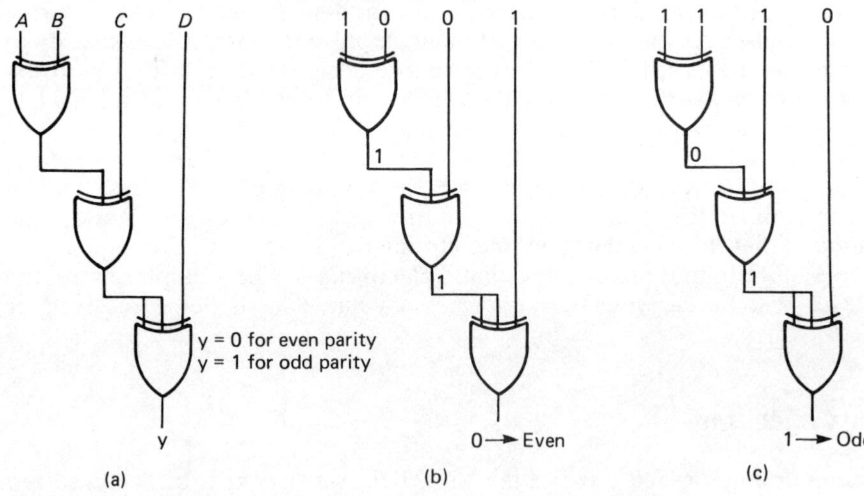

Figure 6.8 Circuit for 4-bit parity checker

Parity checks can only be made when the symbol code has been received, demultiplexed, and stored in parallel form. The circuit of a 4-bit parity checker is shown in *Figure 6.8*. In this example a 4-bit parity check is being carried out on inputs A,B,C,D. We have seen from earlier chapters that the exclusive OR gate will produce the sum of its two inputs. Here the sum of (A) and (B) is added to (C) and this result is then added to (D). If the final sum output from the circuit is a logic

'0' this implies that the parity is even. If a logic '1' results the parity must be odd. For example assume that the inputs A,B,C,D, are equal to 1001.

Figure 6.8 (b) shows this condition set up on the parity checker inputs. A final sum of '0' is produced indicating that the parity is even. A further example is shown in *Figure 6.8 (c)* where A,B,C,D is equal to 1110. The final output is a logic '1' denoting that the parity is odd.

The concept of this circuit can be extended to produce a check on any number of parallel bits, for example with an 8-bit byte such as that used in Teletext 7 exclusive OR gates would be required and assuming that the parity check was correct the final output from the circuit would be a logic '1'. Additional combinational logic gates on the output of the parity checker would then process the data in such a way that if the parity was correct the data would be stored in the decoder RAM, but if the parity was incorrect the decoder would store the code for a space. A blank space would then be displayed on the screen instead of the corrupted character.

6.8 Hamming code correction

The Hamming code was devised by R. W. Hamming of Bell Telephone Laboratories who developed a means of detecting and correcting a single bit error in a received data word. Although the Hamming code makes it possible to detect and correct errors it does unfortunately reduce the number of bits used to relay coded information to the receiver. *Figure 6.9 (a)* shows a normal byte of Teletext data. Bits 1 to 7 are called message bits and are used to generate alphanumeric characters, graphics, etc. Bit 8 is the parity bit which is added on the end of the data word to ensure that the total number of logic '1s' is odd.

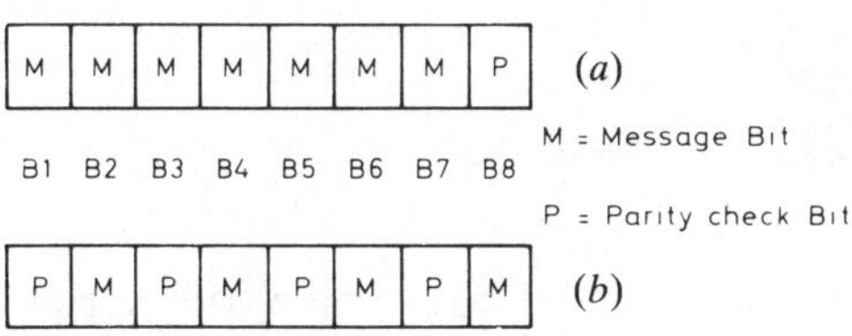

Figure 6.9 Teletext data bytes. (a) normal text data; (b) Hamming coded address data

Code	Bit position and state 1	2	3	4	5	6	7	8
0	1	0	1	0	1	0	0	0
1	0	1	0	0	0	0	0	0
2	1	0	0	1	0	0	1	0
3	0	1	1	1	1	0	1	0
4	0	0	1	0	0	1	1	0
5	1	1	0	0	1	1	1	0
6	0	0	0	1	1	1	0	0
7	1	1	1	1	0	1	0	0
8	0	0	0	0	1	0	1	1
9	1	1	1	0	0	0	1	1
10	0	0	1	1	0	0	0	1
11	1	1	0	1	1	0	0	1
12	1	0	0	0	0	1	0	1
13	0	1	1	0	1	1	0	1
14	1	0	1	1	1	1	1	1
15	0	1	0	1	0	1	1	1

Figure 6.10 Table of Hamming codes

When data is Hamming coded it is transmitted using the format shown in *Figure 6.9 (b)*. It can be seen that parity and data bits alternate so that bits 2,4,6,8 carry data, and bits 1,3,5,7 are parity bits. As there are 4 message bits it is only possible to transmit a maximum of 16 separate codes instead of the 128 codes capable of being transmitted using 7 bits. *Figure 6.10* shows the bit pattern combinations for an 8-bit byte which has been Hamming coded.

To detect a single bit error the electronics in the receiver has to carry out four separate parity checks on the byte of data. Each of these checks is made for an odd parity result, and we will assume that each parity checker will produce a '0' at its output if the parity check is correct.

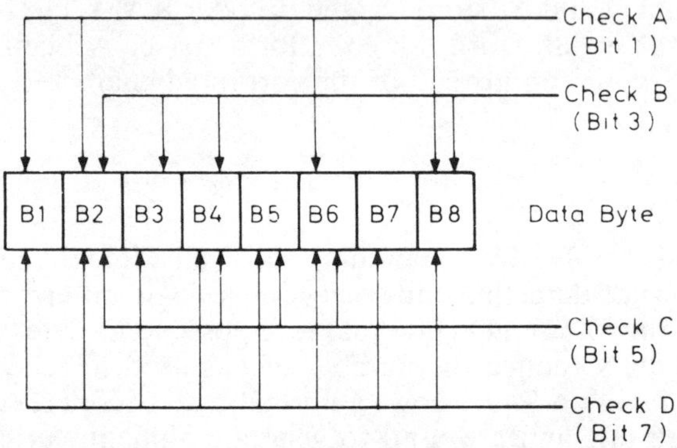

Figure 6.11 Hamming code parity checks

Refer to *Figure 6.11*.
Check A is carried out using bits 1,2,6,8.
Check B is carried out using bits 2,3,4,8.
Check C is carried out using bits 2, 4, 5, 6.
Check D is carried out using bits 1,2,3,4,5,6,7,8.

The four separate parity checks labelled (a) to (d) are made on various combinations of message and parity bits within the byte of data. The result of each parity check will be either a '0' for odd or a '1' for even parity. Examination of these results will enable the receiver to determine if a single or multiple bit error is present. If a single bit error is present the corrupted bit is identified and corrected, but if two or more errors are detected the byte of data is rejected. A table outlining this action is shown in *Figure 6.12*.

If parity checks A,B,C,D, all create a logic '0', indicating that the tests are correct the data is accepted. If check D indicates that the parity is correct but checks A,B,C indicate an error, two or more bits of the byte have been corrupted. In this situation the error cannot be corrected and the byte is rejected. If check D fails, indicating a 1-bit error, then the Hamming code corrector is capable of identifying and correcting the corrupted bit.

The action required is shown in the lower half of *Figure 6.12*. A circuit that carries out this correction process is shown in *Figure 6.13*. The message bits are fed to a set of exclusive OR gates. If the second input to these gates is at a logic '1' the message bit is inverted, but if the second input is at logic '0' the message bit passes through the gate unaffected. Gates G7 to G10 therefore act as either a buffer or an

A	B	C	D	ACTION REQUIRED
0	0	0	0	DATA IS CORRECT
1	0	0	0	REJECT DATA TWO OR MORE ERRORS DETECTED IN BYTE
0	1	0	0	
1	1	0	0	
0	0	1	0	
1	0	1	0	
0	1	1	0	
1	1	1	0	
0	0	0	1	INVERT BIT 7
1	0	0	1	INVERT BIT 1
0	1	0	1	INVERT BIT 3
1	1	0	1	INVERT BIT 8
0	0	1	1	INVERT BIT 5
1	0	1	1	INVERT BIT 6
0	1	1	1	INVERT BIT 4
1	1	1	1	INVERT BIT 2

0 = Parity test O.K.
1 = Error detected

Figure 6.12 Results of parity tests and corrective action required

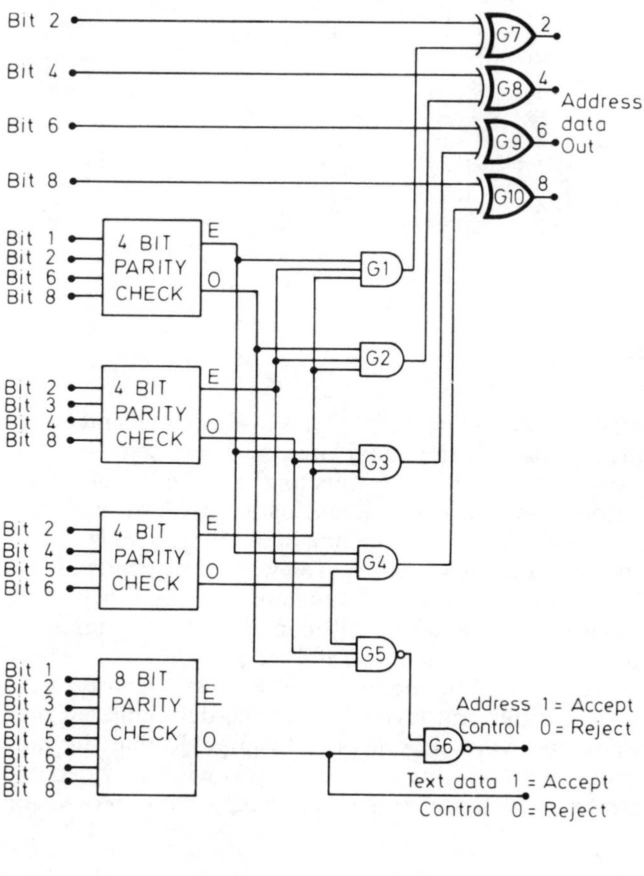

Figure 6.13 Circuit diagram for error detection and correction of Hamming coded date

inverter depending on the logic state of the second input. The error outputs of the 4-bit parity checkers are combined in a set of 4 AND gates which select which bit is to be inverted.

In this example the Hamming code correcting circuit will only take action if any of the *message bits* are corrupted. Correction of the parity bits is not made because these are not used by the receiver. Gates 5,6 and parity checker (D) are used to determine if more than one bit in the received byte is corrupted. If no parity error, or if a single bit error is present, then the two inputs to gate 6 will be either 00,01,10. If the inputs to G6 both rise to a logic '1' then more than one bit in the byte is corrupted and the data outputs from G7 to G10 will be rejected.

6.9 Bandwidth requirements of a Teletext signal

The highest fundamental frequency contained in an NRZ data transmission will be produced when the data alternates between a logic '0' and a logic '1'. This is shown in *Figure 6.14*. If this were a Teletext signal the bit rate would be 6.93 M/Bits

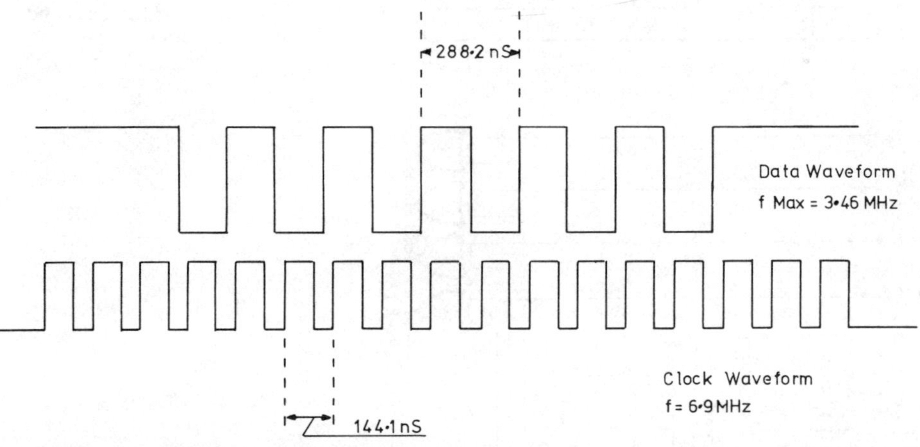

Figure 6.14 Data alternating between logic 0 and 1

producing a timing between one bit and the next of 144.1 nS. This is the amount of time occupied by one bit of the text data. If the signal alternates between a logic '1' and a logic '0' the repetition frequency of the data waveform will be 0.5 of the bit rate, thereforethe highest periodic time in the Teletext data waveform will be 2 times 144.1 nS which equals 288.2 nS giving a frequency of 3.4698126 MHz.

At first thought it may appear that this signal will lie well within the bandwidth of the video spectrum, but we have only considered the fundamental frequency of the signal. A digital signal is a square wave, and this will contain a large number of harmonics which will extend far above the fundamental frequency. If this signal were transmitted the higher frequency components would be lost because of the bandwidth limitations of the transmitter and receiver, and the demodulated data waveform would approximate a sine wave. However, it is possible that the high frequency components generated by a data square wave will mix with the video and sound frequencies producing cross modulation effects, degrading the sound, vision

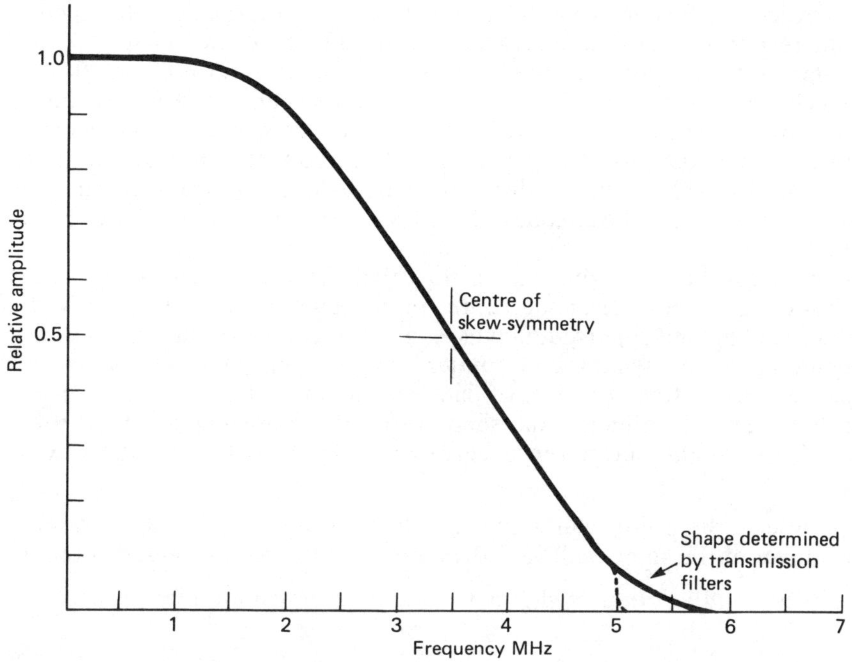

Figure 6.15 Raised cosine response

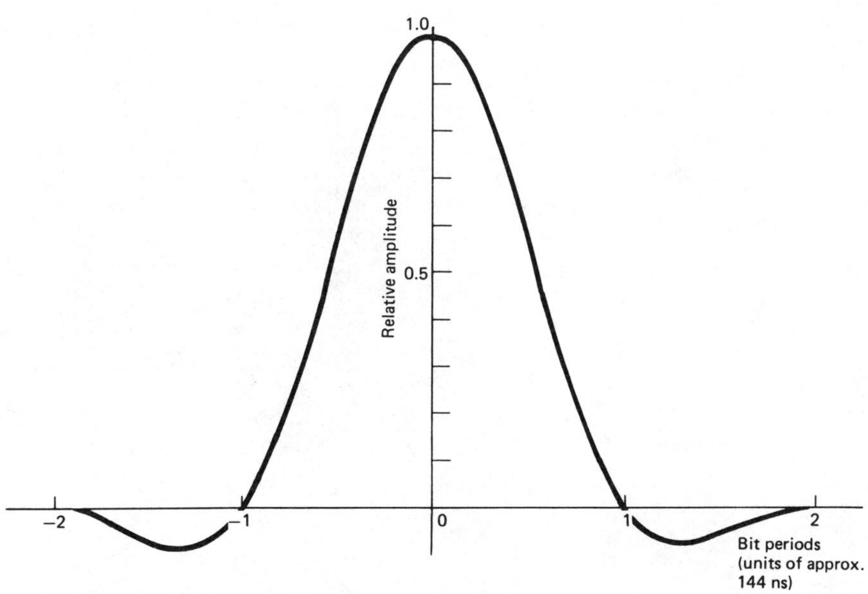

Figure 6.16 Raised cosine characteristic

and data. To overcome this problem the Teletext data is fed through a filter having a raised cosine response. The characteristic of this filter is shown in *Figure 6.15*. and the effect it has on the response of the Teletext data is shown in *Figure 6.16*.

Restricting the bandwidth of a square wave alters its shape, and if this square wave had an equal mark to space ratio the filter output would be predictable. Unfortunately Teletext data uses NRZ coding and consequently the mark to space ratio of the waveform will change as different symbol codes are transmitted. This action causes the fundamental frequency of the waveform to vary up to a frequency of 3.47 MHz.

Furthermore it can be seen from *Figure 6.16* that the filter response produces small overshoots before and after the centre point. This in turn will modify the shape of the preceding and superseding pulses, and the degree of modification will change depending on the symbol code under examination. The overshoots that effect the pattern of the following bits of data will therefore be irregular, but it is safe to say that they will influence the shape of all the following pulses to some extent. The effects of this interference between one data pulse and the next is two-fold:

1. The amplitude of the pulses will vary depending on the data products present.
2. The mean level of the signal will vary depending on the data products present.

These two effects jointly produce what is known as intersymbol interference.

Chapter 7

Microprocessor fundamentals

The introduction of the microprocessor in the early 1970s produced the most significant advances in electronic technology since the advent of the transistor. During their short history microprocessors have found applications in practically every aspect of modern electronic circuits. Their introduction has revolutionized many industries and bought the concept of the home computer into reality.

Microprocessors are finding applications in electronics ranging from control systems for washing machines, cookers etc, through to the more advanced aspects of controlling video cassette recorders, computers, telephone switching systems etc.

To appreciate why the microprocessor has made such a significant impact it is necessary to go back to the fundamentals of any system, whether it is electronic, electro-mechanical or mechanical.

Figure 7.1 shows the basic arrangement of any system. It consists of a number of elements which are working together in a very closely defined manner. These

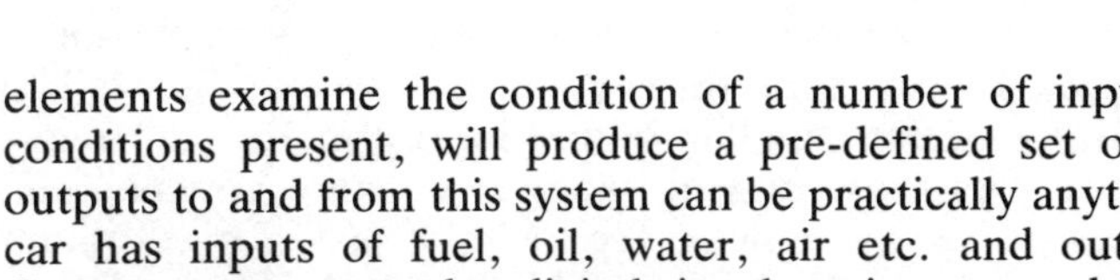

Figure 7.1 Basic system

elements examine the condition of a number of inputs, and depending on the conditions present, will produce a pre-defined set of outputs. The inputs and outputs to and from this system can be practically anything. For example, a motor car has inputs of fuel, oil, water, air etc. and outputs of motion and heat. Computers use complex digital signals as inputs producing outputs in the form of characters on a VDU or text from a printer.

Until the advent of the microprocessor the series of predefined events that took place within a system were determined by the mechanics, or the electronics of the system. This term is known as the hardware. Nearly all electronic systems developed before the advent of the microprocessor were hardware controlled. Hence the electronic circuits and the connections used determined the function of the system. For example a machine used to mechanically insert the components in a printed circuit board would be hardware wired. Once the electronics was designed and produced, the machine could only insert components for a particular printed circuit board. If the design of the printed circuit board was changed, then the electronics or the hardware of the system would have to be changed as well.

A system controlled by a microprocessor will again have the same input and output signals previously described, but the revolutionary aspect of a microprocessor based system is that the sequence of events that takes place is controlled by a programme of instructions. In *Figure 7.2*, the programme determines which inputs will be examined, and what sequence of events will take place to produce the outputs. The hardware, or the electronics of the microprocessor system remains the same, regardless of the system function. This means that component manufacturers can mass produce the ICs used in microprocessing systems very economically, and use these components in a variety of microprocessor based systems.

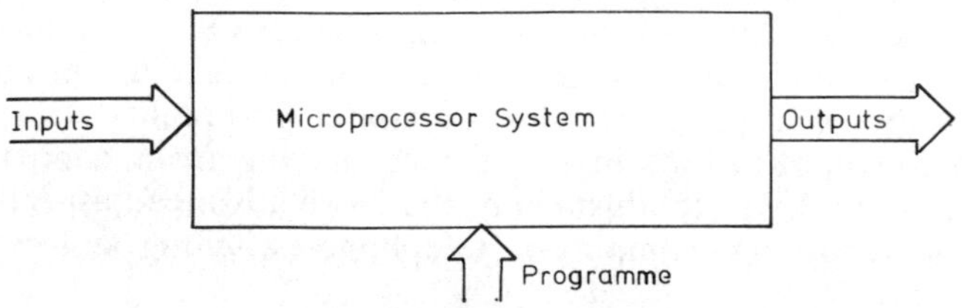

Figure 7.2 Microprocessor controlled system

The precise process which the system eventually carries out will be determined by the coded programme of instructions, and these processes can range from operating an aircraft guidance system, down to controlling an automatic washing machine. It is quite possible that the fundamental elements of the washing machine and the aircraft control system are the same, the only difference being the series of coded instructions that the system is using as a programme.

Microprocessors have found considerable application in many aspects of TV. They first made an appearance in the TV receiver in the form of control systems for digital tuning units. Their applications have now expanded to the field of video cassette recorders where they are used to control all the electromechanical functions of the recorder as well as incorporate many advanced programme selection and safety facilities. Many manufacturers are now using microprocessors in their range of Teletext and Viewdata decoders, where the processor is responsible for the manipulation of data throughout the system as well as controlling cassette tape storage of data, and an output for a printer.

As the cost of microprocessors falls, making them even more commercially viable for the manufacturers of TV equipment, they will be used for additional functions within the receiver such as AGC, ACC and AFC control, HT regulation, timebase control and remote control. It is possible using current technology to produce a single chip microcomputer which could in fact handle all of these functions simultaneously.

7.1 Microprocessor architecture

The term 'architecture' is the name given to the basic structure of a microprocessor based system. It outlines the units within the system and shows how they are interconnected. The microprocessor based system used for TV applications can normally be subdivided into four major sections and these are shown in *Figure 7.3*.

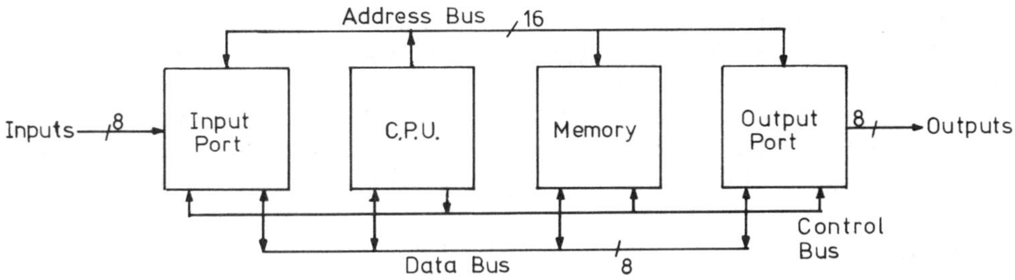

Figure 7.3 Microprocessor architecture

Input/Output ports

Input and Output ports, (I/O ports), are circuits which provide an interface between the microprocessing system and the outside world. Every input and output connected to a microprocessor based system must pass via some form of I/O device.

As the microprocessor system is digitally operated then every input or output signal will be digital, either in serial or parallel form. Any device that interfaces with the system must therefore produce a digital signal either directly, or indirectly via an analogue to digital converter.

Central Processing Unit

The Central Processing Unit (CPU) is the intelligent heart of any microprocessing system. It is an LSI chip containing many tens of thousands of components. It is now mass produced in such large quantities that the unit costs are very low indeed. The CPU contains the electronics which identifies and executes each coded programme instruction, and the majority of CPUs also contain a small internal memory which the programmer can use at his discretion for storing data. All of the microprocessing system is directly controlled by the CPU.

Memory

The memory used in a microprocessing system can be any of those previously described in chapter 4 'Storage Devices'. It is common to find two major types of memory in microprocessing systems—Read Only Memory and Random Access Memory. The ROM normally contains the instructions written by the manufacturer of the system known as firmware. The firmware contains all the instructions required to make the system operate, e.g. how to input serial data or how to output parallel data, how to check for parity errors, how to Hamming code correct etc. The RAM is normally used for storing data generated by the firmware, such as the results of calculations, text data, or possibly instructions entered into the system by the user. As the RAM will be either of the static or dynamic type it will be volatile and will therefore lose its information once the power is removed, unless of course the ICs are provided with a battery backup.

The CPU, memory and I/O ports form the basic building blocks of any microprocessor-controlled system regardless of its application. The components used for each part are mass produced on a very large scale indeed and consequently their combined costs are low. The precise way in which these elements operate in conjunction with each other is determined purely by the programme of instructions stored in the memory. Simply changing the contents of a memory can change the entire function of the whole system.

Communication between each of the individual elements of a microprocessing system takes place using three data highways. These are known as the Address Bus, Data Bus and Control Bus.

Address bus

In chapter 4 the concept of an address bus for memory was outlined, and the theory applied in that chapter is the same for the address system of a microprocessor. The address bus on most microprocessor systems is 16 bits wide, i.e. it contains 16 conductors each one carrying a single bit of data. It is therefore possible to address a total of 65 536 different circuits. Each circuit element will have its own unique address which is generated by the CPU and placed on the address bus when that circuit element is to be used. The CPU is the only device within the whole system can generate an address, and consequently every circuit in the system is under the control of the CPU via the programme.

Input and output ports will also have an address in addition to the storage elements of the memory. It is therefore possible for the CPU to select which input or output port is to be used in exactly the same way as it selects which memory cell is to be used in the RAM. The diagram illustrated in *Figure 7.3* has only one input and one output port in order to keep the diagram as simple as possible. In practice several input/output ports will be used and it is possible to use a maximum of 256 ports in a practical system.

Data bus

The data bus is normally 8 bits wide and is used to transport data between the individual elements of the microprocessing system. For example, if the CPU was reading a memory location, the data from that memory location would be placed on the data bus and transported to one of the internal stores of the CPU.

Only two circuits will use the data bus at any time, these will be the CPU and the circuit whose address is on the address bus. If the CPU and two or more circuits access the data bus simultaneously then a data conflict will occur, and this results in the corruption of the data on the bus.

Control bus

The control bus is a group of conductors, which carry control signals between the CPU, memory, and the input/output ports. These control signals enable the CPU to tell the addressed circuit what is required, such as read or write. For example, if the CPU was to write data to a memory location the following sequence of events would take place:

1. The address of the memory location would be placed on the address bus to identify the memory cell.
2. The data to be written to that location would be placed on the data bus by the CPU.
3. The control bus would carry a memory write signal from the CPU to the memory cell causing the addressed memory cell to store the information present on the data bus.

Other control bits on the control bus operate input and output circuits, UARTs, etc. The number of lines used on the control bus is determined by the type of CPU and differs between systems.

System timing

A Central Processor Unit will execute a series of coded instructions stored in memory sequentially, and a clock oscillator is used to govern the timing of the whole system. Clock frequencies of 4 MHz are common for microprocessors and instructions will vary in timing from about 1 to 10 microseconds. Programme execution speeds are therefore directly related to the clock frequency of the CPU. A high clock frequency ensures rapid programme execution.

Carrying out logic functions using a microprocessor is very slow when compared with the alternative of using custom, made hardware with combinational logic circuits. Hence the use of a microprocessor to simulate combinational logic circuits is only a practical proposition when the handling speed of the data is not important. In the chapter on data transmission it was mentioned that Viewdata is transmitted at very slow data speeds of 75 and 1200 baud. As the microprocessor is capable of manipulating data throughout the system at speeds far in excess of this then it is practical to use a processor to control the reception of Viewdata.

The Teletext system uses a very high baud rate of 6.9 Mbits/sec, and as the clock frequency of processors is typically 4 MHz the use of the microprocessor to manipulate Teletext data is not a practical proposition until the cost of the faster CPU chips falls dramatically.

7.2 Practical microprocessor system

A more practical microprocessor system is shown in *Figure 7.4*. Again it can be seen that the system contains the basic elements previously outlined, the CPU, memory, and input/output ports. Here all the ports are housed in a single input/output unit. One input port will input 8 bits of data from a peripheral to the system and one output port will output 8 bits of data from the system to the peripheral. The remaining two ports can transfer data in either direction e.g. 4-input bits and 4-output bits or any combination thereof.

A clock oscillator has been included in this diagram to indicate that the series of events carried out by the system is governed by a central timing source. In some systems this clock oscillator will be separate from the CPU and in the more advanced designs it may be an integral part of the CPU chip.

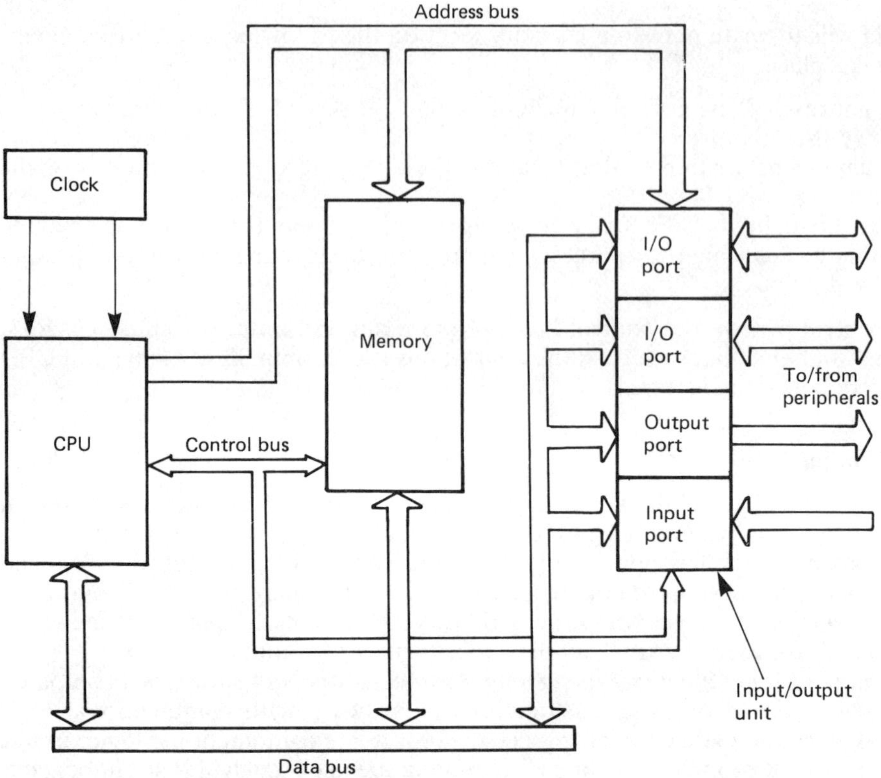

Figure 7.4 A practical microprocessor system

7.3 Instruction execution

A series of coded instructions, stored in the memory, is executed sequentially by the CPU. These instructions are fed from the memory to the CPU via the data bus, and consequently as the data bus is 8 bits wide each instruction will contain 8 bits of data. This byte of data is known as the Op-Code. If an Op-Code is contained in one byte of data then a maximum of 256 instructions can be coded in binary form. Most microprocessors take advantage of nearly all the possible bit combinations and therefore have instruction sets of about 240 commands.

More advanced CPUs have Op-Codes of two bytes, but microprocessors of this complexity are not required for television since a simple 8-bit processor is sufficiently powerful.

Instruction sequence

Each instruction executed by the CPU will have three distinct phases. These are:

1. Fetch
2. Decode
3. Execute

The fetch phase

At the beginning of each instruction the microprocessor will place a 16-bit code on the address bus, which will address the memory location that contains the instruction. At the same time a memory read signal will be placed on the control bus so that the addressed memory location containing the instruction is read. The data from this memory location is then placed on the data bus by the memory and will enter the CPU where it is stored.

Decode phase

When the data from the memory is stored in the CPU it can be interrogated using a small microprogramme to identify the instruction which the CPU is to execute.

Execute phase

Once the instruction has been decoded, the microprocessor will know exactly what sequence of events to initiate in order to execute the instruction. The execution phase may involve just an internal CPU operation, e.g. add two bytes together, or the instruction may require additional memory or 1/0 read or write phases. For example the CPU may be required to output data to an output port, in which case the address of the output port would be placed on the address bus, an output port write signal would be placed on the control bus and the data to be written to the output port would be placed on the data bus.

It can be seen from the above description that the fetch and decode phases will be the same for any instruction, but the execution phase will vary in complexity and timing depending on the nature of the instruction. Some instructions will take longer to execute than others, and consequently the length of time it takes to execute a programme depends entirely on the complexity of the instructions contained in the programme and the speed of the clock frequency to the CPU.

7.4 Internal architecture of the CPU

Detailed CPU operation, and the control of these operations under the direction of the machine code programme is way beyond the scope of this book, but it is hoped that this brief outline of the major CPU components will give the reader an appreciation of the more fundamental internal operations of a microprocessor chip.

The design of a CPU will differ considerably between one manufacturer and another. It is therefore impossible to give the reader a complete picture of CPU architecture because of these considerable differences. A choice has to be made as to which microprocessor chip to use as an example. The author has chosen the Intel 8080/8085 processor for inclusion at this stage for the following reasons:

1. It is an industry standard and therefore widely used in many microprocessor based systems.
2. Its architecture tends to be widely used in modified form by many other manufacturers.
3. It is possibly one of the easiest microprocessors to understand for the beginner.
4. Single chip CPUs described in chapter 10 'Viewdata Decoders' use similar architecture.

The internal block diagram of the 8080/8085 microprocessor is shown in *Figure 7.5*, and for the purposes of this description it will be subdivided into three major areas. These are:

1. Instruction decoding and control.
2. General purpose and 16-bit registers.
3. Arithmetic logic unit and associated circuits.

Instruction decoding and control

The 8080/8085 is an 8-bit processor and therefore the data bus in an 8080/8085 based system will be 8 bits wide. This means that the maximum number of instructions the processor will be able to execute from a single 8-bit Op-Code is 256. All the instructions for the 8080/8085 have one Op-Code and the processor is capable of decoding and executing 244 instructions.

We have seen that the first operation any CPU has to carry out in order to execute an instruction is to fetch the Op-Code from the memory. This is achieved by placing the address of the memory location which contains the Op-Code on the address bus, and reading that memory location. The data that is read is placed on

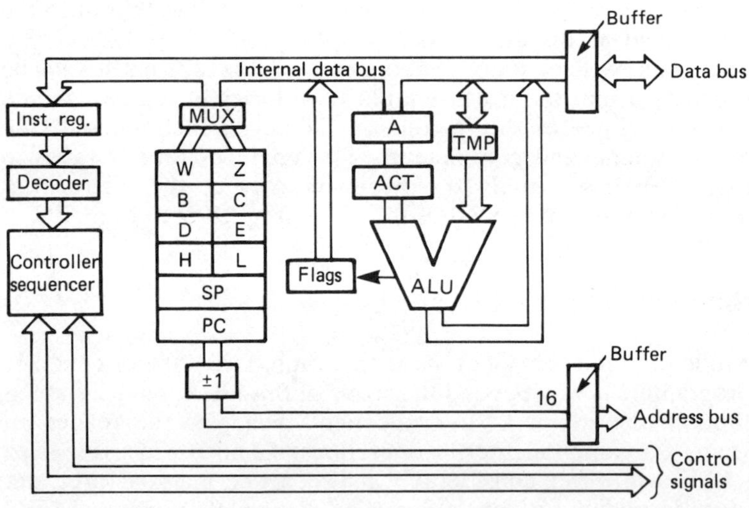

Figure 7.5 Internal structure of CPU

the data bus and this is passed to the CPU. By referring to *Figure 7.5* it can be seen that the CPU has its own internal data bus which is connected to a number of stages. One of these is the Instruction Register, (INST REG), which is used to store the Op-Code representing the instruction which the processor is about to execute.

Once the Op-Code is stored in the instruction register it is interrogated by the decoder so that the desired operation to be carried out by the CPU can be identified. When the instruction has been identified the controller sequencer, under the control of its own microprogramme, will generate the necessary sequence of control signals to execute the decoded instruction. This sequence of events may be an internal operation such as adding or subtracting two bytes of data, or it may require further memory read or write operations for the more complex instructions.

The instruction register, decoder and controller sequence are therefore used purely to receive and identify the Op-Code, and then initiate the necessary control sequences to execute the decoded instruction.

General purpose and 16-bit registers

In the centre of *Figure 7.5* is a series of 8- and 16-bit registers. The 8-bit registers are labelled (W,Z)(B,C)(D,E)(H,L) and the 16-bit registers are labelled (SP), and (PC).

The 8-bit registers will receive or send data via the 8-bit data bus, and only one register can access the data bus at a time. The registers W and Z are used by the microprocessor itself to aid in the execution of some instructions. The remaining registers (B,C,D,E,H,L) are under programme control and can be selected individually or in pairs by coded instructions used by the programmer. Registers B,C,D,E are normally used when it is more convenient to store temporary bytes of data inside the microprocessor rather than go to the lengths of storing the data in RAM, e.g. saving the results of a calculation or saving a byte of data that the computer received from an input port.

Registers H and L can be used as 8-bit registers in the same way as B,C,D, and E, or they can be used as a 16-bit register pair. In the register pair mode of operation the H and L registers are used as a memory pointer, for instructions which require the microprocessor to access a memory location, for example if the programmer wanted to write data into memory location 0800 (Hex) then the H and L register pair would contain the memory address to be used. The H register would store 08 and the L register 00. They are called H and L because the H register stores the High order memory address byte and the L register stores the Low order memory address byte.

Two 16-bit registers are used solely for memory addressing. The contents of the stack pointer address a memory location which may be used for the rapid storage of data. In some modes of operation it is used by the CPU automatically, and in other modes it may be used by the programmer.

The programme counter contains the address of the next instruction stored in the memory. As instructions are executed the programme counter is automatically incremented by the controller sequence so that the CPU knows the location of the next instruction in the programme. It is therefore the programme counter that creates the memory address that is placed on the address bus during the execution of the majority of the CPU instructions. During some of these instructions however the 16-bit stack pointer or the 16-bit register pair H and L may also place their contents on the address bus in order to access memory locations which store data rather than programme instructions.

Arithmetic Logic Unit

The Arithmetic Logic Unit (ALU) is, as its name implies, responsible for carrying out all the arithmetic or logic instructions which the CPU is capable of executing.

These instructions are quite limited, and the following list gives an indication of what is normally possible.

1. Add.
2. Add With Carry.
3. Subtract.
4. Subtract With Carry.
5. Increment.
6. Decrement.
7. Compare.
8. Logic AND.
9. Logic OR.
10. Logic XOR.
11. Shift Left.
12. Shift Right.
13. Complement.

Instructions such as add, subtract AND, OR, etc, require two bytes of data to be present at the input to the ALU. One of these bytes will always come from the accumulator, via the temporary accumulator, (shown as A and ACT in *Figure 7.5*), and the second byte will come from a temporary register, (shown as TMP in *Figure 7.5*). The TMP register receives its data from a store or register via the internal data bus of the CPU. This means that for any ALU operation one byte must be in the accumulator and the other can come from any of the 8-bit internal general purpose registers or from memory.

The result of an ALU operation is fed back to the data bus and is always stored in the accumulator. In the 8080/8085 and most other 8-bit CPUs, there are no facilities for storing the result of an ALU operation directly in memory or in any of the other general purpose registers.

The ALU also feeds data to the flag register located on the left hand side of the ALU in *Figure 7.5*. The flag register stores the results of special events that have occurred during the operation of the ALU. For example if the data FF (Hex) stored in the accumulator was added to OF (Hex) in the B register the result would produce a carry from bit 7 to bit 8 resulting in a 9 bit answer.

Hex		*Binary*
FF	=	11111111
+OF	=	+00001111
10E	=	100001110

It has been previously mentioned that the result of the ALU is stored in the accumulator, which is an 8-bit register. Consequently the most significant bit of the answer would be lost. To overcome this problem the carry is stored in one of the 8 bits of the flag register, which can then be interrogated to find out if a carry was generated or not. The flag register stores a number of these special conditions generated by the ALU and the more common of these are:

(a) Zero.
(b) Parity.
(c) Carry.
(d) Auxiliary Carry.
(e) Sign.

These special flag bits give the microprocessor its real power of making decisions. All the decision making processes carried out in the 8080/8085 CPU take place as the result of the condition of one or more flags. For example the programmer can subtract two numbers and then test to see if the result is zero by interrogating the zero flag bit in the flag register. Depending on the flag bit being set to a logic '1' or cleared to a logic '0' the CPU can execute separate programmes carrying out a different sequence of events.

7.5 Input and Output ports

Input and Output (I/O) ports are the interface between the microprocessing system and the outside world. They enable bytes of data to be transferred from the microprocessing system to peripherals, and from the peripheral to the system. All of these operations are carried out under the control of the programme and consequently each I/O port will have an address which enables the CPU to identify it as one of the many elements operating within the system.

Interfacing the CPU with peripherals such as keyboards, VDUs, disc storage systems, etc is one of the most complex aspects of any microprocessing system in

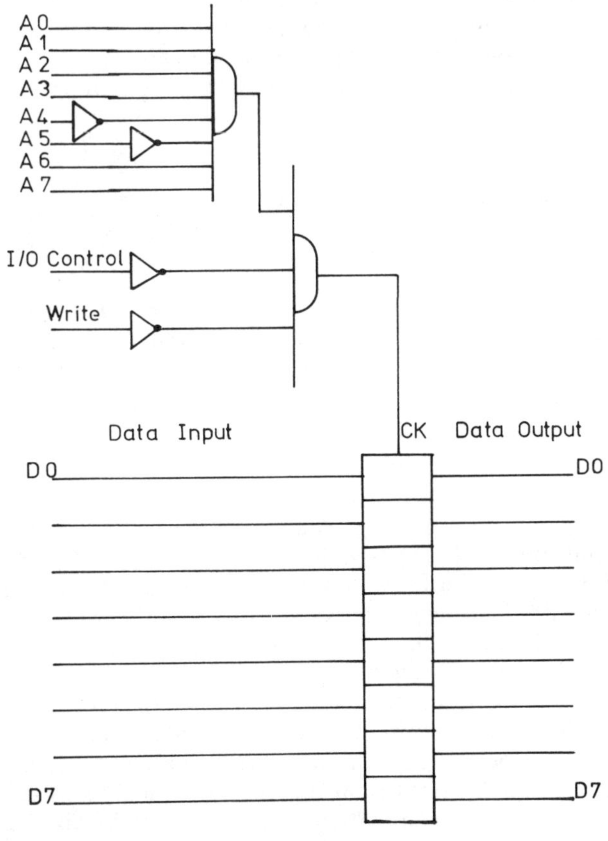

Figure 7.6 o/p port using D-type flip-flops

both hardware and software terms. A large proportion of hardware and software will be involved with I/O techniques, and some of the more advanced I/O systems such as handling the transfer of data between RAM and disc store are very complex indeed.

In its simplest form an output port is a series of eight D-type flip-flops, each one connected to its respective bit of the data bus. These flip-flops are clocked so that data is transferred from the data bus to the output of the flip-flops under the control of the CPU. As each output port has its own memory address then the clock input must be derived from decoded address and control signals which will ensure that the output port will only operate when it has been correctly addressed.

A simplified arrangement of an output port is shown in *Figure 7.6*. In this example, the port address is derived from the least significant 8 bits of the address bus and has a port address of CF (Hex). The address decoding circuitry will therefore interrogate the least significant 8 bits of the address bus and produce the desired output when these 8 bits produce CF (Hex). This address decoder will produce a logic '1' whenever CF (Hex) is present on the address bus.

Many memory address locations will also use the code CF (Hex) in the least significant 8 bits of their address, e.g. memory addresses 003C, 013C, 023C, etc. The CPU must therefore identify if the address on the address bus is designed to be used by memory circuits or for use by I/O circuits. This is achieved by a special control signal which will fall to a logic '0' whenever an input or output port is being used. (Note the majority of microprocessor control signals are active low). The CPU will also generate a read or write signal to identify whether the addressed port is being written to or read from. In this example the port is an output port and consequently we will require to clock the D-type flip-flops only when the following conditions exist:

1. The least significant 8 bits of the address bus = CF (Hex).
2. The I/O control line is at logic '0'.
3. The write line is at logic '0'.

When these three conditions exist we know that output port 3F (Hex) is being written to, and data on the data bus will be transferred to the output of this port. *Figure 7.6* shows the additional control and address line decoding required to generate the clock signal for the flip-flops.

Input ports

An input port operates in a similar way to an output port apart from the inclusion of a tri-state buffer between the outputs of the D-type flip-flops and the data bus. These are required because whenever the input port is not being used the outputs from the flip-flops which connect with the data bus must go tri-state. If they do not, a data conflict will occur and they will modify the pattern of 0s and 1s present on the data bus during other memory or I/O operations.

In the example shown in *Figure 7.7* the same port address of CF (Hex) has been used, but this time the control decoding circuits are decoding an I/O and read operation in order to generate the clock signal for the port. When the correct combination of control and address signals are present the input port will clock the input data to the Q output of each flip-flop and then transfer this data via the tri-state buffers to the data bus. Once this has been done the address and control

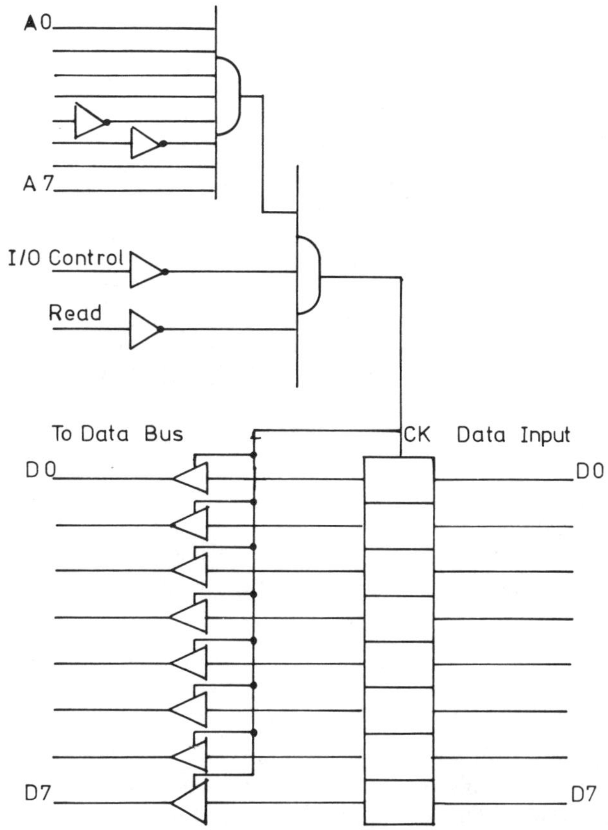

Figure 7.7 i/p port using D types and tri-state buffers

signals will change causing the input port to become inactive. Under these conditions the D-type flip-flops will not be clocked and the tri-state buffers will create an open circuit between the flip-flops and the data bus. This action releases the data bus for use by other circuits within the microprocessing system.

Programmable I/O ports

Manufacturers of microprocessor chips such as Intel, Zilog, Rockwell and Motorola etc. also produce support ICs for use with their central processor chips. In chapter 5 'Display Devices' we met one of these support ICs in the form of the Universal Asynchronous Receiver Transmitter (UART). Another dedicated support chip for a microprocessor system is the Programmable Input/Output Port or PIO.

This integrated circuit is specifically manufactured for use with a particular microprocessor chip. It normally has two parallel ports contained in the one IC package, and these can be programmed to be either input or output ports, keeping the system as flexible as possible. Each port of the PIO is bit selectable i.e. each data bit of the port can be selected to be either an input or an output. This gives the programmer the following choices:

1. 8-bit parallel input port.
2. 8-bit parallel output port.
3. Some of the bits are input bits and some are output bits.

In the last example it would be possible to programme the PIO so that bits 0 to 3 become input bits and bits 4 to 7 become output bits, creating two 4-bit I/O ports from a single 8-bit port.

Before a PIO can be used with a microprocessor system it must be programmed to establish which of the above modes is to be used. This small programme is executed by the CPU before any input of output operations take place, and it is known as the PIO initialization programme. The initialization programme is executed for each of the two ports to be used. Once the PIO has been initialized the system can then use the port in the desired mode. As the PIO is a programmable device it is possible to change the function of the I/O port during the execution of a programme, giving the user greater flexibility if it is required.

Most microprocessor based systems use a PIO chip in preference to the simple input techniques previously described because of the following reasons:

1. They give greater flexibility to the system.
2. They interface directly with the system with very little additional hardware.
3. Input/output handling can take place with the minimum of software.

7.6 The dedicated microcomputer chip

We have seen that the basic components of a microprocessing system are the CPU, RAM, ROM and I/O ports. It is not surprising that manufacturers have reduced the complete microprocessing system to a single chip, or the single chip microcomputer. Using modern IC techniques, manufacturers are now able to implement all the major functions of a complete microprocessing system within the constraints of a single chip, and a typical block diagram of such a microcomputer is shown in *Figure 7.8*.

It can be seen that the integrated circuit contains its own small area of Random Access Memory which is used for the temporary storage of data. It also has a 1 K byte ROM. This will contain the programme of instructions which causes the complete system to carry out the desired process. Because the programme is contained in the same chip as the CPU, I/O ports etc the complete system will only be able to carry out the coded instructions contained in this ROM. The user of the system will not be able to use the microcomputer for any other function apart from that implemented by the programmed ROM.

This technology is therefore only used where a manufacturer has an application for a large number of microprocessing systems carrying out exactly the same function, e.g. controlling washing machines, petrol pumps, fuel injection systems or games machines, etc. Applications for the single chip microcomputer are found in TV in the form of control systems for Video cassette recorders and this has been expanded to include their use in Teletext/Viewdata decoders, and video disc players.

Once the equipment manufacturer of an electronic system has the programme written for the microcomputer chip, it is sent to the IC manufacturer, who uses the programme to determine the pattern of the ROM connections within the microcomputer. Once the ROM mask connections have been calculated it is used together with the remainder of the mask in the final manufacturing process of the chip.

Initially only a few of the chips will be made and then tested by the equipment manufacturer before large orders are placed. If software errors are encountered the programme will have to be corrected before another mask pattern is calculated for the ROM inside the microcomputer. Once the equipment manufacturer is convinced that the microcomputer system is working perfectly then the integrated circuit can be mass produced by the IC manufacturer at a very economical rate.

Equipment manufacturers are increasingly using this technique because it reduces the component count of the system, and is significantly cheaper than producing a discrete component microprocessing system using separate ROM, RAM, CPU and I/O chips.

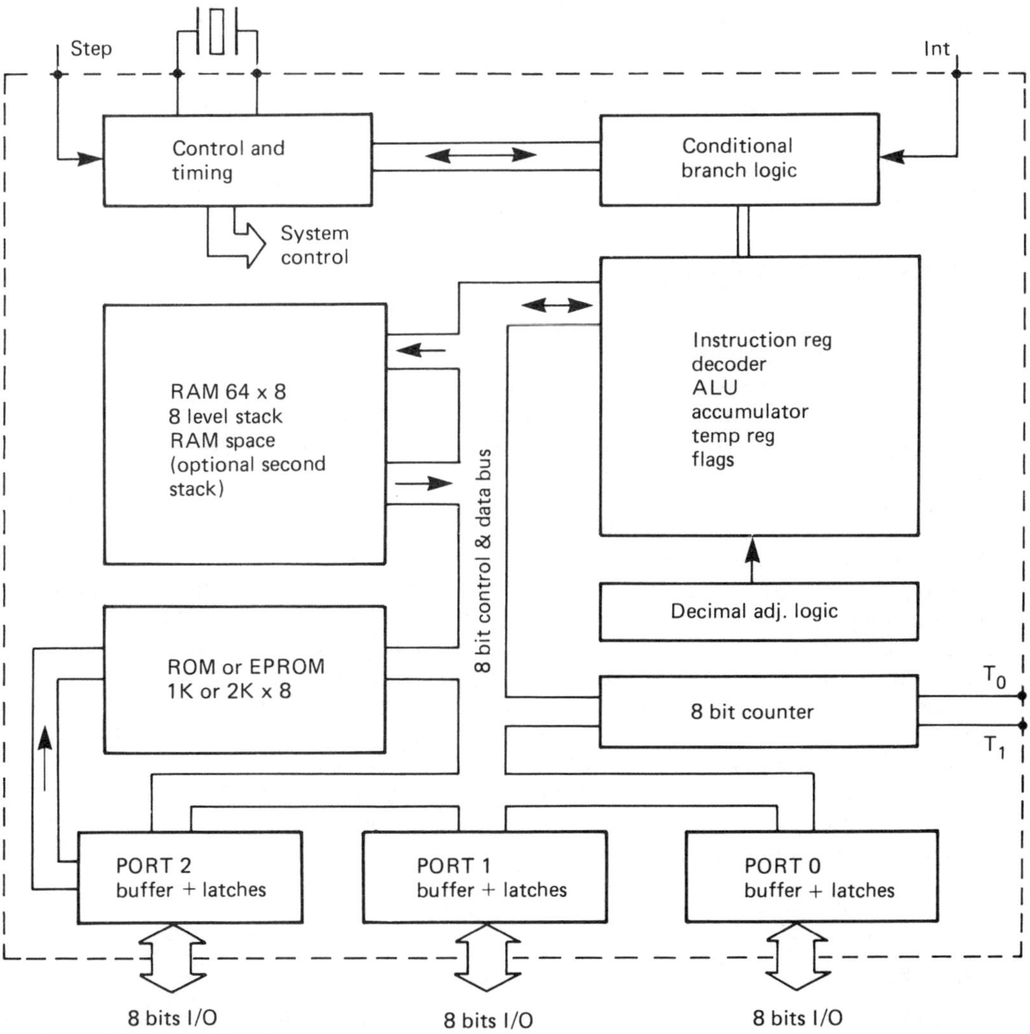

Figure 7.8 Single chip microcomputer

At the bottom of *Figure 7.8* there are three input/output ports. These allow a total of twenty-four separate digital connections to be made between the microcomputer and the rest of the system. Each of these I/O ports is programmable giving the programmer maximum flexibility in software and external hardware design. One of the first things that the microcomputer chip will do when power is applied is to run the initialization programmes to set up the required I/O configuration on these ports.

To the right of the RAM and the ROM is the Central Processing Unit which contains its own ALU and flag register. At the top of the diagram is the clock and timing circuits together with the instruction decoder and controller. All of these blocks carry out exactly the same functions as those previously described for the 8080/8085 CPU.

Most single chip microcomputers have the facilities for extending the address data and control signals to the IC package pins. This facility provides the user of the microcomputer chip with the option of using additional RAM or ROM if required.

As all the electronics of the microcomputing system is implemented on a single chip of silicon, the speed of the system can be increased because all the circuits are in close proximity, and little interfacing is required between them. The diagram shown is a simplified version of the Intel 8048 single chip microcomputer, which uses a 6 MHz clock frequency compared with the 4 MHz clock frequencies normally encountered with the single CPU chips such as the 8080/8085 or Z80 processors.

7.7 Example of a simple programme

Programming a microprocessor is not a situation met by the TV service engineer and consequently, as with other aspects of this chapter it will not be dealt with in any great detail. It is important, however, that the engineer has an appreciation of the sort of data manipulation processes that take place within a typical microcomputing system, as it is these processes which will be controlling circuits and systems which the engineer will have to fault find.

The following description is of a simple sequence of programme instructions which will give the reader an appreciation of the microprocessor. Let us assume that the microprocessor has to input a 7-bit data word, which has been generated by a keyboard, via an input port. At the input port there will be a 7-bit binary code representing the letter or number of the key that was depressed. The microprocessor system is required to add an even parity bit to this character code to create an 8-bit symbol code, and then output the complete symbol code to an output port. This sort of operation would be required of a Viewdata decoder where the operator of the system is able to transmit figures and commands to the data base, and these will be 8-bit symbol codes using even parity.

The first step is to devise an algorithm for the programme. An algorithm is a step by step solution to a problem and it can be expressed in any form because no computer can understand a natural language. The algorithm then has to be converted into a series of programme instructions which the microprocessor can understand. The algorithm of this simple exercise is shown below.

Step one

The first step is to input the 7-bit word of data from the input port. As the data is only 7 bits wide and we are to use an 8-bit microprocessing system then the most significant bit of the input port will be connected to logic '0' (Bit 7). This first step will store the data generated by the keyboard in the CPU.

Step two

The second step requires the CPU to check the parity of this byte of data. It has been seen in the previous description of the CPU that this can only be done by the Arithmetic Logic Unit, which operates on two bytes of data, leaving the result in the accumulator. What is required therefore is a process which will cause the ALU to set the parity flag in the flag register without altering the byte of data from the keyboard. This can be achieved by a logic OR between the data generated by the keyboard and 00000000 in binary. The result of this process will leave the keyboard data unaffected but set the parity flag in the flag register.

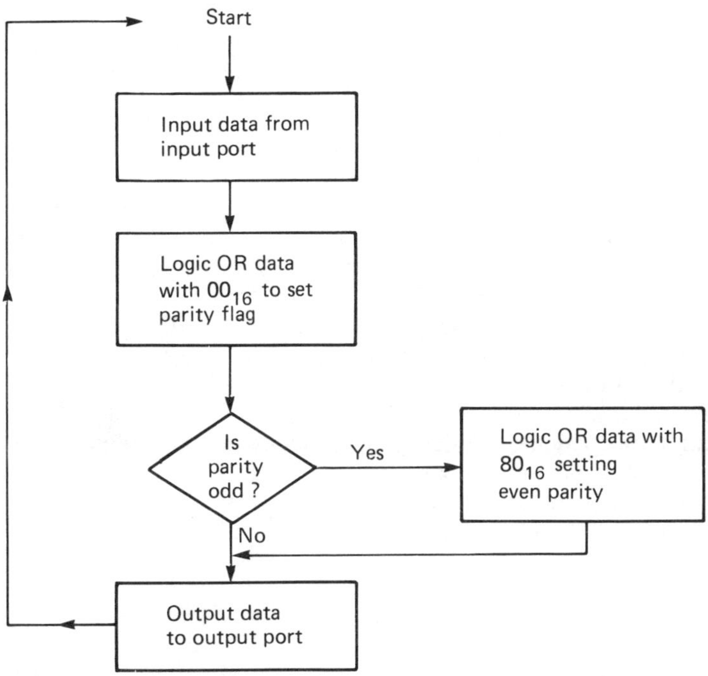

Figure 7.9 Programme flow chart

Step three

If the parity flag after step two indicates that odd parity exists then we have to set bit 7 to a logic '1' as described in step four, but if parity is even we have to leave bit 7 at logic '0' and jump to step five.

Step four

If the parity flag indicates odd parity the keyboard data will be logic ORed with 10000000 in binary so that the keyboard data remains unaffected but bit 7 is set to a logic '1'.

Step five

Now that we have an 8-bit byte of data with correct parity it is output to the required port.

Step six

When new data appears at the input port the programme will jump back to step one.

Having decided that this algorithm has a chance of working, the programmer creates a flow chart which lays out the step by step sequence of programme instructions which will fulfil the requirements of the algorithm, and this flow chart is shown in *Figure 7.9*. Again the feasibility of each operation is checked as the flow chart is created, and any errors that are found in the algorithm are corrected.

On completion of the flow chart each box is converted into its corresponding series of machine code instructions which the CPU will then execute. The programme is entered into a computer development system which simulates the operation of the process under development and is then tested. Any errors that occur have to be removed, and the algorithm/flow chart modified accordingly until the programme runs smoothly.

It is normally the last stage of this process which takes the longest period of time. It is quite common for a programmer to find that certain elements of the algorithm have been overlooked or underestimated, and that the programme does not run successfully when first entered into the development system. If this is the case then the programmer has to step through the programme one instruction at a time, making sure that the desired results are being produced. This process of checking and correcting a programme is very time consuming, and is reflected in the high cost of original software.

Manufacturers of equipment which use microprocessors as part of their control system would have invested a considerable sum of money into the development of the system software, and it is quite common to find that software development constitutes at least 50% of the total system development costs.

Chapter 8

The Teletext system

Teletext is a generalized term given to the transmission and reception of digitally coded text which is inserted during the field blanking period of a TV video waveform. This system of transmission was pioneered in the early nineteen seventies by the British Broadcasting Corporation (BBC) and the Independent Broadcasting Authority (IBA).

ORACLE (Optical Reception of Announcements by Coded Line Electronics) was the name given to the IBA system which started test transmissions in April 1973 and CEEFAX (See Facts) was the name given to the BBC system which started its test transmissions in July 1973. Initially the two systems were non-compatible but in 1974 the BBC and the IBA adopted a joint standard now known as Teletext, which essentially means text at a distance.

During these early trials specially adapted receivers were used to evaluate the potential and performance of the Teletext system. These receivers contained 'Teletext decoders' which used a large amount of custom built combinational and sequential logic electronics. The early decoders were large and complex, with some containing more than one hundred integrated circuits. This made them very expensive and they were not commercially viable for the domestic market.

As soon as these trials were proven the major manufacturers of integrated circuits invested large amounts of time and money into the development of large scale integrated circuits (LSI) which would reduce the component count, and hence the cost of a Teletext decoder. Texas Instruments was the first manufacturer to produce a dedicated Teletext decoder module using LSI technology.

In 1978 Mullard produced their first decoder using four custom made LSIs and six smaller integrated circuits. Since the introduction of the first commercial decoders both manufacturers have extended their facilities to include the Viewdata system (described in chapter 10).

Because of the introduction of these compact modules the concept of Teletext has been commercially realised. With further advancements in the technology of the integrated circuit the size, complexity and cost of these modules will fall considerably until a time is reached where the Teletext receiver is as commonplace as the conventional colour receiver.

8.1 The Teletext display

Before examining the specification of the Teletext system in detail it is necessary to familiarize the reader with the major display facilities which Teletext offers. Current LSI Teletext decoders have a very wide range of display modes such as boxing the real time clock into the 'off air' picture, and enabling the viewer to display the top or bottom halves of a page separately. The following examples are designed to explain Teletext facilities, and it has been decided to leave out many of the more advanced display modes which comprehensive decoders provide.

Display modes

A Teletext receiver will normally have four display modes, these being:

(a) The 'off air' picture alone.
(b) Teletext pages alone.
(c) Teletext superimposed on top of the 'off air' picture.
(d) Teletext boxed into the 'off air' picture.

Picture mode

In the picture mode, the television operates as a conventional receiver and no text is displayed on the screen.

Teletext mode

In the Teletext mode, pages of text and graphics are displayed using a format similar to that shown in *Figure 8.1*. Most pages are displayed alone on the screen.

Superimpose mode

Text is superimposed on top of the 'off air' picture, so that the receiver displays a colour picture and monochrome text simultaneously.

Boxed mode

A boxed display is used almost exclusively for News Flash pages and subtitles for the hard of hearing. In this mode the 'off air' picture is displayed and the text is boxed into the picture. In the case of the News Flash the boxed area normally appears in the lower portion of the picture while subtitles are positioned anywhere in the picture below the page header row.

8.2 Teletext pages

Pages are normally transmitted in numerical sequence at an approximate rate of eight per second. Each television channel has the ability to transmit a maximum of 800 pages, which are subdivided into 8 magazines each having a maximum of 100 pages. Simple calculations will show that with 4 lines of text transmitted per field (worst possible case), it will take 96 seconds or 1.6 minutes to receive a page of text.

Figure 8.1 CEEFAX Index page

Access times of this length are unacceptable, and to offset this problem the broadcasting authorities have reduced the number of pages they transmit, with the more widely used pages being transmitted more frequently than the others, e.g. a magazine of 100 news and current affairs pages could be inserted twice in a 400 page sequence in the following order:

0 to 99	Low priority text
100 to 199	News and current affairs
200 to 299	Low priority text
100 to 199	News and current affairs
300 to 399	Low priority text
	Repeat transmission sequence.

This action will decrease the access time for the news and current affairs pages but will increase the access time for the others.

Another way of decreasing the access time is to increase the number of lines in the field blanking period which contain Teletext data. At the present time four lines per field blanking period are used and if, for example, this were to be increased to eight lines then access times would be halved. This concept is discussed in more detail later in this chapter under the heading 'The Teletext specification'.

Rotating sub pages

In some situations it is desirable to sub divide a page allocation into a number of sub pages, e.g. page 199 may have 4 sub pages; 199/1, 199/2, 199/3, and 199/4. These sub pages are transmitted in sequence on the occurrence of the page number. The first time page 199 is received page 199/1 is displayed. The second time page 199 is received page 199/2 will be displayed etc. this facility enables far more that the maximum of 800 pages to be transmitted, but of course the access time to any one of the sub pages is considerable. The introduction of sub pages does not effect the access time of other pages in the magazine.

A single page of text can be sub-divided into two major sections:

(a) The header row.
(b) The page text.

The header row

The header row appears at the top of every page and indicates to the viewer:

(a) The name of the Teletext system being received. (CEEFAX or ORACLE).
(b) The page number being displayed.
(c) The day, date and month.
(d) The real time clock.

(*See Figure 8.1*)
Many decoders insert to the left of the header row the page number that the viewer has selected. This enables the viewer to confirm that the page being displayed is in fact the page that was selected.

Rolling header row

Most decoders have a rolling header facility which comes into operation immediately after a new page of text has been selected. A rolling header displays the header rows being received on a real-time basis so that the viewer can watch the sequence of pages as they are received. When the received page number corresponds with the page number selected the rolling header stops, and the requested page header is displayed in its place.

The header row can only display a maximum of 32 characters (not including those inserted by the decoder), because the other character codes received are used for non display purposes by the decoder. (See 8.4 'Teletext specification').

The page of text

The area of the page containing text information consists of 23 rows each containing 40 characters. These characters may be displayed as alphanumeric (upper and lower case characters or numbers), or graphics. Each character can be displayed in white or in any of the primary or complementary colours, e.g. red, green, blue, yellow, cyan and magenta. Using the index page from the BBC CEEFAX Magazine shown in *Figure 8.1*, it can be seen that the rows of text immediately below the page header row contain graphic display characters giving the name of the magazine. Below this the rest of the display consists mainly of alphanumeric text in a variety of colours.

Double height characters

Alphanumerics are normally of a fixed height, pre-determined by the height of each row of text. Decoders also have a double height facility which enables alphanumeric characters to be displayed over the height of two rows of text and not just one. An example of the double height facility is shown on the Engineering Test page in *Figure 8.2*.

Reveal

Some pages contain question and answer information which makes use of the reveal feature. This is a facility under the control of the viewer, where a page is initially displayed with certain predetermined portions of the text missing from the display. This enables the decoder to display questions, and when the viewer selects the reveal function the decoder inserts the answers.

Flash

A facility that cannot be shown using photographs is Flash. Here portions of the text display, which can be either alphanumeric or graphic, can be made to flash on and off at a rate which is pre-determined by circuits in the Teletext decoder. This facility is very useful when the viewers' attention must be drawn to a portion of the page.

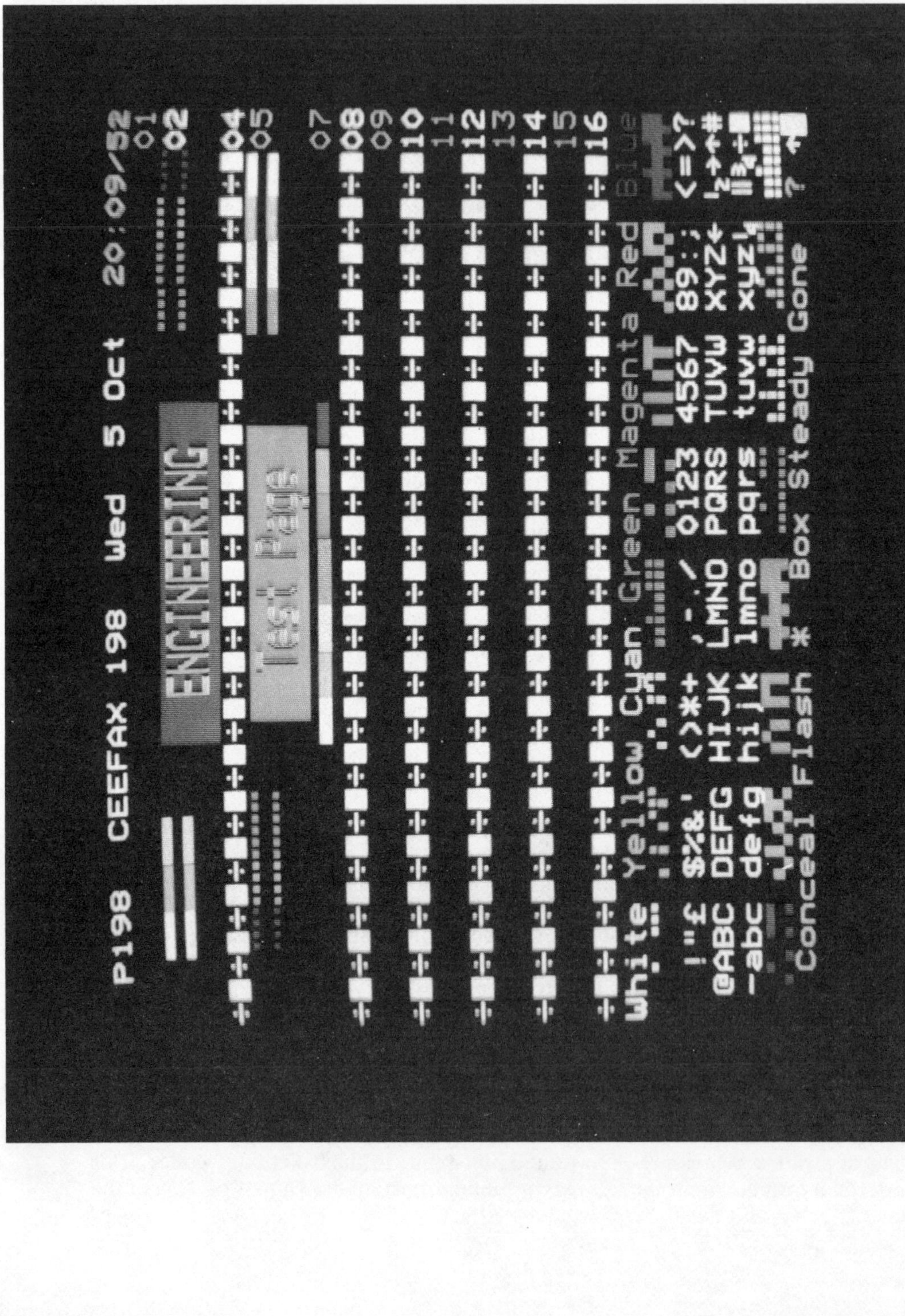

Figure 8.2 CEEFAX Engineering Test page

8.3 The engineering test page

Engineering test pages are transmitted by both the BBC and the IBA to aid the TV engineer in fault finding and assessing the performance of a Teletext receiver. Test pages differ in presentation between the broadcasting companies. Examples given in this section are taken from magazines transmitted by the BBC.

Clock cracker

Accurate synchronization of the data clock oscillator in the decoder is imperative if error-free decoding is to be achieved. The first two bytes of data received on a line of text are known as the clock run in, and they consist of alternating logic '1s' and '0s' which are used to synchronize a clock oscillator at the beginning of the line. After the clock run in oscillator synchronism is provided by the transitions between logic levels that take place during the transmission of the text.

The character code of the text is designed to have at least one transition in the logic levels during each byte of data, and the worst possible case which can be encountered is where there is only one transition between logic levels during each byte of data. As Non Return to Zero (NRZ) coding is used this situation can only arise during the transmission of two characters, these being the divide and the block characters. Referring to *Figure 8.3* it can be seen that the divide character has a code of 01111111 and the block character has a code of 11111110. Both of these codes have been written with the addition of an odd parity bit and with the LSB first.

The clock cracker pattern is illustrated in the engineering test page *Figure 8.2*, shows that each row of text consists of alternating divide and block characters. During the transmission of these two bytes of data there will only be one transition in the logic levels during each byte (0111111111111110). This is the worst possible situation for the data clock oscillator as it receives the smallest possible number of logic transitions per byte of data. If the data oscillator is to lose synchronism it will be during the transmission of the clock cracker and the errors will be at their worst towards the end of the line. This test is able to give an indication of the oscillator stability and can be used to set the decoder's data oscillator frequency.

The combined engineering test page

Lines 01 to 07 form the identification of the page with the words 'Engineering Test Page' being displayed using double height characters on a coloured background. Both contiguous and separated graphics are used in lines 02 and 05.

Lines 08 to 16 display the clock cracker pattern described in the previous section. Each line is identified by a red number at the far right hand side of the picture.

Lines 17 to 22 contain the complete range of alphanumeric characters together with examples of separated and contiguous graphics. Many of the earlier Teletext decoders do not have a separated graphic mode, hence all the graphic characters in this area of the page will be displayed in the contiguous mode. Row 17 consists of coloured alphanumeric characters forming the names of the three primary and three complementary colours which are available in the Teletext system. As with a standard colour bar pattern they are arranged in descending order of luminance.

Line 23 enables the engineer to check that the reveal, box and flash circuits of the decoder are operating correctly. The photograph has been taken with the reveal feature selected and the flash characters displayed.

b4	b3	b2	b1	Row \ Col	0 (b7 b6 b5 = 0 0 0)	1 (0 0 1)	2 (0 1 0)	2a (0 1 0)	3 (0 1 1)	3a (0 1 1)	4 (1 0 0)	5 (1 0 1)	6 (1 1 0)	6a (1 1 0)	7 (1 1 1)	7a (1 1 1)
0	0	0	0	0	NUL①	DLE①			0		@	P	—		p	
0	0	0	1	1	Alpha^n Red	Graphics Red	!		1		A	Q	a		q	
0	0	1	0	2	Alpha^n Green	Graphics Green	"		2		B	R	b		r	
0	0	1	1	3	Alpha^n Yellow	Graphics Yellow	£		3		C	S	c		s	
0	1	0	0	4	Alpha^n Blue	Graphics Blue	$		4		D	T	d		t	
0	1	0	1	5	Alpha^n Magenta	Graphics Magenta	%		5		E	U	e		u	
0	1	1	0	6	Alpha^n Cyan	Graphics Cyan	&		6		F	V	f		v	
0	1	1	1	7	Alpha^n White②	Graphics White	'		7		G	W	g		w	
1	0	0	0	8	Flash	Conceal Display	(		8		H	X	h		x	
1	0	0	1	9	Steady②	Contiguous② Graphics	)		9		I	Y	i		y	
1	0	1	0	10	End Box②	Separated Graphics	*		:		J	Z	j		z	
1	0	1	1	11	Start Box	ESC①	+		;		K	←	k		¼	
1	1	0	0	12	Normal② Height	Black② Background	,		<		L	½	l		‖	
1	1	0	1	13	Double Height	New Background	-		=		M	→	m		¾	
1	1	1	0	14	SO①	Hold Graphics	.		>		N	↑	n		÷	
1	1	1	1	15	SI①	Release② Graphics	/		?		O	#	o		■	

① These control characters are reserved for compatability with other data codes

② These control characters are presumed before each row begins

Codes may be referred to by their column and row e.g. 2/5 refers to %

□ Character rectangle

Black represents display colour

White represents background

Figure 8.3 Teletext character codes

8.4 The Teletext specification

The first Teletext specification was published in September 1974 jointly by the BBC, IBA and BREMA (British Radio & Electronic Manufacturers' Association)[1]. Since then the specification has been updated on two occasions, first in January 1976[2], and in September 1976[3]. The September 1976 publication of the Broadcast Teletext Specification contained the essential features of the then existing system plus new bytes of data which gave improved graphics and other facilities. Since the September 1976 publication, the Broadcast Teletext Specification has changed very little and therefore the following description is based on this publication.

The Teletext data is contained during the field blanking period of both the odd and even fields, and its position in field blanking is shown in *Figure 8.4*. The first section of the field blanking period contains the equalizing and field synchronizing pulses which the receiver uses to trigger the field timebase. It takes several line periods for the electron beam to return from the bottom of one field scan to the top of the next, and to prevent the danger of video information appearing on the screen during this flyback period a number of lines are transmitted after the field synchronizing information to allow the new field scan to 'settle'.

The colour burst is not transmitted during the equalizing and field pulse sequence, hence the subcarrier oscillator in the decoder of a colour receiver will lose synchronism during this period. Before a new field of colour information can be displayed on the screen the receiver's subcarrier oscillator must be accurately locked to the colour burst, and this is achieved by including a colour burst on most of the blank lines transmitted during the field blanking period.

During some transmissions it is possible that additional data will appear during the field blanking period. This information is used internally by the broadcasting companies to identify the source of the accompanying picture together with network status.

Lines 19 and 20 on the even field and lines 332 and 333 on the odd field contain video test waveforms which enable the broadcast engineer to monitor the performance of video lines and transmitters. They can be of some use to the television service engineer in assessing the performance of the receivers IF and video amplifiers. Before the Vertical Interval Test Signals (VITS) are four lines which contain the Teletext data, these are numbered 15, 16, 17 and 18 on the even field, and 328, 329, 330 and 331 on the odd field. It is this information which is extracted from the video waveform and used by the Teletext decoder to display the pages of text. Initially just two lines were used for data transmission, but in 1981 this was expanded to four, and could be further increased if necessary.

Each line of data occupies the same space as a standard picture (52 micro-seconds), and has a colour burst on the back porch, see *Figure 8.5*. Data is transmitted in serial form using NRZ techniques, with the pulses shaped by a complex filter to enable the data to fit inside the pass band of the 625-line television system. A level of binary logic '0' corresponds to the normal black level of the picture plus or minus 2%, where as a binary logic '1' represents a video level of 66% plus or minus 6%. (Peak white = 100%).

The complex nature of the waveform and the use of NRZ transmission causes the data waveform to contain overshoots so that the peak to peak data amplitude will exceed the basic data amplitude. A bit rate of 6.9375 Mbits/sec. has been chosen which is 444 times the nominal television line frequency. This enables the

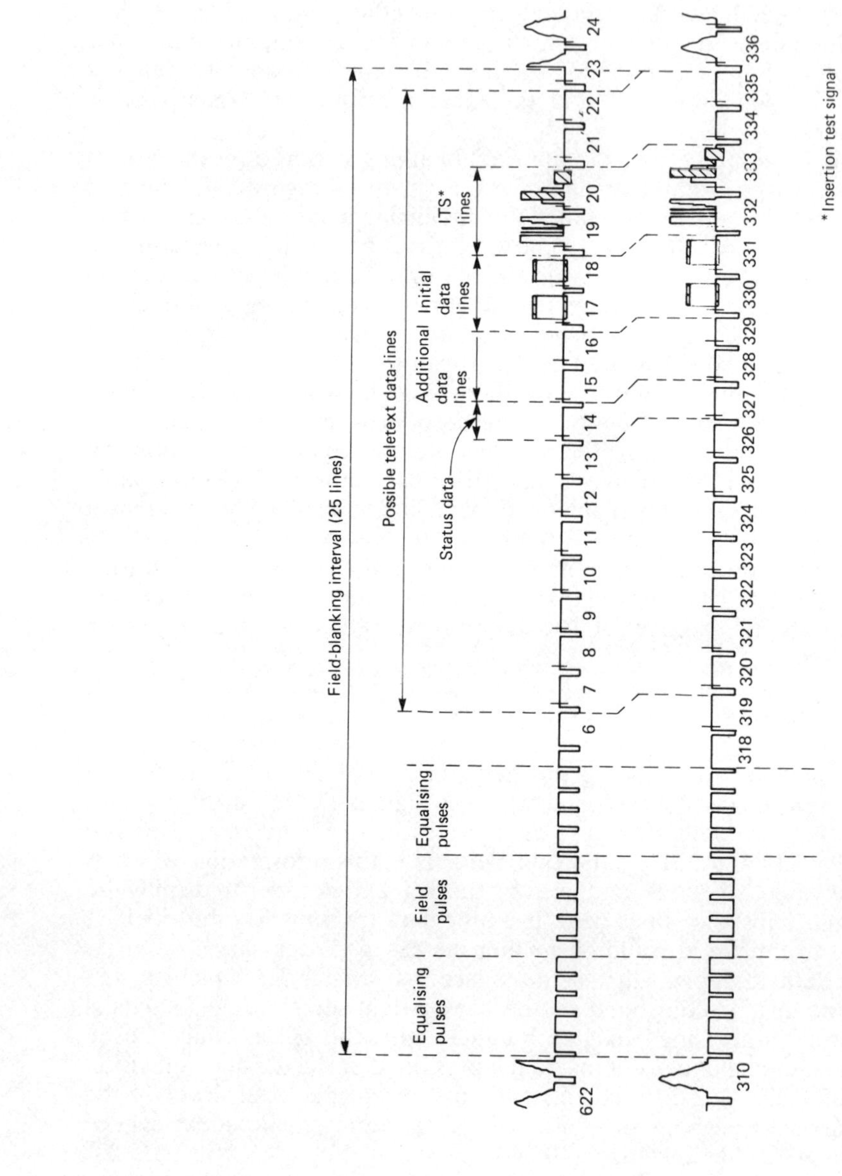

Figure 8.4 Field blanking period with Teletext

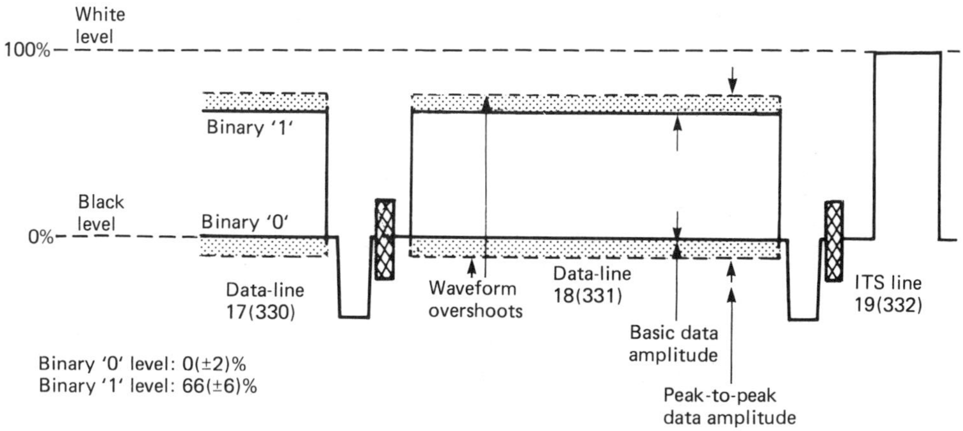

Figure 8.5 Two-lines of Teletext data

transmission of 360 bits of data, divided into 45 separate 8-bit bytes, during the active picture time of each line. Seven bits of the byte are used to identify the character code to be displayed, while the eighth bit enables odd parity detection on each character. The exception to this rule is in the first three bytes of data on each line which all have even parity.

The clock run-in

A line of Teletext data can be identified by the first two bytes of data which consist of a clock run-in sequence of alternating logic states producing a pattern of 1010101010101010. Refer to *Figure 8.6*. This clock run in has a frequency of 3.46875 MHz which is exactly half the data bit rate. It is used to synchronize a bit sampling clock oscillator in the Teletext decoder and can be compared in function to the colour burst which is used to synchronize the subcarrier oscillator in the PAL colour decoder.

Once the data clock oscillator in the decoder is locked, oscillator synchronism is maintained during the rest of the Teletext line by the transitions in logic levels that occur during the transmission of text. Data coding of the text characters is arranged so that there is a minimum of one transition during each byte of data.

Framing code

Following the two bytes of clock run in is a unique single byte of data known as the framing code. This code follows the pattern of 11100100. It is used to synchronize the decoding of each byte of data and tells the decoder the precise point at which each byte of text starts. Because of the very careful choice of framing code pattern, decoders are able to detect and correct a single bit error should the framing code be corrupted by noise or interference. Immediately after the last bit of the framing code is received and identified the Teletext decoder 'knows' that the next bit of data received is the first bit of the first byte of text.

Magazine and row address

The next two bytes are the first bytes of text which the decoder can use to produce a display, and they are known as the magazine and row address group. Each of these

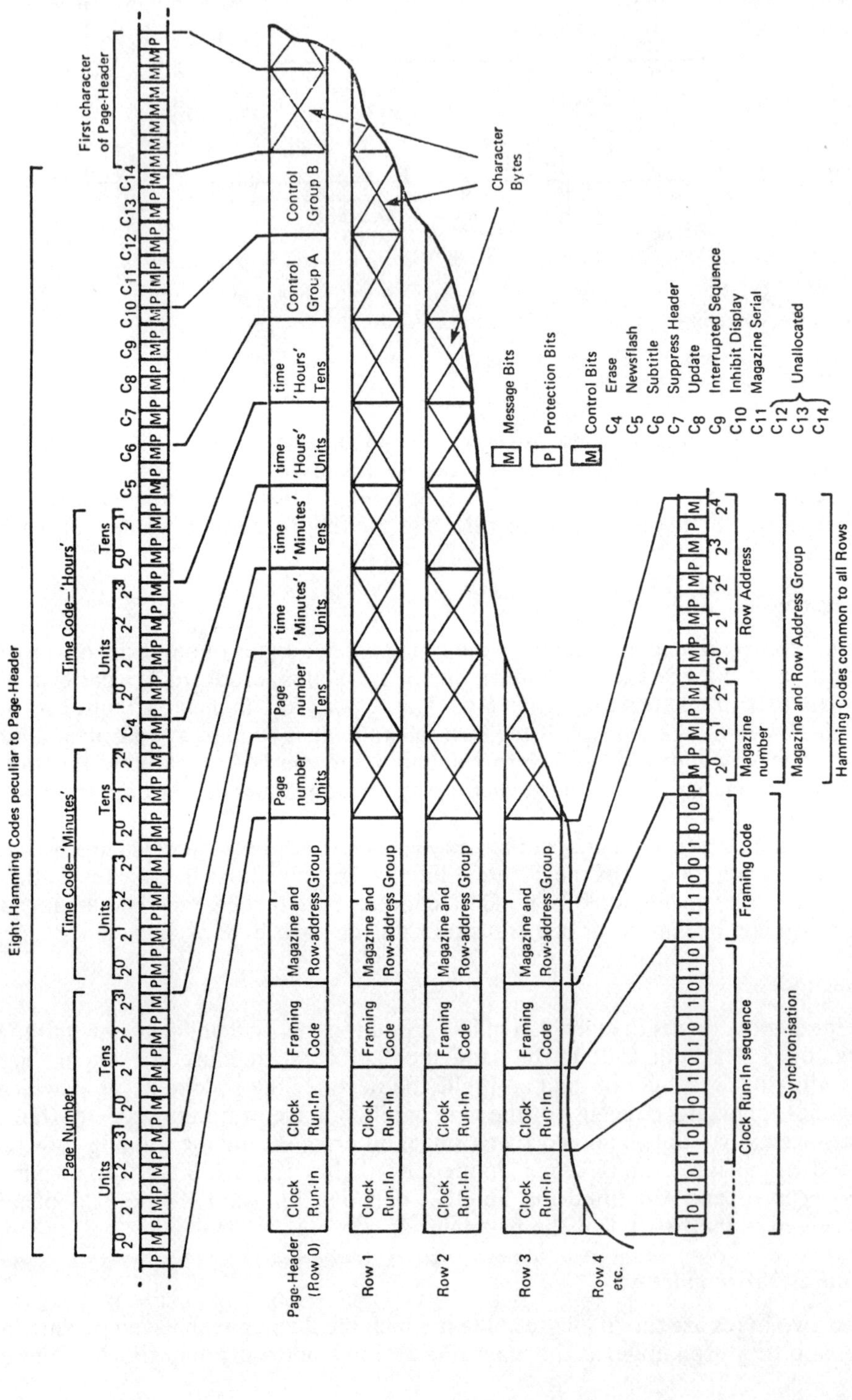

Figure 8.6 Detail of the Teletext rows

bytes is Hamming code corrected and contains four message bits and four parity bits which enable the detection and correction of a single bit error. (See chapter 6 for a more detailed description). The first three message bits indicate the magazine number which can be from 0 (000) to 7 (111), giving a maximum of eight separate magazines. Message bits four to eight identify the row address which can be any number from 0 to 31. A page of normal text contains a maximum of 24 rows and hence row addresses in the range of 0 to 23 will be the only rows decoded. Row addresses in the range of 24 to 31 can be transmitted but will be ignored by a conventional decoder.

The function of the first five bytes is the same for each row, they do not produce a text display, and are summarized below.

Function	*No. of bytes*
Clock run in	2
Framing code	1
Mag. and Row address	2

Every one of these five bytes is required by the decoder before text can be accurately decoded.

The page header row

Teletext lines having a row address of 0 are known as page header rows. They are always transmitted on lines 18 or 331, and have an additional 8 Hamming coded bytes of data after the magazine and row address group. The format of these additional eight bytes is shown in *Figure 8.6* and the message bits contained in the bytes identify:

1. The page number of the magazine being transmitted.
2. The time, coded in hours tens and units, and minutes tens and units.
3. Control functions which provide special decoding facilities.

Page numbering

Each page of text is numbered just once, in the page header row. This obviates the need to have the page number coded into every line of text and therefore enables more bytes in each row to be used for the text display. Bytes six and seven of the header row have message bits interleaved with parity bits containing the Binary Coded Decimal (BCD) page number. These numbers follow standard data transmission practice in that the least significant byte is transmitted first, e.g. the first four message bits contain the code of the page units and the last four message bits contain the code for the page tens. This gives each magazine a maximum of 100 pages (0 to 99).

Time-coded bytes

Bytes eight to eleven of the page header row contain time-coded data with the least significant digit transmitted first. Hence the first four message bits represent the minutes units, the second group of three message bits are the minutes tens, the third group of four is the hours units and the final two bits represent hours tens. All are transmitted in BCD form. Message bits one and two of the hours tens byte are used for the hours tens coding, e.g. numbers 0 to 3. Message bits 3 and 4 of this byte and message bit 4 of the page tens byte, are control bits.

The hours and minutes of the time code are not necessary related to the real-time clock which is transmitted at the end of the header row, although for most of the time the bytes do follow the real-time clock. This enables the viewer to select a predetermined page of text at a predetermined time of the day.

A second facility which time coding offers is the detection of up to 3200 different versions of a single page, e.g. page 100 can be transmitted with 3200 different numbers contained in the time coded bytes of data. Note that the hours tens figure may have any value between 0 and 3 and the minutes tens may have any value between 0 and 7.

Control groups

Control bits are required to indicate to the decoder that special display functions are required on the following page of text. Eight control bits are used and they are numbered from C4 to C11. Control bits 12 to 14 inclusive are spare. When a control function is not being used the corresponding control bit is cleared to a logic '0'. A control bit set to a logic '1' indicates to the decoder the need to modify its normal operation.

Bit C4 clear page

When this bit is set to a logic '1' it indicates to the decoder that the page about to be received is significantly different to the previous transmission of the same magazine and page number. When this bit is detected the Teletext decoder will completely erase any stored text during the following field scan, so that the following lines of text will be written into a memory which has been pre-stored with logic '0s'. It is therefore impossible for any previously stored data to be displayed on the screen.

Bit C5 news flash

This bit identifies the transmission of a news flash page and when it is set to a logic '1' the decoder switches the receiver display back to the 'off air' picture and then boxes in any text that appears on the Teletext page. The header row is automatically suppressed from the display.

Bit C6 subtitle

This control bit identifies the transmission of subtitles and when it is set to a logic '1' the decoder again switches the display back to the 'off air' picture and boxes any text. As with the news flash the header row is suppressed.

Bit C7 suppress header

When this control bit is set to a logic '1' the page header row is suppressed completely. This facility is used mainly with pages of graphic material where the presence of the header row would spoil the overall presentation of the page.

Bit C8 update indicator

This control bit is set to a logic '1' when part or all of the following page contains updated information on the previous transmission of the magazine and page number. It is possible that the following page of text may contain only one or two rows of updated rows of text, so the decoder can slot them into the correct place in the store. This prevents the need for a completely new page to be transmitted and therefore decreases the access time to the magazine.

Bit C9 interrupt sequence

When subtitle pages are transmitted, e.g. as an aid to the hard of hearing, it is necessary to interrupt the sequence of page transmission in order to synchronize the change of a subtitle to the change of an 'off air' picture. When this is required the interrupt sequence control bit is set to a logic '1' and the page number in the rolling header is suppressed. This avoids discontinuity of the displayed page numbers in the rolling header.

Bit C10 inhibit display

This control bit is set to a logic '1' when the contents of the following page of text cannot be usefully interpreted as Teletext data, e.g. with the use of experimental pages containing computer data etc. During this mode of operation the data is written into the decoder memory but is not displayed on the television screen.

Bit C11 magazine serial

When this control bit is set to a logic '1' the header row displays a rolling header even though a single page may have been selected and stored by the decoder.

Bit C12 to C14

These control bits are unallocated at the time of writing and enable additional control facilities to be added to the system for future use.

Page header text

After the transmission of control group B there are 24 bytes of text which contain the name of the Teletext magazine being received, i.e. CEEFAX or ORACLE together with a page number, day, date and month. The format of these 24 characters is not fixed and can be altered by the editor of the magazines. Two possible examples are shown in *Figure 8.7*.

Following these 24 characters are the final eight bytes of data which produce a display of a real time clock. This block of eight characters is continuously displayed in the top right hand corner of the page to give the viewer an accurate time check. Some Teletext decoders enable the real time display to be boxed into the off air picture for a short period of time, to provide a time check during the picture mode.

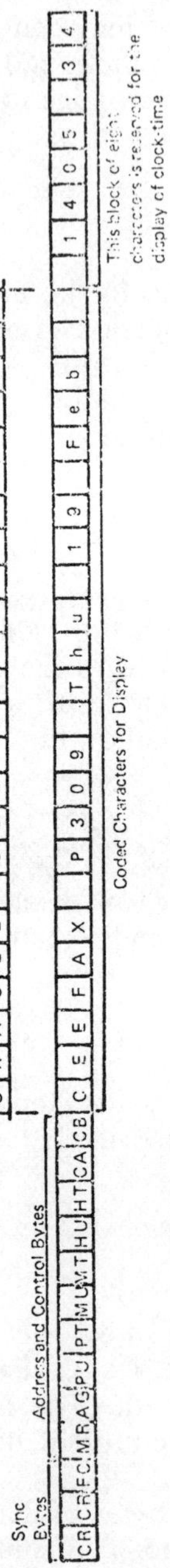

Figure 8.7 Two examples of header row presentation

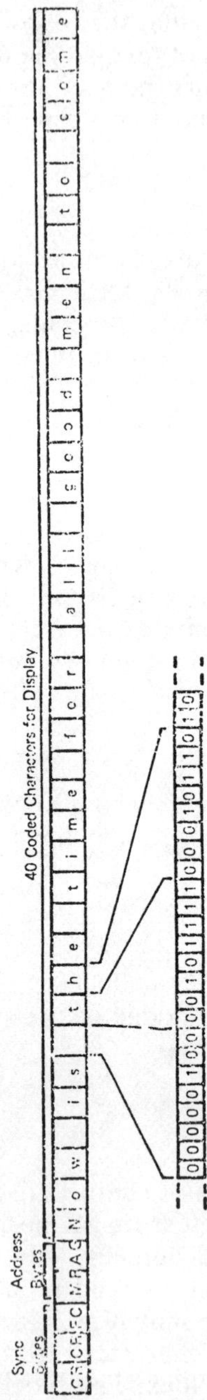

Figure 8.8 Detail of display rows 1–23

The page display

A total of 45 bytes are transmitted per row of text, the first five being used for the clock run in, framing code and magazine and row address group. This leaves 40 bytes of data which can be used for text display. Each byte contains a 7-bit character code which represents either a display character or a control character. Display characters are used to form the graphics or text to be displayed on the screen, while control characters inform the decoder of changes in the display mode, e.g. a change from alphanumerics to graphics. It should be noted that the control characters do not produce display information. A typical format for rows 1 to 23 is shown in *Figure 8.8*.

Whenever a new row of text is decoded the display circuits in the decoder assume that certain display modes are to be used. These are as follows:

1. Alphanumerics

The Teletext decoder will automatically assume that the first byte of text received after the magazine and row address group will be an alphanumeric character.

2. Contiguous graphics

When the decoder switches to the graphics mode, after having received the appropriate control character, the graphic shape that will be displayed will fill the character rectangle.

3. Display colour

The RGB outputs from the Teletext decoder will all contain video information thus producing a white character or graphic display.

4. Black background

The surrounding background to the alphanumeric or graphic display will be black.

5. Reveal

The reveal circuits of the decoder will be set to display all the text received.

6. Steady

The circuits of the decoder which enable a flashing display will be set so that the RGB output from the Teletext decoder provides a steady (non-flashing) display.

7. Unboxed

All text transmitted will be displayed unboxed, e.g. Not inserted to the off air TV picture.

8. Normal height

All text will be displayed using normal height characters.

9. Release graphics

All control characters transmitted during the graphics mode will be displayed on the screen as a blank character space.

8.5 Teletext character codes

Figure 8.3 shows the Teletext display and control characters together with their respective binary codes. Seven bits of the binary code are shown in the table, the eighth bit being an odd parity bit.

Control characters are shown in columns 0 and 1 with their equivalent binary code shown to the left and above each column, e.g. the control character for alphanumerics red is in column 0, row 1 and its binary code is 1000000. (Least significant bit is transmitted first). It is important to note that bits 6 and 7 of a control character are always at logic '0'. This enables the decoder to determine if the character code received contains control or display information.

Columns 2 to 7 contain the display characters, with the columns suffixed by an (a) showing the corresponding graphics display for the alphanumeric character, e.g. if the decoder's display circuits are set for alphanumerics and a character code of 1000011 is received a lower case 'a' will be displayed. (Column 6 Row 1). However, if the decoder is set to the graphics mode and the same character code is transmitted the graphic symbol (top left and bottom right square lit) will be displayed. This 'doubling up' of displays for a single character code enables a very large range of graphic and alphanumeric characters to be coded using a 7-bit code. Seven bits of data can give a maximum of 128 unique states and these are used to produce the following:

1. 96 alphanumeric characters. (listed in columns 2,3,4,5,6 and 7 of *Figure 8.3*).
2. 27 control characters used to alter the display modes. (listed in columns 0 and 1 of *Figure 8.3*).
3. 64 contiguous graphic characters (which can be used with 32 blast through alphanumeric characters from columns 4 and 5).
4. 64 separated graphic characters corresponding to the contiguous graphic characters listed in columns 2a, 3a, 6a and 7a in *Figure 8.3* (which can be used with 32 blast through alphanumeric characters from columns 4 and 5).

References

1. *Specification of Standards for Information Transmission by Digitally-coded Signals in the Field-blanking Interval of 625-line Television Systems,* Published jointly by the BBC, IBA, BREMA (September 1974)
2. *Specification of Standards for Broadcast Teletext Signals,* Published jointly by the BBC, IBA, BREMA (January 1976)
3. *Broadcast Teletext Specification*, Published jointly by the BBC, IBA, BREMA (September 1976)

Chapter 9

Teletext decoders

When Teletext was first introduced in 1973 the decoders used to process the data were purpose-built devices developed by broadcasting companies and IC manufacturers. The decoders used a large number of 74 series integrated circuits, together with RAM and ROM chips that were already available. A complete decoder used on average about 100 integrated circuits.

IC manufacturers such as Mullard developed their own Teletext decoders and made these available to receiver manufacturers for inclusion in their 'up market' range of televisions. Decoders using these techniques were naturally very expensive, and well beyond the range of the average TV user. Early Teletext receivers were priced at approximately £1200 to £1400.

Home enthusiasts who had a good working knowledge of both digital techniques and receiver design were able to build their own decoders. The first series of constructional articles on a Teletext decoder was produced by *Wireless World* in 1975[1]. This decoder was built into the receiver and had manual controls for all of its functions. Even a home constructed decoder at this point in time cost over £200 to build.

Later on, in 1977, the magazine *Television* also produced a series of constructional articles[2]. This design differed to the one produced by *Wireless World* in many respects. It was an add-on unit to an existing TV receiver, and hence did not require interfacing with the receiver electronics. It had its own UHF tuner and IF amplifier which were responsible for receiving the off air signal. The requested page of Teletext was extracted, decoded and then used to generate a monochrome UHF output that was fed to the TV receiver aerial input. This system considerably reduced the interfacing problems met with the *Wireless World* decoder, but many of the advanced features of Teletext such as the colour display, box and flash modes etc, were sacrificed to keep the cost and complexity down to a reasonable level. Even so a complete kit of decoder parts and the cabinet cost in the region of £200.

The Teletext system only became a commercially reality when integrated circuit manufacturers introduced decoders using dedicated LSI chips. These decoders also made use of recently introduced remote control techniques so that all the decoder functions could be controlled by a small hand held transmitter. Remote control contributed to the commercially feasibility of Teletext because the TV user was able to select either TV or text, page numbers etc, without having to go to the receiver to input the information. Ease of operation is a major consideration in the design of Teletext control systems.

Development costs in these early stages were very high and manufacturers had to recoup these costs over a relatively short period of time. Initial decoder prices were therefore high, keeping the average remote controlled Teletext receiver at about £1000. However, as sales increased, IC manufacturing techniques improved, and initial development costs were recouped, the price of the decoder fell rapidly. The additional cost of remote control and Teletext is now measured in terms of extra tens instead of hundreds of pounds, and this has been reflected in the lowering of rental costs.

As all Teletext decoders currently available in TV receivers use dedicated LSI chip sets, a description of the individual components of decoders is unnecessary and superfluous to current technical needs. In this chapter the Teletext decoder will be explained in block schematic form, using simple circuits as an example of the techniques used in the more important areas of the decoder. Once decoder operation has been covered a number of commercially available Teletext circuits will be examined to give the reader an appreciation of current techniques and future technology trends.

A Teletext decoder has a series of basic functions to fulfil. These are categorized below:

1. To extract the lines of Teletext data from the video waveform.
2. To convert the serial data being received into parallel bytes of data which can be processed in the decoder.
3. Compare the page number and magazine number of the transmitted data with the page and magazine number requested by the user and identify the selected page when it is received.
4. Store the requested page in the decoder Random Access Memory.
5. Read this memory at a rate synchronized to the receiver's timebases, and produce the text and graphics for the CRT.

The operation of the Teletext decoder will be explained step-by-step using the sequence of events listed above. The first of these is to remove the lines of Teletext data from the video signal.

9.1 Data separator

Figure 9.1 shows a Teletext decoder in block diagram form and in this example the complexities of remote control have not been included. The decoder has a video input at the top left hand side, and the RGB text leaves the decoder at the bottom right hand-corner.

We have seen from chapter 8 that the Teletext signal is currently transmitted during four lines of the field blanking period, and the first task that the decoder has is to extract this data from the composite colour signal. All Teletext data is transmitted using pulse shaped NRZ coding. The first stage of the decoder therefore has two tasks:

1. To separate the data from the video.
2. To square up the data to produce a clean digital square wave.

At first it may seem logical to approach these two functions in the order outlined above but it is, in fact, common practice for the data separator to tackle these two tasks in reverse order. In terms of processing it is easier to square up the complete

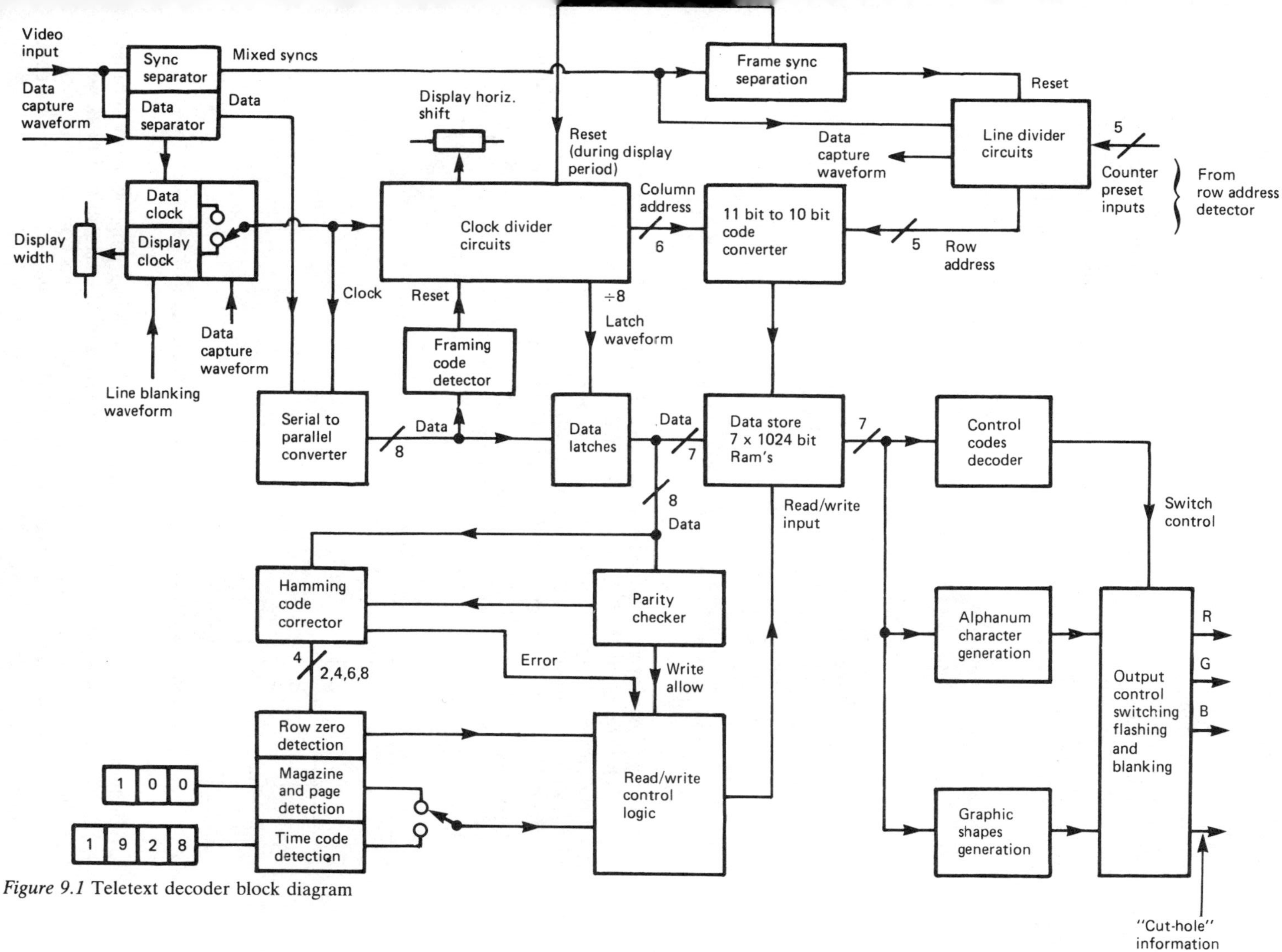

Figure 9.1 Teletext decoder block diagram

video waveform first and then extract the digital signal representing the Teletext data.

The first task carried out by the data separator is to slice the video waveform half way between black level and peak white. This process is shown in *Figure 9.2* and its action will square both the Teletext data and the picture information.

If the data slicer is to cut the Teletext data at a 50% level it must be able to either measure the peak to peak amplitude of the data, and then set the slice level at a mid point, or be able to guarantee that the peak to peak amplitude of the data will be fixed, regardless of signal conditions.

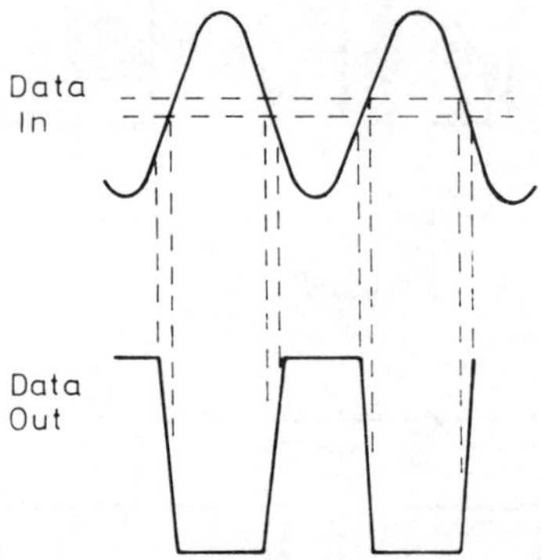

Figure 9.2 Waveforms for a simple slicer

In its simpliest form a data slicer can be produced using a comparator such as the 710 IC. The composite video signal is applied to one of the comparator inputs and a fixed d.c. level is applied to the other. Whenever the video input rises above the slice level set by the d.c. voltage the data output from the 710 will rise to a logic '1'. Refer to *Figure 9.3*. When the video level drops below the voltage set by the slice level the data output will fall to a logic '0'. The main disadvantage of this form of slicer is that if the input video signal changes in amplitude then the Teletext data will be sliced at the incorrect level and consequently invalid data may be produced.

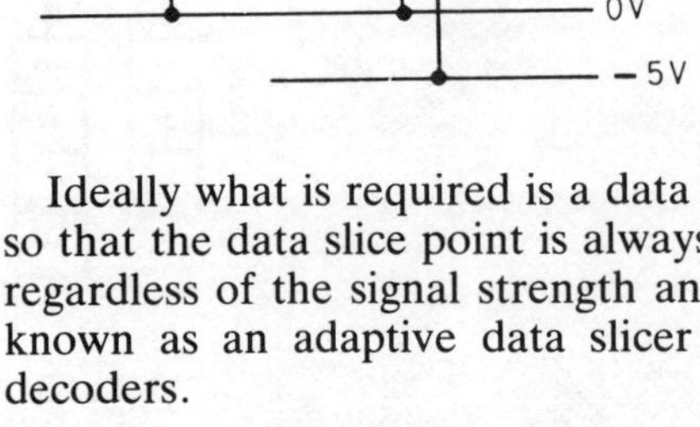

Figure 9.3 Practical data slicer using a 710

Ideally what is required is a data slicer that adapts to varying video amplitudes, so that the data slice point is always midway between black level and peak white, regardless of the signal strength and video amplitude. This form of data slicer is known as an adaptive data slicer and is used exclusively in modern Teletext decoders.

An adaptive data slicer is shown in *Figure 9.4*. The incoming video waveform is applied to the comparator as previously described, and also to two detectors. One detector produces an output voltage which is proportional to the average positive peaks of the video waveform, and the other produces a voltage which is proportional to the average black level. These two voltages are fed to a resistance network, from which a reference slice level is obtained for the comparator.

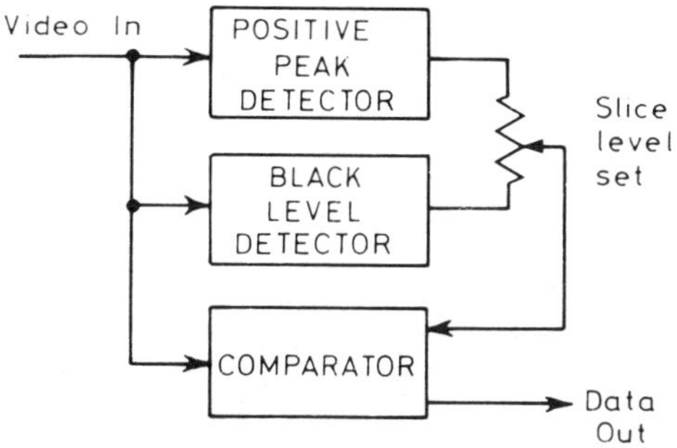

Figure 9.4 Block diagram for an adaptive data slicer

If the mean video amplitude now changes, the reference slice voltage will also change in sympathy, ensuring that the video is still sliced at the correct point. The digital signal leaving the comparator will change between a logic '0' and a logic '1' level in phase with the Teletext data input, and can be processed using conventional logic circuitry from this point onwards.

The Teletext data gate

Once the data slicer has converted the complete television waveform into a true digital signal the lines of Teletext can be extracted using a data gate.

Initially Teletext data was transmitted on lines 17 and 18 of the even field and 330 and 331 of the odd field, but in 1981 the transmission of text was extended to cover lines 15,16,17,18 of the even field and 328,329,330,331 on the odd field. It is possible that in future years a further expansion of the number of lines containing data will take place. Consequently any Teletext decoder should be capable of extracting data during the complete field blanking period. The total possible Teletext data lines are illustrated in *Figure 8.4* (chapter 8) where it can be seen that on the even field lines 7 to 22 could be used and on the odd field lines 320 to 335 could be used.

It is common practice for broadcasting companies to transmit additional information during the field blanking period known as ITS signals (Interval Test Signals). These normally consist of two lines of analogue test waveforms (pulse and bar, grey scale etc.), and a digital signal containing status information.

A Teletext data gate will open immediately after the field sync pulse period and close at the end of line 22 on the even field and line 335 on the odd field, extracting both the Teletext data and the ITS signals. Several circuits have been devised, and one possible configuration is shown in *Figure 9.5*.

A monostable multivibrator is triggered by field sync pulses which have been separated from the incoming video waveform. The timing of the monostable is such that when triggered its output will fall to a logic '0' for the duration of the possible Teletext data lines and will rise back to a logic '1' sometime during line 22 of the even field or line 335 of the odd field. The monostable therefore provides an approximate Teletext data gating waveform, but the edges of this negative pulse will not be accurately positioned because of the indeterminate nature of its timing components.

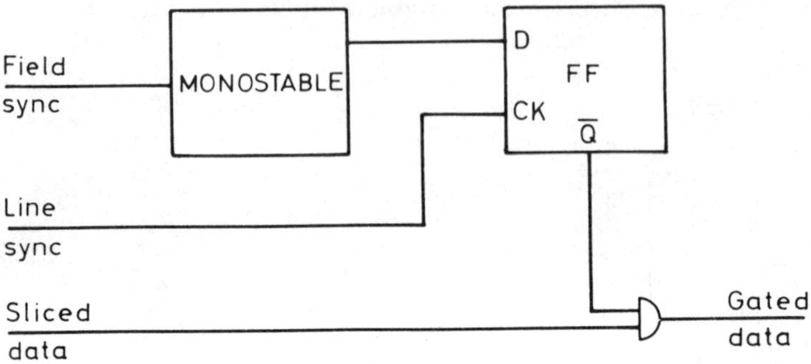

Figure 9.5 Monostable type data gate

To overcome this problem, a D-type flip-flop is used to accurately position the start and stop timings of the data gate. The approximate timing pulse produced by the monostable is fed to the D input of the flip-flop, which is clocked by line sync pulses. This arrangement ensures that the data gate waveform produced at its $\overline{Q}$ output will only change at the beginning of a TV line. The monostable is triggered into its unstable state sometime between line 6/319 and line 7/320 causing the D input to the flip-flop to fall to a logic '0'. The flip-flop will only be triggered on the next line sync pulse, causing the data gate waveform to rise to a logic '1' at the beginning of line 7/320. Similarly the monostable will time out sometime during line 22/335, causing the D-input to the flip-flop to rise back to a logic '1'. However the data gate waveform will not change until the flip-flop is clocked at the beginning of line 23/336, causing it to fall back to a logic '0'.

This accurately timed signal is applied together with the sliced data to a two input AND gate, which switches through all information contained in the field blanking period. This information will include not only the Teletext data but also the ITS lines and any status information. They will not affect the decoder's operation because neither have a clock run in or framing code which the decoder requires before it can decode the serial data. Gated data from this circuit is fed in two directions:

1. To the clock oscillator which provides bit synchronization for the decoder.
2. To a serial to parallel converter which deserializes the text.

9.2 Data clock oscillator

The data bit rate is 444 times line frequency and the most accurate way to synchronize this oscillator to the incoming data is shown in *Figure 9.6*. A voltage controlled oscillator forms the heart of the circuit, and runs at a nominal frequency of 6.9375 MHz. This output is divided by 444 and then phase compared with line sync. Any error voltage developed is used to control the frequency of the oscillator.

Providing the design of the oscillator ensures a stable output frequency, then even when Teletext data is not transmitted the oscillator should be synchronized to the incoming data. There will however be a phase difference between the incoming data and the output of the oscillator due to a variety of elements in the transmission path and reception circuits. Although the oscillator may be at the correct nominal frequency its output may be phase shifted when compared with the input data, and will therefore produce inaccurate text decoding.

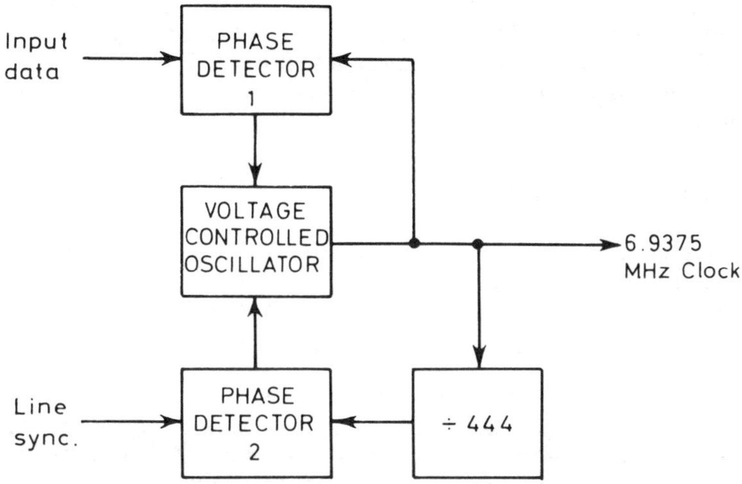

Figure 9.6 Clock generator using voltage – controlled oscillator (VCO)

To overcome this problem a second phase detector is used to alter the phase of the VCO. This detector compares the input data with the direct output of the oscillator, and any small phase differences between the two are used to modify the control voltage from phase detector 2. The result will therefore be a very accurately controlled bit rate clock.

This method provides the ideal solution to the problem because the VCO is continuously running at the nominal data frequency, and large changes in control voltage do not occur. It does have the disadvantage of using a considerable amount of circuitry, especially in the divide by 444 stage. Consequently this circuit would only be used in broadcast quality decoders or in the dedicated LSI chips used for producing clock and sliced data output.

A simpler solution is to use an LC oscillator with a high Q parallel tuned circuit to determine the frequency. Refer to *Figure 9.7*. Sliced data is fed to the base of the

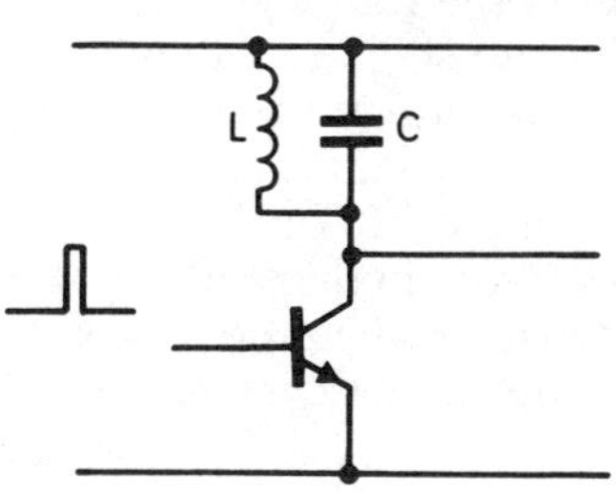

Figure 9.7 Simple L–C clock generator circuit

transistor and shocks the circuit into oscillation. The high Q tuned circuit in the transistor collector will produce a damped oscillation of fixed frequency and decaying amplitude. Data is used to shock this oscillator continuously throughout the Teletext line, ensuring that oscillation, and synchronism is maintained. This form of data locked oscillator was fully described in chapter 3 'Sequential logic circuits'. (Refer to *Figures 3.23 and 3.24*).

Whichever of the above oscillators is used it must be phase locked to the Teletext data before characters can be decoded. This synchronization is provided by the two bytes of clock run in. A series of alternating logic '1s' and '0s' are transmitted for the first two bytes of any Teletext line. These bytes provide a reference signal which will phase lock the oscillator before the transmission of text itself. From the end of these two bytes the clock oscillator will rely on the transitions in the data to maintain synchronism.

9.3 Serial-to-parallel conversion of text

The second output of the data separator provides an input for the serial to parallel converter shown in *Figure 9.1*. This circuit can take the form of an 8-bit shift register feeding a series of D-type flip-flops connected as shown in *Figure 9.8*. Also included in this diagram is the framing code detector.

Sliced data from the data separator is applied to the serial input of the 8-bit shift register, and the register is clocked by the output waveform of the synchronized data clock oscillator. Every time a new bit of data appears at the serial input, the shift register is clocked, shifting data through one bit at a time. At this point the decoder has no means of knowing when the parallel output is correct, because it is not able to differentiate between the end of one 8-bit byte and the beginning of the next. A framing code is transmitted to provide the decoder with this byte synchronization reference.

9.4 Framing code detector

The framing code is a byte of data that immediately follows the clock run in, and has a binary code of 11100100. The decoder identifies this bit combination and then resets a variety of circuits prior to the reception of character bytes.

When the clock run-in moves through the shift register, the output will be a series of alternating '0s' and '1s'. As soon as the framing code arrives this bit pattern will gradually change until the framing code is correctly loaded. At this point the framing code detector will identify the code 11100100 and reset the decoder.

Figure 9.9 shows the combination of bit patterns present at the output of the shift register as the framing code is clocked through. A vertical column of steps and the number of bits matched to the framing code are shown on the right.

Step one shows the clock run in bits stored in the shift register immediately prior to the reception of the first bit of the framing code. Step two shows the first bit of the framing code loaded into the MSB position of the shift register. At this stage only four bits of the shift register will match the desired bit pattern of the framing code. On step three the second bit of the framing code is present in the shift register and again four bits match. This process continues until step nine when the complete framing code is present on the eight parallel outputs of the shift register, and all eight bits match.

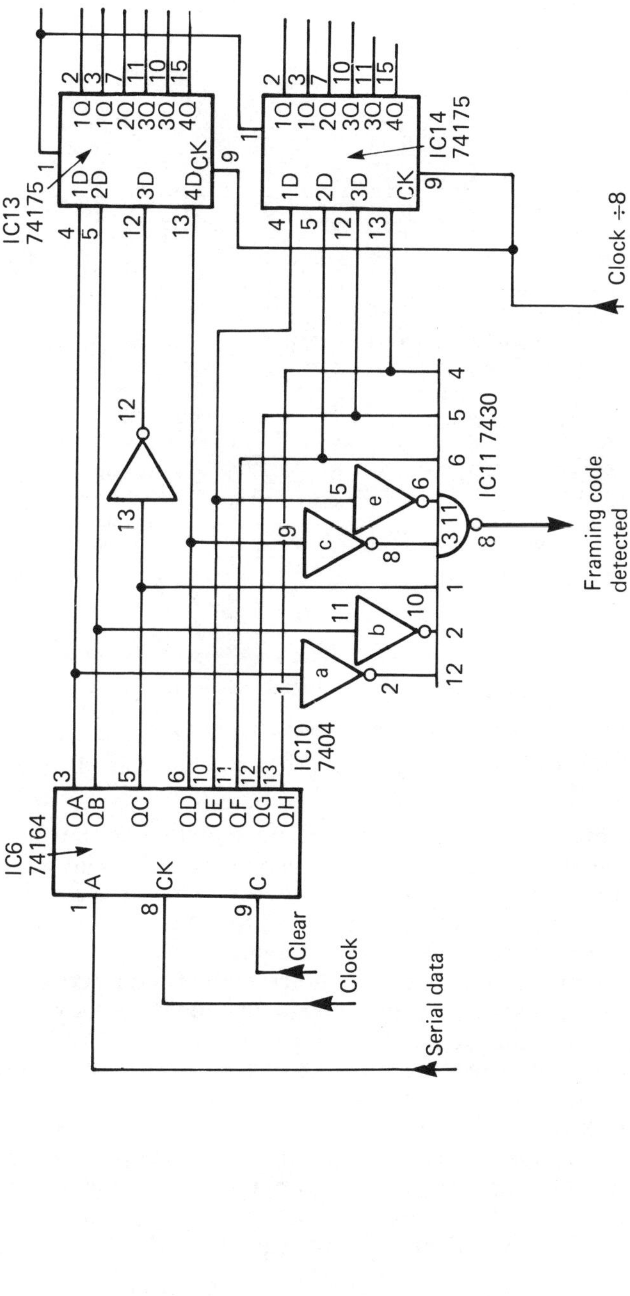

Figure 9.8 Serial-to-parallel converter and framing code detector

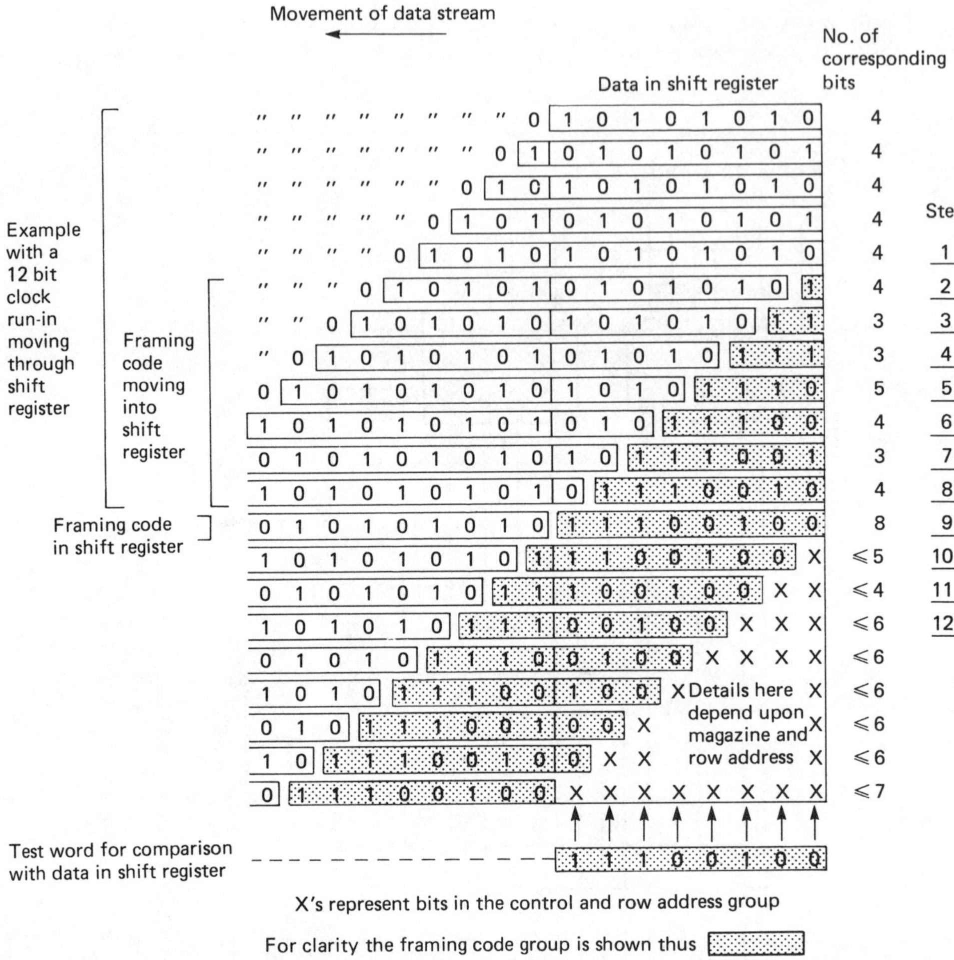

Figure 9.9 Pattern of bits in the input shift register as the framing code passes through it

Examination of the number of bits which are matched as the framing code is clocked through the shift register will show that there will be at least a three bit error in the output prior to the reception of all eight bits of the framing code. This will ensure that even if the framing code is corrupted there is little possibility of the framing code detector producing an invalid output.

Considerable thought has gone into the choice of the framing code bit pattern, so that there is the smallest possibility of the framing code detector producing a false output. After the framing code has been transmitted at steps 10,11, and 12 the number of bits that match drop to a maximum of 5, 4, and 6 respectively. Again the detector is unlikely to produce a false output.

The simplest of all the framing code detectors has been included in *Figure 9.8*. This technique follows the examples given in earlier chapters for identifying a combination of bits. The logic '1s' of the framing code are passed directly to the input of a NAND gate, while the logic '0s' are inverted by a series of NOT gates and then applied to the NAND gate. Consequently when the framing code is correctly loaded in the shift register the output of the NAND gate will fall to a logic '0' identifying the presence of the correctly positioned framing code. This output is then used to reset a number of circuits within the decoder.

It has been shown in *Figure 9.9* that a maximum of five matched bits will be present prior to the correct loading of the framing code in the shift register. If there is a single bit error the detector will not produce a false output, but unfortunately it will not produce a correct output either. To overcome this problem circuits have been devised which identify and correct single bit errors, resulting in a correct framing code detect output even when a single bit is corrupted.

One function of the framing code detector output is to reset a divide-by-8 counter which is clocked by the synchronized data oscillator. When this counter is reset it will count successive clock pulses and produce an output on every eighth pulse. This output will therefore be a data BYTE clock frequency, indicating when a byte of valid data is present at the output of the shift register. This data byte frequency is at 867 kHz, and is used to clock the D-type flip-flops at the output of the circuit in *Figure 9.8*. Every time a valid byte of data is present at the output of the shift register these D-type flip-flops will store the data. During the next seven clock periods the shift register will clock through the next byte of information, but it will only be transferred to the data latch during the eighth clock pulse.

At the output of the latch there will be a stationary byte of data for 1.1 microseconds, which can be processed by the rest of the decoder circuitry. This is the first time a parallel byte of data has been generated for use by the decoder electronics, and all further processing of this data must take place within the next 1.1 microseconds, before the following byte is latched in to the D-type flip-flops.

The output from the latches provides the remaining circuits of the decoder with the data they require. Bits 1 to 7 contain the character codes, and these are fed to the decoder memory. The parity bit is not used in the generation of character information hence, bit 8 is not stored.

All eight bits of data are however used by the data processing areas of the decoder such as the Hamming code corrector, parity checker, row detector, magazine detector, etc. These circuits are used to generate the read/write control signals for the decoder RAM.

9.5 Parity checking

The parity circuits will carry out a check on all eight bits of data, and produce a write allow signal for the read/write control logic when odd parity is detected. Should even-parity be detected the read/write control logic will be inhibited, preventing the corrupted character code from being stored. A parity check on all eight bits is also required as part of the function of the Hamming code corrector.

9.6 Hamming code corrector

The Magazine and Row Address group which follows the Framing Code and many of the bytes contained in the header row are Hamming coded. Bits 2,4,6,8 of these bytes contain the message information, while bits 1,3,5,7 are parity bits. A description of the Hamming code corrector was given in chapter 6 'Data transmission', and the reader should refer to section 6.8 for detailed information on the circuit. Message bits leaving the Hamming code corrector are used for a variety of tasks:

1. Detect the magazine and row address.
2. Detect the header row.
3. Detect the page number in the header row.
4. Detect the time coded information in the header row.
5. Detect the control bits in the header row.

9.7 Magazine and Row Address Detector

The first of these tasks is to identify the magazine number, (which is message bits 2,4,6 of byte 4), and row address, (which is message bit 8 of byte 4 and message bits 2,4,6,8 of byte 5). These bytes of data are illustrated in *Figure 8.6*. A circuit showing one way of detecting these message bits is shown in *Figure 9.10*.

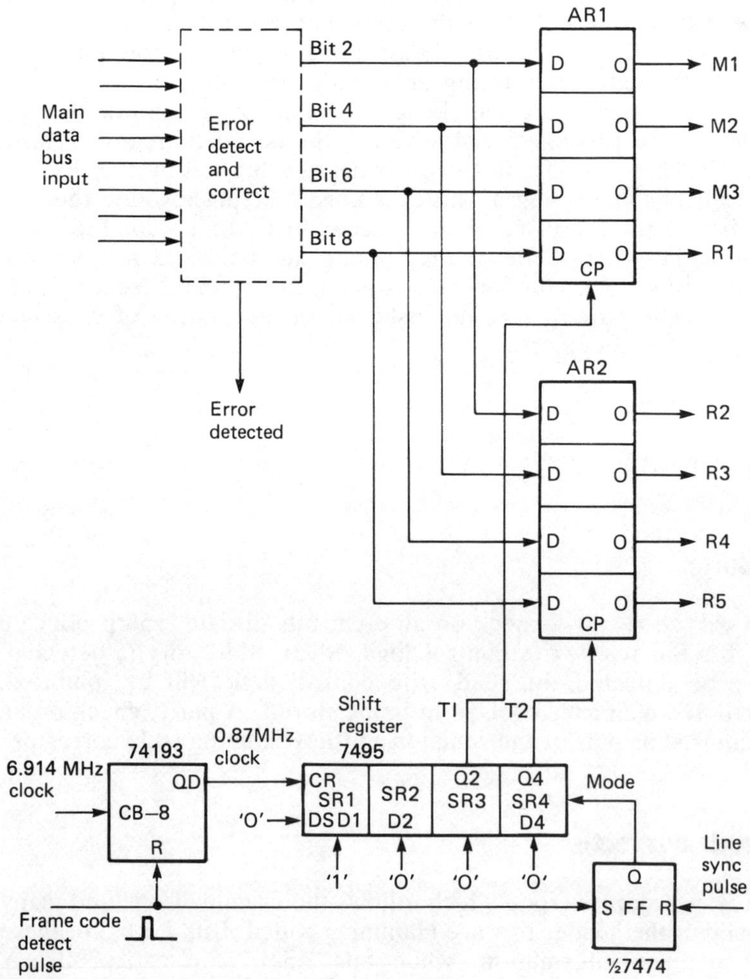

Figure 9.10 Magazine and row address detector

The main 8-bit data bus is shown entering the Hamming code correction circuit on the left hand side of the diagram, with bits 2,4,6,8 leaving the error corrector. These message bits are fed to the D inputs of 8 D type flip-flops arranged in 2 groups of 4. The top group is used to store the message bits of byte 4 and the bottom group is used to store the message bits of byte 5. In order to carry out this storing process the decoder must be able to count the number of bytes received after the reception of the framing code, and clock latch AR1 (address and row byte 1), and latch AR2 (address and row byte 2) at the correct times. This is achieved as follows.

A 4-bit parallel in/serial out shift register labelled SR1,2,3,4 is used to count the number of bytes of data received after reception of the framing code. The mode input to this shift register is derived from an SR flip-flop which is reset by line sync pulses, and set by the framing code detect pulse. In the reset mode this flip-flop will load in the parallel bit pattern 1000 as shown on the diagram.

On arrival of a framing code detect pulse, the Q output of the SR flip-flop will be set, causing the shift register to operate in the serial mode. The framing code detect pulse is also used to reset the divide by 8 byte counter previously described, so that from the moment this counter is reset the shift register receives its first clock pulse, shifting the logic '1' stored in SR1 into SR2. The bit pattern contained in the shift register when the framing code has been detected will therefore be 0100 (left to right). The divide by 8 counter is clocked by the data oscillator and will count the next 8 bits of serial data received. On the 8th clock pulse its QD output will change shifting the logic '1' in SR2 one place to the right into SR3. The code in the shift register will now be 0010, and the data on the bus will be the first byte of the magazine and row address group.

After Hamming code correction, message bits 2,4,6,8 are clocked into the D-type latch AR1 by clock pulse T1 from the shift register. Magazine bits 1,2,3 and row address bit 1 will now be stored in AR1. While this process is being carried out the next byte of data is being received, and its bits counted by the divide-by-8 counter. When the complete byte of data is present the divide-by-8 counter will again clock the shift register, shifting the logic '1' stored in SR3 into SR4. When this takes place the second word of the magazine and row address group will be present on the data bus, and its message bits are clocked into latch AR2 by clock pulse T2.

As soon as the magazine and row address codes have been transmitted they will be stored in the eight D-type flip-flops. Output M1,2,3 will indicate the magazine number which is coded in binary form (M1=LSB). This magazine number is the most significant digit of the page number. For example, assume the magazine number was 011 in binary, which is magazine number 3. This will form the most significant digit of a page number between 300 and 399.

Outputs R1 to R5 represent the horizontal row number of the text being transmitted, with R1=LSB. As there are 24 rows of text these binary codes will range from 00000 to 10111 (0–23).

These latched bits of data can now be used by the decoder for two functions:

1. To generate the row address for the RAM if the received line of data is to be stored. This row address will indicate to the memory circuits where to store the following 40 characters, so that when the memory is read during the display period the horizontal row of text will appear in the correct position on the screen.

2. If the horizontal row number is 0 it indicates that the header row of a new page is being transmitted. This being the case the decoder must take further action to store:
   (a) The page units and tens (bytes 6 and 7).
   (b) The minutes units and tens, hours units and tens of time coded information (bytes 8,9,10,11), together with control bits C4 to C6.
   (c) The control bits C7 to C14 (bytes 12 and 13).

All the above data is Hamming coded.

The next task of the decoder therefore is to identify the condition R1 to R5= 00000. This is simply achieved by complementing each R output and feeding the result to a 5 input NAND gate, causing the output from the gate to fall to a logic '0' when the header row is being transmitted. Gates G1 to G6 of *Figure 9.11* form the row 0 detection circuit.

9.8 Page number detection

If the output of G6 in the row 0 detection circuit falls to a logic '0', indicating that a header row is being transmitted, the decoder must store the next eight bytes of data, all of which are Hamming coded. The techniques used to achieve this are very similar to those previously described for storing the magazine and row address group. *Figure 9.11* shows the storing process of the page units and tens bytes. The remaining bytes of the header row are stored using identical techniques.

At the top of the circuit diagram is the data bus, Hamming code corrector and two groups of four D-type flip-flops labelled page register 1 (PR1) and page register 2 (PR2). When a header row is being received (PR1) must be clocked when byte 6 is present on the data bus, and (PR2) clocked when byte 7 is present. Correct timing of the circuit is achieved by the shift register (TR1 to TR4), which is arranged in a very similar way to the previous example. When the header row is detected by gates G1 to G6 the mode input to the timing register will fall to a logic '0'. This will load in the parallel bit pattern 1000. This data will be stored in the timing register while the second byte of the row address group is present on the data bus, and only when the row address is 00000.

On the next transition of the byte counter the timing register will be clocked, producing the code 0100 in the register. Output Q2 will now rise to a logic '1' causing message bits 2,4,6,8, from the Hamming code corrector to be latched into PR1. Page units bits 1,2,3,4, will therefore be stored.

During the next data byte counter output the contents in the timing register will be shifted one place to the right again, producing the code 0010. PR2 is now clocked causing the page tens bits PT1 to PT4 to be stored.

Subsequent pulses from the byte counter will shift the contents of the timing register further places to the right. Similar circuits to the one illustrated in *Figure 9.11* will then be used to store minutes units, minutes tens, hours units, hours tens, and the control bytes A and B. After transmission of control byte C2, the decoder will have all the necessary data to control the storage of the new page of text.

Data contained in the header row can now be interrogated to find out if the magazine and page numbers requested by the user match those contained in the header row. This comparison has to take place digit by digit. If a match on all the digits is made than an accept data signal is generated by the decoder telling the read/write control logic to write the data into the decoder RAM.

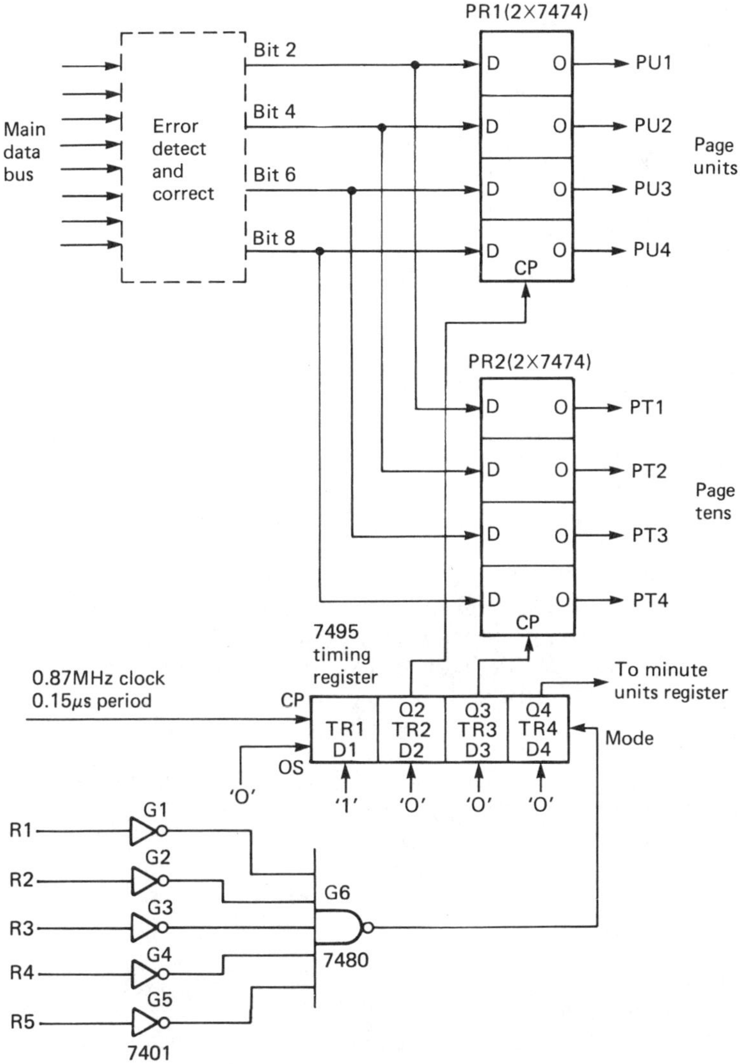

Figure 9.11 Page number detector circuit

Modern Teletext decoders are all remote controlled, and consequently the magazine and page number requested by the user will be stored in the decoder. The comparison process takes place using combinational logic gates and the circuitry used is quite extensive. To simplify the description of this process one method of basic code comparison will be described in circuit form, and the total concept explained in block diagram form.

In a decoder using remote control techniques, the selected four bit page units code will be stored in a memory, and then fed to a set of 4 XOR gates as shown in *Figure 9.12*. The received page code from the PR1 array in *Figure 9.11* is applied to the second set of inputs. Each XOR gate makes a single bit comparison between the received and selected page codes. If the two bits are the same the XOR gate will

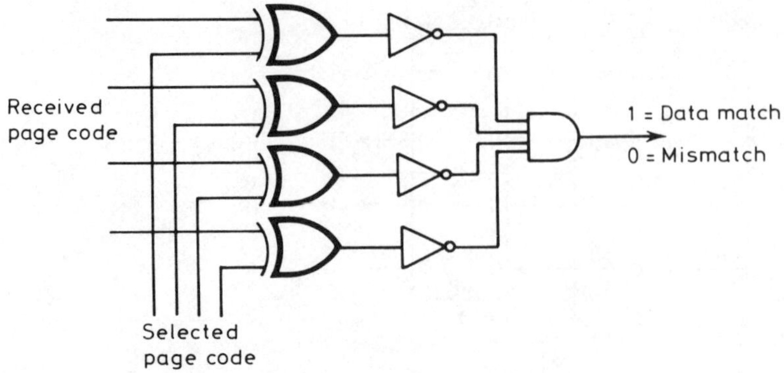

Figure 9.12 Basic code comparison logic using Exclusive OR gates

produce a logic '0', and if they are different it will produce a logic '1'. Each output of the XOR gates is complemented by a NOT gate and then fed to a four input AND gate. If all bits match there will be four logic '1s' to the input of this gate, producing a logic '1' out. If there is a miss-match the AND gate output will fall to a logic '0'.

This theory is extended to the other four bits of the page number, and the three bits of the magazine number. Consider *Figure 9.13*; the received eight bits of the page code is compared in the top comparator with the eight bits of the selected page code. If the two groups of 8-bits match a logic '1' from the comparator is latched into a page match register. A similar process takes place for the three bits of the

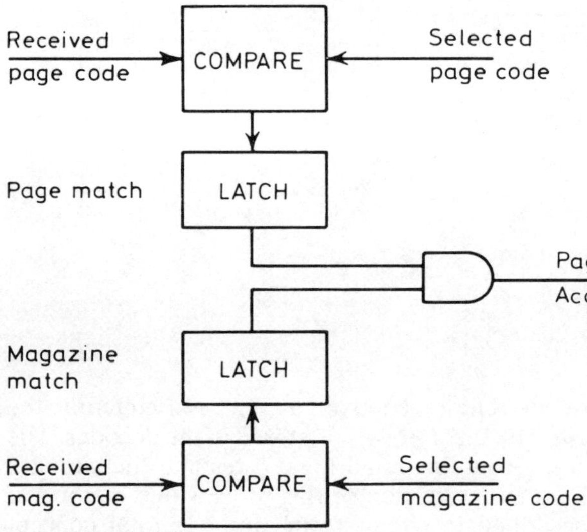

Figure 9.13 Page and magazine comparison logic

magazine code received and the three bits of the magazine code selected. If a match is made a logic '1' is now stored in a magazine latch. The outputs from the page and magazine latches are ANDED together to produce a page accept signal which is then fed to the memory read/write control logic.

9.9 Time code detection

Bytes 8,9,10,11, of the page header row contain time coded bits. These are a 4-bit minutes unit, 3-bit minutes tens, 4-bit hours units and 2-bit hours tens, transmitted in this order. If the decoder is operating in the time code mode then these bytes have to be interrogated to determine if the time code selected by the user matches the time code of the page transmitted. If a match is obtained then the read/write control logic for the memory will store the page, but if a match is not made the read/write control logic remains in the read mode.

All decoders have the facility to override time code detection so that the page selected will be written into the decoder RAM regardless of the time code data.

The detection circuits for the time-coded bytes of the header row will be a modified form of the circuits previously described for magazine and page detection. In the previous example the data being compared consisted of a 3-bit magazine number, a 4-bit page tens number, and a 4-bit page units number. In this application 4 BCD time-coded digits extracted from the header row will be compared with BCD time-coded digits entered by the user.

The row 0 detection, magazine and page detection, and time code detection circuits all generate signals which are used by the read/write control logic. In addition to these three signals are further inputs from the parity checker and Hamming code corrector. Other gating signals enable the real time clock transmitted in each header row to be written into the memory, and control bit C4 to erase the memory prior to the reception of a new page. (For simplification these two are not shown in *Figure 9.1*.)

9.10 Read/write control logic

The read/write control logic for the decoder RAM is a group of combinational logic circuits which determine whether data is to be written to the memory or read from it. A variety of inputs are used to determine the mode of operation, some of which have been outlined above. A simplified circuit schematic showing how they are gated is illustrated in *Figure 9.14*.

OR gate 1 combines three separate signals which determine if data is to be written to or read from the memory. These three inputs are:

1. *Page Erase*. This is carried out when the page stored in memory is significantly different from the page about to be written to the memory. This process is initiated by control bit C4 being set to a logic '1' in the header row. (See chapter 8 'The Teletext system'.)
2. *Page Write*. A logic '1' will enable the decoder to store the requested page of data when it is transmitted.
3. *Real Time Clock*. This input supplies the memory with a write signal only when the real time clock is transmitted. (Characters 33 to 40 of the header row.)

Page erase

We have seen from chapter 8 that the header row contains control bits numbered from C4 to C14. C4, which is the erase page control bit, is the last bit transmitted in the time coded minutes tens word. Refer to *Figure 8.6* (chapter 8); this control bit is set to a logic '1' when the information on the page about to be received is significantly different from that in the previous version of the same page. When C4 is set to a logic '1' the header row will always be followed by a page erasure interval. This is an active television field period which will occur between the initial page header row and further rows of the same page. This field period allows the receiver to erase all the data in the memory prior to the reception of the new page of data.

In order to carry out this process the memory read/write line from the control logic has to produce a memory write operation for one complete field scan. As the memory is addressed during the field period the code for a blank space is written into every location of the store.

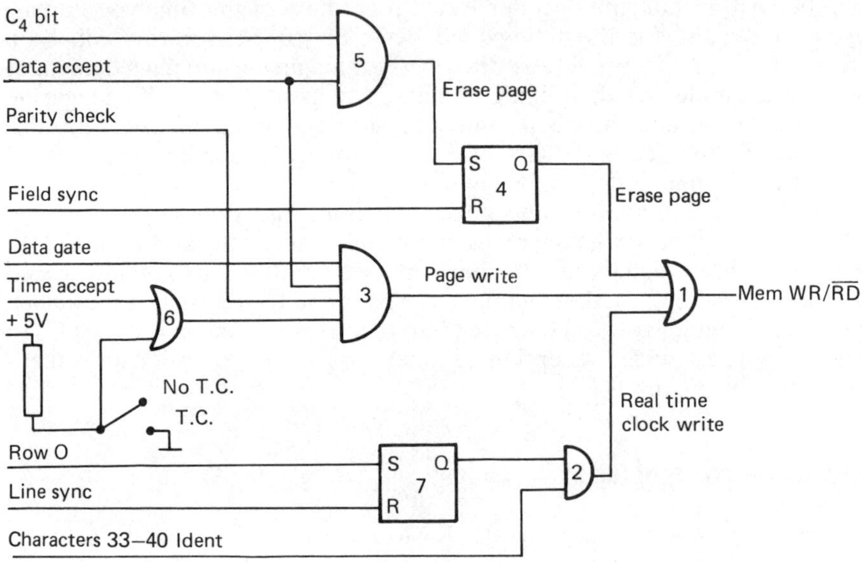

Figure 9.14 A simple memory read/write logic circuit

The data accept signal from the magazine and page detection circuits previously described is logic ANDED with control bit C4 in gate 5. If the page stored in memory is to be erased, the S input of flip-flop 4 will be set to a logic '1', resulting in a logic '1' at the output of OR gate 1 for the complete field period. The Q output of flip-flop 4 is reset to a logic '0' on the next occurrence of the field sync pulse. This procedure will result in the memory being erased during the field scan period immediately following the header row.

Page read/write

Normal operation of the page read/write circuit is controlled by AND gate 3, which has four inputs. Input 1 is derived from the output of the Teletext gate previously described in section 9.1. This signal rises to a logic '1' during the field blanking period when the decoder can expect to receive Teletext data. After the field blanking period the waveform falls back to a logic '0', causing the output of AND gate 3 to fall to a logic '0'. This will ensure that the memory can only be written to during the field blanking period, and during the field scan the memory will be read by the display electronics.

Input 2 of the AND gate comes from the magazine and page detection circuits, and this will rise to a logic '1' only when the transmitted magazine and page numbers correspond to thoses requested by the user.

The third input comes from the parity detection circuit. If parity is correct, this input will rise to a logic '1' allowing the byte of data present on the decoder data bus to be written into the memory. Should a parity check fail this input will fall to a logic '0' preventing the byte of data from being stored.

The fourth input to AND gate 3 is produced by the time code detection circuits, and when this input is at a logic '1' it indicates that either the time code number selected by the user matches that contained in the header row, or that the time code detection facility is not being used, OR gate 6 provides the manual/automatic operation of the time code facility.

When all four inputs to AND gate 3 rise to a logic '1' the output of the gate will also rise to a logic '1' indicating that:

1. Valid text is present on the decoder data bus.
2. The text present on the data bus has been requested by the user.
3. The text has the correct parity.
4. The time coded bytes present in the header row match those requested by the user, or the time code facility is not being used.

All four signals together, will result in a memory write operation.

Real-time clock display

In the top right-hand corner of every page of Teletext is a 6-digit clock which occupies bytes 33 to 40 of the header row. This clock is displayed on a real-time basis and is updated every second during the transmission of the header row. For the decoder to display this clock on a real-time basis, the clock characters have to written into the memory every time the header row is transmitted. On subsequent field scans an updated clock display will be produced.

Flip-flop 7 and AND gate 2 generate the write signal for the memory only during bytes 33 to 40 of the header row. The Q output of flip-flop 7 is set to a logic '1' every time row 0 is detected, and is reset to a logic '0' every time a line sync period occurs. This Q output is coupled to one input of AND gate 2. A second input to this AND gate is derived from the clock divider circuits shown in the block diagram of *Figure 9.1.*. This signal will rise to a logic '1' at the beginning of character 33 and back to a logic '0' at the end of character 40. When this signal is logic ANDED with the Q output of flip-flop 7 the resulting waveform will rise to a logic '1' during characters 33 to 40 of the header row. This is then combined with the two other signals at the input of OR gate 1 so that a memory write signal will be produced by the circuit every time the data for the real time clock is present on the data bus of the decoder.

This circuit of the read/write control logic has been considerably simplified so that the reader has an easy introduction to some of the logic operations required to generate the memory read/write signal. In a practical decoder this circuitry will be significantly more complex if all the teletext facilities are incorporated. For example control bit C7 of the header row is used to suppress the header row itself. If bit C7 is set to a logic '1' the header row must not be written into the memory, and this will obviously require additional logic to that shown.

Similarly when control bit C10 is set to a logic '1' the display will be inhibited. This is used when the contents of the page cannot be usefully interpreted as a

Teletext transmission, e.g. experimental pages of Telesoftware, etc. Again in a practical decoder it would be necessary to detect bit C10 and inhibit the memory write signal if required. Control bit C11 is set to a logic '1' when the transmission sequence of magazine pages is such that it is preferable to display all page headers as rolling headers. Consequently if C11 was set to a logic '1' the memory read/write control logic would provide a memory write signal every time a page header row was transmitted.

Additional facilities such as those outlined above can easily be incorporated in the memory read/write logic, but unnecessarily complicate the basic description required here.

9.11 Random Access Memory (RAM)

Memory organization will differ between one decoder and another, and the storage format will determine the design of the memory addressing circuits. Any Teletext decoder memory must conform to the basic specification outlined below, regardless of memory organization.

1. As there is a total of 40 characters per row of text, and 24 rows of text per page then the minimum number of memory locations required will be $40 \times 24 = 960$ words.
2. As the parity bit is not required to be stored each memory location will be a minimum of 7 bits wide.
3. As data is transmitted at 6.9 Mega-bits per second the decoder memory access time should be considerably less than 1 microsecond as this is the nominal period between displayed characters.

Earlier decoders used 7 1K by 1-bit static RAMs organized as shown in *Figure 9.15*. Each static RAM IC is used to store one bit of the data. All 7 memory chips share a common 5 bit row address, 5 bit column address, and read/write control signals. Although standard 1 K static RAMs are readily available, the design of the addressing circuits is somewhat complicated by the memory organization.

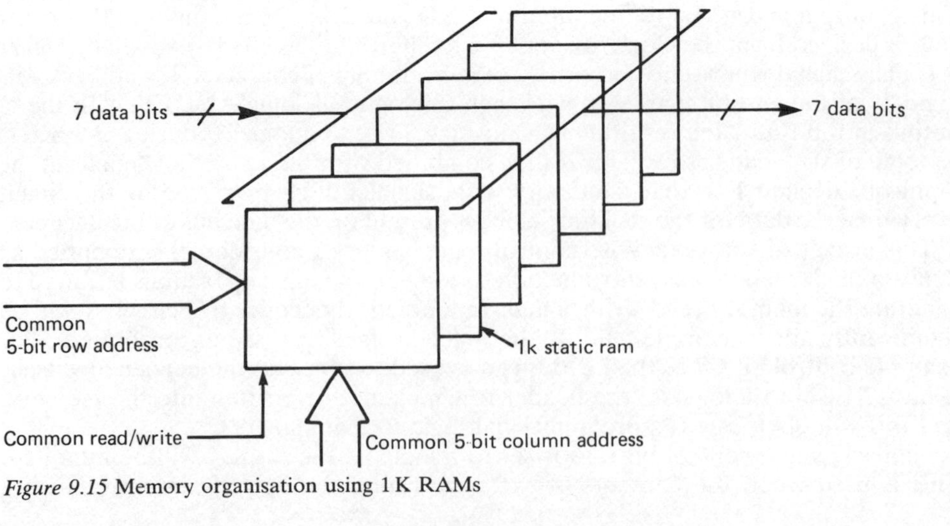

Figure 9.15 Memory organisation using 1 K RAMs

A 1 K (1024 bits) memory is made up from a 32 by 32 matrix addressed by a 10 bit code as shown in *Figure 9.16*. A five bit column address, feeding a 1 of 32 decoder, will select one of the vertical columns, while a five row address selects a horizontal row. The memory cell located at the cross point of the vertical column and the horizontal row is the addressed memory cell.

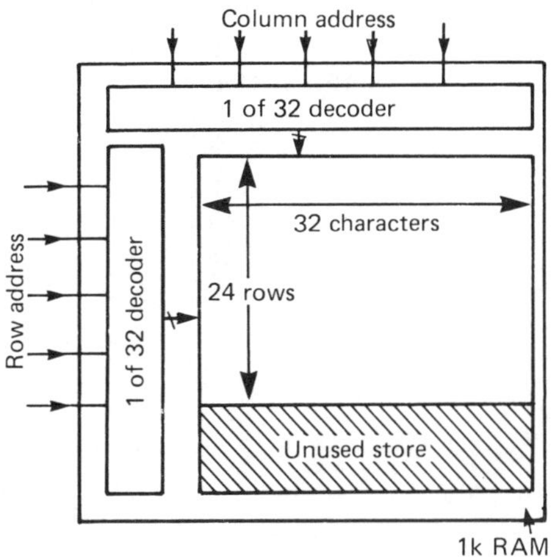

Figure 9.16 Addressing of 1 K RAM

If each line of text contained only 32 characters, instead of 40, this mode of addressing would be ideal, because the addresses derived in the decoder could be applied directly to the memory. Unfortunately a line of text contains 40 characters and this complicates the design of the memory address circuitry. The technique most widely adopted is shown in *Figure 9.17*.

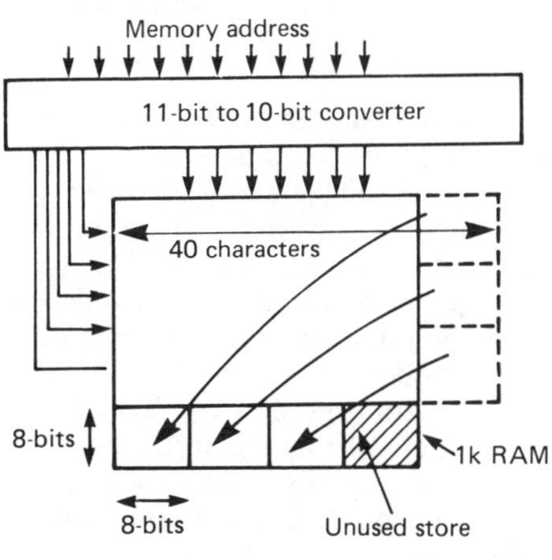

Figure 9.17 Addressing of a Teletext 1 K RAM

Six bits are required to address 40 characters and five bits to address 24 rows. This results in 6 column and 5 row address bits being fed to the memory. An 11 bit to 10 bit address code converter is required to interface these inputs to the 32 by 32 bit memory matrix. The first 24 horizontal rows of memory store character numbers 1 to 32 of each line of Teletext, and the remaining characters of each row

(33 to 40) are stored in horizontal rows 25 to 32 as shown in the diagram. The 11 to 10 bit address code converter counts the column address signal, and detects characters 33 to 40. When these occur the address to the memory matrix is changed, to store the remaining 8 characters of the row in the spare RAM locations.

The task of the 11- to 10-bit converter is to correctly arrange the stored information inside the memory as shown in *Figure 9.18*. It will consist of a combinational logic circuit which will use the row and column address signals generated by the Teletext decoder to directly address the matrix up until the transmission of character 32. When character 33 is present on the decoder data bus.

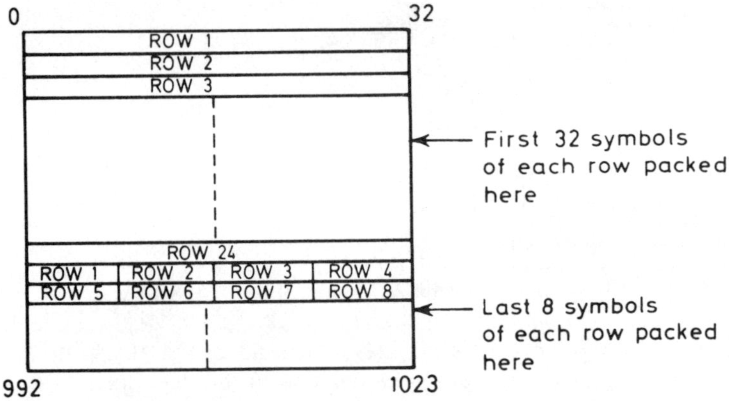

Figure 9.18 Memory packing

the converter will modify the memory address to the matrix and store the remaining 8 character codes for that horizontal row in the unused portion of the matrix. A simplified diagram showing the theory behind this code conversion process is shown in *Figure 9.19*.

Assume a new line of text is being stored starting with the first character of data. The row address inputs on lines 1,2,4,8,16, will have the binary code of the horizontal row transmitted. This is extracted from the magazine and row address group of the Teletext line. Initially the column address on inputs 1,2,4,8,16,32, will be 000000. The first character of Teletext data will therefore be stored in memory location 0 of the horizontal row whose number is set up on the row address inputs. As each new character of Teletext is received the column address is incremented to point to the next memory location in the row.

This process continues up to, and including, the transmission of character 32 which is stored in column 31 (address code of 011111). During the transmission of Teletext character 33 the column address will change to 100000. At this point the MSB of the column address will rise to a logic '1', causing a changeover gate to switch the address lines, producing the following results:

1. The 8 and 16 lines of the row address both rise to a logic '1' creating a row address of between 25 and 32.
2. The 8 and 16 bits of the row address are now switched so that they replace the 8 and 16 bits of the column address.

The result of this switching action is to store the last 8 character symbols of each row in an unused lower area of RAM as shown in *Figure 9.18*.

In practice, of course, each of the change-over switches will be replaced by a combinational logic circuit, and it has been seen from chapter 2 that it is an easy process to construct a single pole 2-way changeover switch using logic gates. The circuitry required for just one of these switches is shown in *Figure 9.20*.

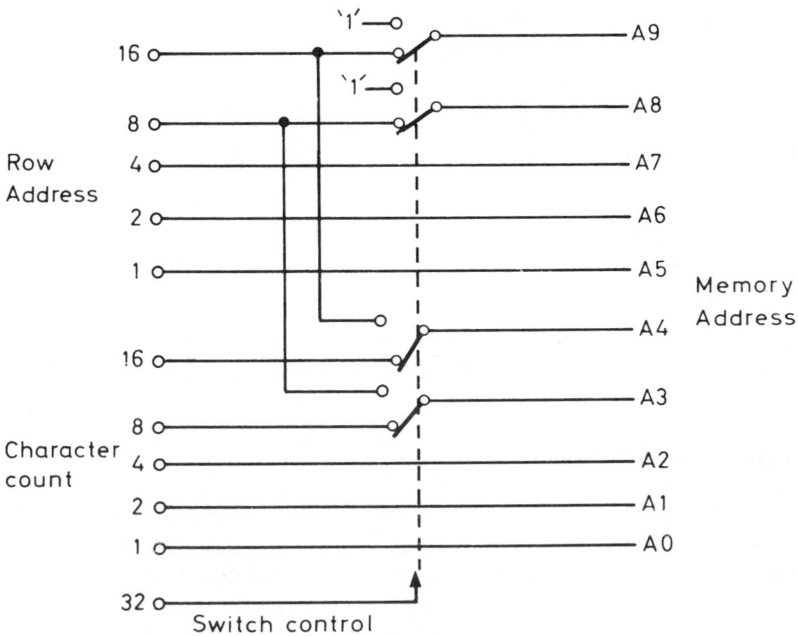

Figure 9.19 Address control using a switch

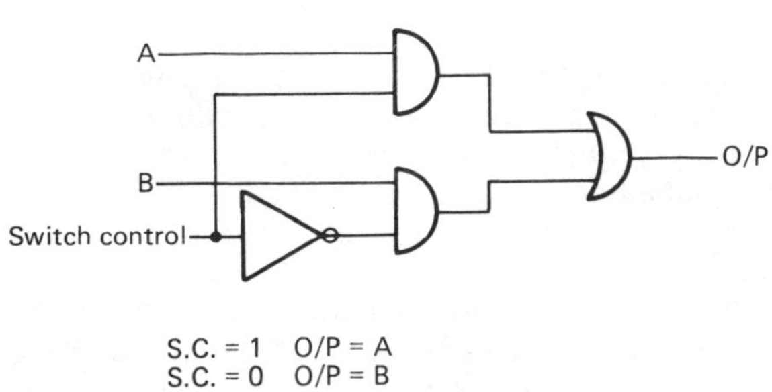

Figure 9.20 SPDT changeover logic diagram

A Teletext memory following the techniques that have just been described is not ideal for several reasons. First of all it is apparent that the memory addressing is unnecessarily complex because the standard 1 K by 1 bit IC is made up by a 32 by 32 matrix. If a memory IC having a 40 by 24 matrix was commercially available then this could be used without the need for the 11- to 10-bit address code converter.

Secondly, the memory organization outlined uses random access techniques and this will enable any single character to be accessed. simply by placing its horizontal row and vertical column position on the address inputs to the code converter. This facility is unnecessary because each row of text is written to, and read from, the memory in serial form. Hence, only serial access is required to the characters stored. Random access of horizontal rows is still necessary because the rows may be transmitted in any order, and some may even be omitted.

The ideal memory for a Teletext decoder will be row addressable, using 5 bits and have a 7-bit input output bus. A vertical column address is not required, and this will be replaced by a simple clock pulse which causes the address to increment to the next location every cycle. The total memory required will be $960 \times 7 = 6720$ bits and this would be easily fitted into a single IC package. The ideal Teletext memory chip would therefore have the following interconnections:

7 Data lines sharing both input and output.
1 Memory read/write control line.
1 Chip Enable control line.
5 Row address lines.
1 Clock input.
0 and +5 V supplies.

As IC technology progresses it is likely that this form of memory will be included in one of the LSI decoder chips, possibly the IC responsible for the data acquisition.

9.12 Memory address generation

We have seen that a Teletext memory requires a 6-bit column address and a 5-bit row address. The row and column address signals have to be generated in such a way that during a memory write operation they are both controlled by the incoming Teletext data, and during the display period, when the memory is being read, the address signals will be controlled and timed from the receiver's timebases.

Column address

The column address identifies the location of each character within a horizontal row of text, and as there are 40 characters in a horizontal row the column address will be 6 bits wide. The block diagram in *Figure 9.1* shows the column address being generated by a section of the decoder labelled clock divider circuits, and this block contains binary counters which are used to generate the necessary timing waveforms required by the decoder.

When data is being written into the memory the 6-bit column address has to be incremented every time a new character is received so that it will be stored in the next horizontal position of the addressed row in the memory. The clock divider stage is shown in simplified form in *Figure 9.21*.

When the framing code has been detected the divide-by-8 counter is reset to zero, and the divide-by-40 counter is reset two bytes later, after transmission of the magazine and row address group. The divide-by-8 counter produces a character clock, synchronized to the Teletext byte rate, and is used to indicate to other decoder circuits when a complete byte of Teletext data is latched on the data bus. The character clock waveform is also fed to the divide-by-40 counter which produces a 6-bit binary count of the number of characters received after the magazine and row address group, starting from 000000 and rising to 100111. This

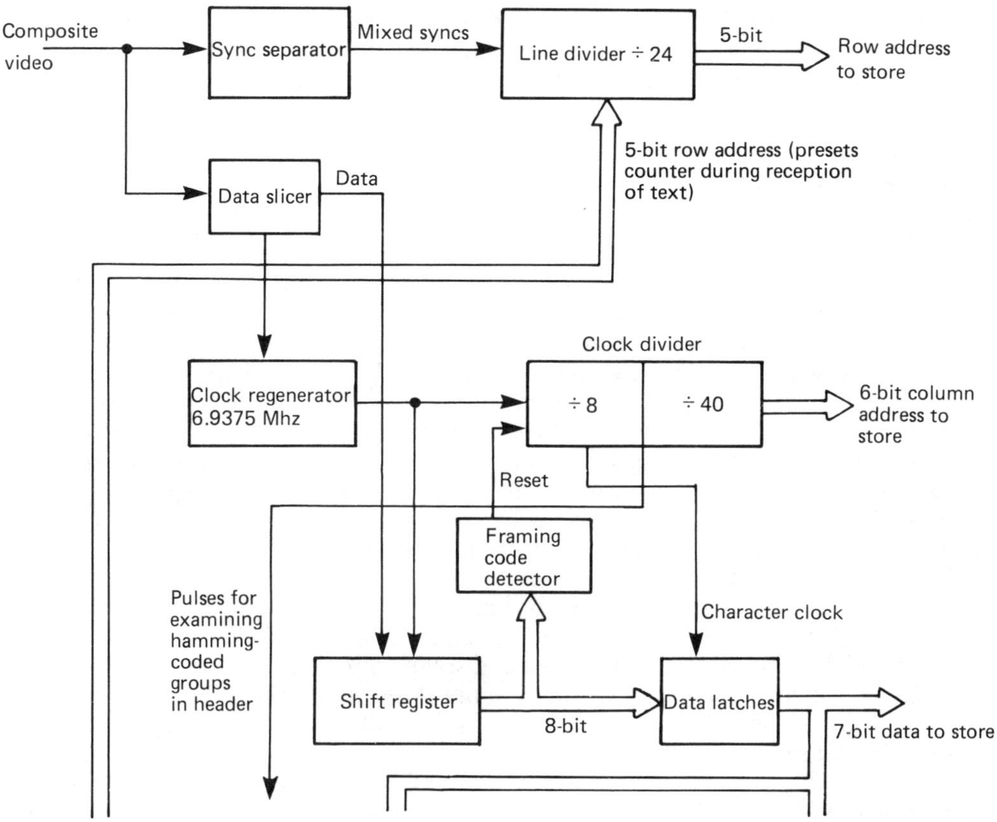

Figure 9.21 Simplified form of clock divider stage

6-bit column address will indicate to the memory the horizontal position of the character in the row of text. (000000 being the first character transmitted and 100111 for the last character.)

During the display period, when the memory is being read, the data clock oscillator is replaced by a display clock oscillator. Refer to *Figure 9.1*. Its design will vary between one decoder and another, but it is normally crystal controlled and its frequency may be variable over a limited range in order to change the width of the display. This oscillator is also inhibited during the line blanking so that the divider circuits are not clocked during the line flyback period. A completely separate oscillator therefore produces the clock waveform for the divider during the display mode.

The switching between these two oscillators is controlled by the data capture waveform used to gate the Teletext data from the field blanking period. When this waveform rises to a logic '1', indicating that teletext data is about to be received, the switch feeds the output from the data clock oscillator to the clock divider circuits. But when the waveform falls back to a logic '0', for the display period, the display clock output is switched to the divider.

During the display period a line blanking pulse is used to reset the divider circuits at the beginning of each line. The display clock oscillator output is then counted in exactly the same way as before, producing a binary coded column address which will count from 000000 to 100111 during the 52 microsecond active picture period. This column address is applied to the memory via the 11- to 10-bit code converter so that each character in the horizontal row is read from the memory during the line scan period.

Row Address Generation

We have seen that the magazine and row address group contains the horizontal row number of the data being transmitted. The 5-bit Teletext row address code is used by the decoder to create the memory row address. This technique allows each row of the memory to be addressed on a random basis, so that if one horizontal row within the page contains no characters or graphics it may be omitted from the transmission sequence. The memory address can therefore 'skip' to the next row being transmitted, speeding up the access time of other pages in the magazine.

In the block diagram of the decoder in *Figure 9.1* the line divider circuits in the top right hand corner are shown with five inputs from the row address detector. These inputs are applied directly to the 11- to 10-bit code converter during the period of time occupied by the data capture waveform. In this mode the row address for the memory is controlled directly by the magazine and row address group of the Teletext data.

Throughout the display period the horizontal row address for the memory has a completely different format. The stored character code has to be read repetitively over a number of lines, so that the correct display pattern for each character can be created. A horizontal row of text stored in the RAM may have to be read up to ten times so that the decoder display circuits can recreate the video waveform for the stored characters.

Assume that the characters displayed on the CRT occupy a height of ten lines. Each horizontal row of data stored in the decoder RAM will therefore have to be read ten times, once for each line of the character scan. The row address present at the input of the 11- to 10-bit code converter during the display mode will therefore be reset to 00000 at the beginning of the field scan and then incremented up to 10111 on every 10th line of the picture.

All the characters in a page of text will occupy 240 lines of the field scan. Each field scan is 312.5 lines leaving $312 - 240 = 72$ lines of the field scan unused. Every row of the text is spaced by one line hence 23 of the 72 unused lines are allocated for row spacing leaving 49 redundant lines. These are split between the top and the bottom of the picture so that the page of text is centred vertically in the field scan. On the next field the interlaced lines of the characters are added (10 odd lines + 10 even lines), giving each character a height of 20 lines with a 2 line spacing.

Display circuits

Before describing the electronics used to convert the character codes stored in the decoder RAM into the alphanumeric and graphic shapes which appear on the CRT it is necessary to understand the fundamentals of the display process.

We have seen that a page of text consists of 24 horizontal rows, each containing 40 characters. Each of these characters will be positioned inside a character rectangle as shown in *Figure 9.22*. In this example the character rectangle occupies 10 horizontal scanning lines and approximately 1/40th of the picture width. The boundaries of each character rectangle indicate the maximum size of the display produced by any graphic or alphanumeric shape. Graphic shapes occupy the whole of this character rectangle, while alphanumeric characters occupy a smaller space known as the ROM rectangle, in this case formed by an 8 by 5 dot matrix. The range of the alphanumeric characters which appear in this 8 by 5 rectangle are generated by a Read Only Memory. The ROM rectangle is smaller than the

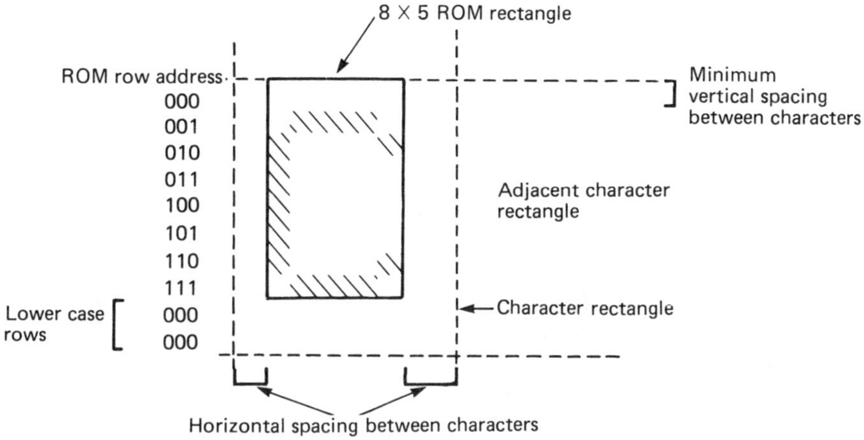

Figure 9.22 Character rectangle

character rectangle because spacing must be provided in between horizontal rows of text and in between individual characters on the same row. This facility is not required by graphics because in most applications of the graphic shapes it is necessary to follow one graphic rectangle immediately by another without spacing.

Each horizontal spacing in the ROM rectangle represents the periodic time of one cycle of the display clock oscillator, while each vertical spacing represents the width of one horizontal scanning line. The complete character rectangle will therefore occupy 8 display clock periods in width and 10 lines in height. The reduced width of the ROM rectangle when compared with the character rectangle allows 3 clock periods to be inserted in between individual characters (1 before and 2 after). The height of the ROM rectangle is reduced by 2 lines when compared with the character rectangle in order to allow for lower case character tails, e.g. j, g, q, p.

In this example the vertical spacing between rows will be a minimum of 1 line and a maximum of 3. Each character or graphic shape has its own character code which will be stored in the decoder RAM. These codes have to be read from the RAM and used to display the shapes within the character rectangle during the field scan period. Generation of the alphanumeric characters is done with the aid of a Read Only Memory, while generation of the graphics characters is carried out by a group of sequential and combinational logic circuits.

9.13 Principles of character generation

A basic character generating circuit is shown in *Figure 9.23*. At the top of the diagram is the decoder RAM, with its row and column address inputs to the left hand side and a read/write input to the right hand side. During a memory read, the character code stored in the addressed RAM location will appear on the 7-bit data output. This character code is used to address a ROM location which contains the pattern of '0s' and '1s' that form the video waveform.

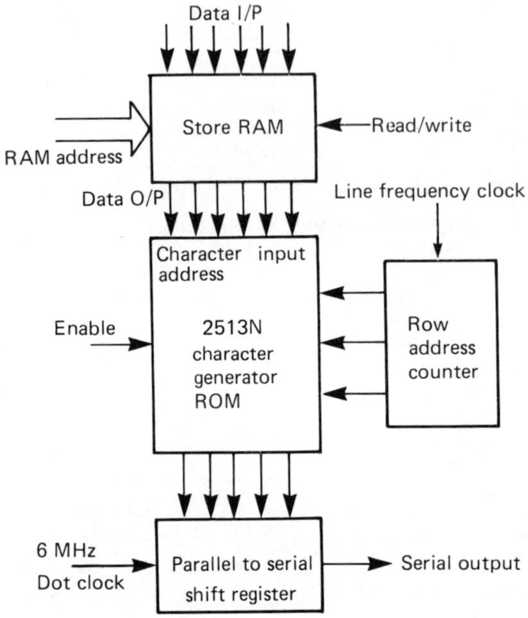

Figure 9.23 A basic character generating circuit

The ROM will require additional address inputs indicating which horizontal line of the addressed character is being scanned, so that it can generate the correct pattern of '0s' and '1s' to be used for that line of the character. To achieve this a row address counter provides the character generator ROM with a further three address inputs, making a total of 10 address lines. The address inputs to the character generator ROM are therefore subdivided into two sections:

1. The 7-bit character code from the decoder RAM.
2. The 3-bit row address code indicating the line of the character being scanned.

The memory location which has been addressed by these 10 inputs is five bits wide, and these five bits determine the illumination pattern for the addressed horizontal row of the character rectangle. Consider *Figure 9.24*, which shows how the ROM would be addressed over an 8-line period to generate the letters 'C, A, N'. The 7-bit character codes read from the RAM are shown across the top of the diagram, while the binary codes from the row address counter are shown in the left hand vertical column. On row address 000 the output from the character generator ROM will be 00000 for each character, therefore the top horizontal row of each rectangle will be blank. On the next horizontal scan the row address is incremented to 001 and the character code for the letter 'C' is again read from the decoder RAM and used to address the ROM.

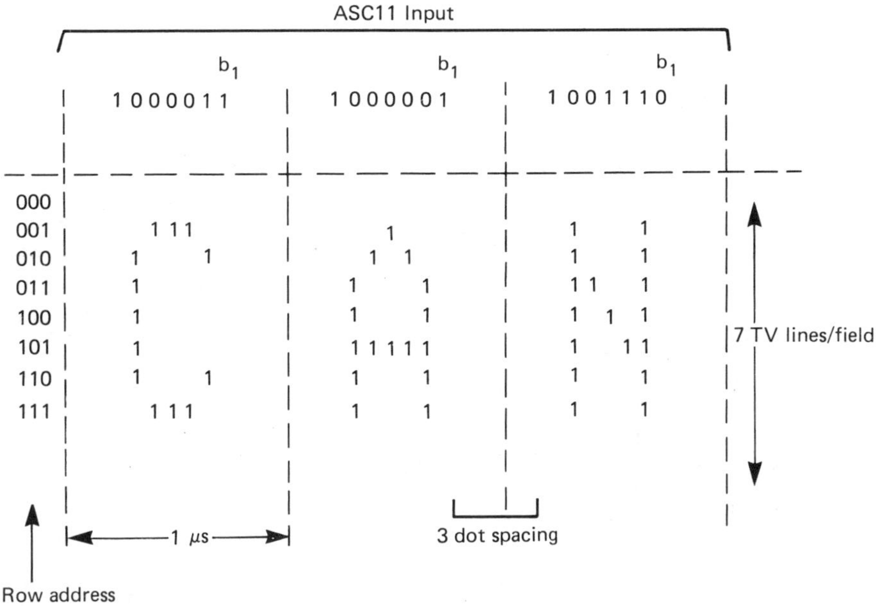

Figure 9.24 Teletext ROM addressing

The data stored at this ROM address is 01110, and will produce the video information for the top line of the letter 'C'. While this top line of the first ROM rectangle is being scanned the character code input to the ROM is changed for the next character in the horizontal row, in this case the letter 'A'. When scanning of the first letter is complete the output from the ROM will be 00100 providing the top horizontal row of the second letter. This data is then used to produce the lop line of the letter 'A'. While this information is being scanned the character code input to the ROM changes again, this time to the code for the letter 'N'. This new address will produce the output of 10001 from the ROM, forming the top line for the letter 'N'.

This process is repeated for all 40 characters in row 001 completing the scanning process for the top line of each character in the row. During line flyback the row address input to the ROM is incremented to 010. The row address to the RAM remains the same for 10-line periods, but the RAM column address is reset back to 000000. At the beginning of the new line the character code for the letter 'C' reappears at the RAM output, causing the ROM, with its new row address to produce an output of 10001 for the second horizontal row of the letter 'C'. The process is now repeated producing 01010 for the second row of the letter 'A', 10001 for the second row of the letter 'N', etc. It can be seen from the diagram that over a period of 8 scanning lines the letters are created on the screen.

Parallel-to-serial text conversion

The output from the character generator ROM is a five-bit parallel data pattern. When this data pattern is interrogated from left to right it will give the beam intensity for the CRT. Logic '0' switching the beam off and a logic '1' switching the beam on. A parallel in serial out shift register is used to convert the five-bit pattern of '0s' and '1s' into a serial digital signal which is to form the video waveform for the CRT.

In the simplified diagram of *Figure 9.25* the parallel to serial shift register is shown receiving five inputs from the character generator ROM. When ROM has been correctly addressed by the RAM, and row address counter, the correct pattern of data will appear at its output. This data is loaded into the parallel to serial shift register, and then clocked out, one bit at a time, under the control of the display clock oscillator. While the contents of the shift register are being clocked out the address to the decoder RAM is changed so that the character generator ROM receives the address for the next character to be displayed. As soon as the shift register is empty, it is reloaded with the bit pattern for the new character, and the process repeated.

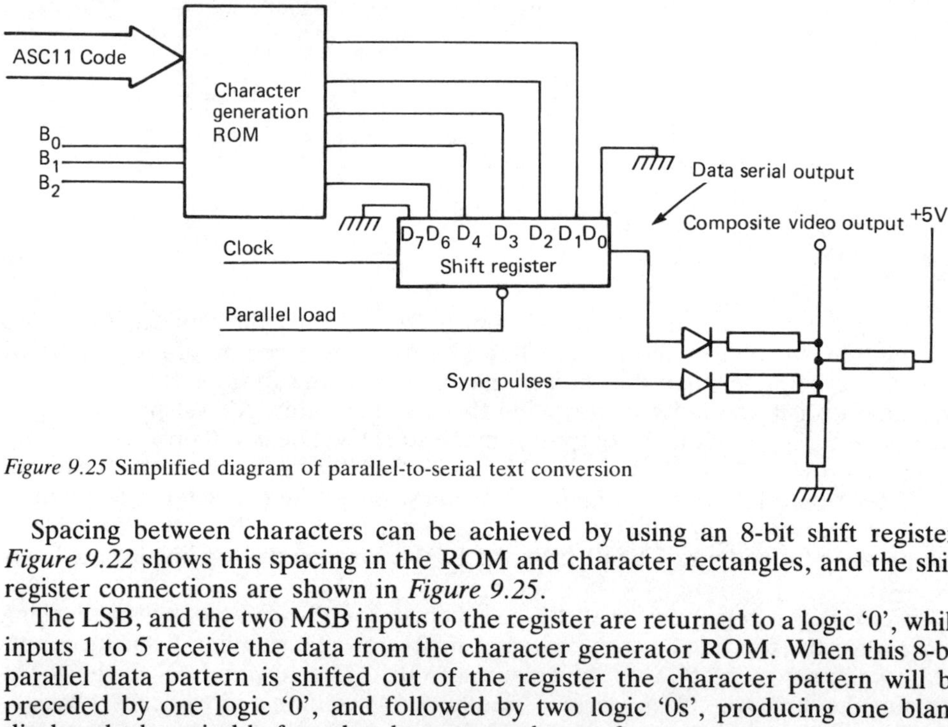

Figure 9.25 Simplified diagram of parallel-to-serial text conversion

Spacing between characters can be achieved by using an 8-bit shift register. *Figure 9.22* shows this spacing in the ROM and character rectangles, and the shift register connections are shown in *Figure 9.25*.

The LSB, and the two MSB inputs to the register are returned to a logic '0', while inputs 1 to 5 receive the data from the character generator ROM. When this 8-bit parallel data pattern is shifted out of the register the character pattern will be preceded by one logic '0', and followed by two logic '0s', producing one blank display clock period before the character and two after.

Dedicated Teletext ROMs

The character set of a Teletext decoder has been specifically designed by the broadcasting authorities for use in this country. A dedicated Teletext ROM will have the following specification:

1. It uses a 9 by 5 dot matrix, with all the the dots within the matrix being used.
2. It has a true Teletext font. (See *Figure 9.26*)
3. By increasing the ROM rectangle from the previous 8 by 5 matrix to a 9 by 5 matrix the extra rows produced can be used for the descending tails of some of the lower case characters such as 'g' or 'j', and these extra rows can also be used for some numerals such as 1/2 etc.

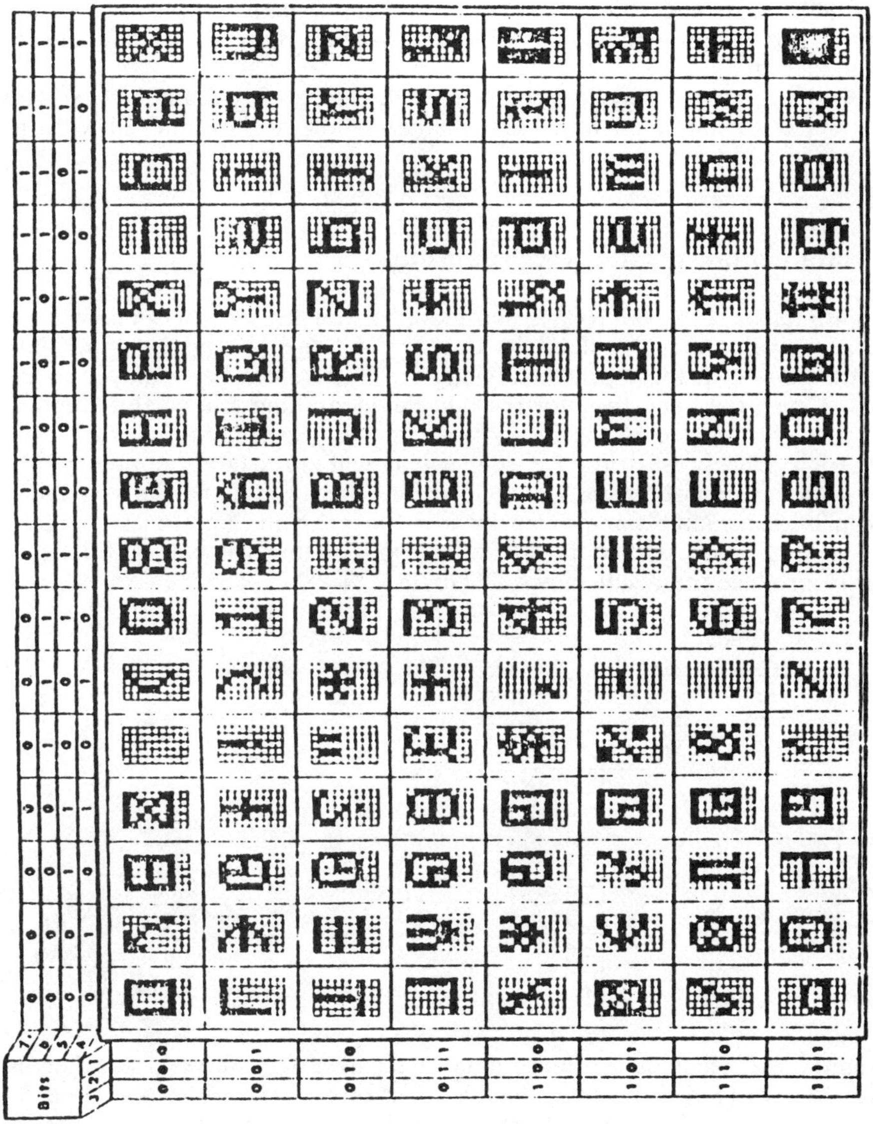

Figure 9.26 Teletext character format

4. The memory has a capacity of 128 characters each one of 45 bits making a total of 5760 bits.
5. Although the ROM rectangle is 9 by 5 most characters occupy a space of 7 by 5 dots within the rectangle.
6. As the character rectangle has 9 horizontal rows the ROM will require a 4-bit row address and a 7-bit character address, making a total of 11 address lines.
7. As the ROM rectangle does not contain blank horizontal rows or vertical columns, vertical and horizontal character spacing must be carried out by external circuitry.
8. The ROM will have a very low character access time of typically 180 nS. This allows character rounding, which is explained below, to be achieved with the minimum of additional circuitry.

9.14 Fundamentals of character rounding

In the examples of character ROMs considered so far the data output from the ROM for any given character will be the same for both the odd and the even fields. Each matrix square in the character rectangle will therefore be displayed twice, once for the even field and once for the odd field, resulting in a pair of scanning lines creating every matrix of the character rectangle. (See *Figure 9.27 (a)*).

This pairing process will produce coarse diagonals in a large number of the characters, the upper case letter 'Y' being used here as an example. This character format can be considerably improved by using a technique known as character rounding. This is a process of adding half dot periods on the diagonal sections of the character to produce the result shown in *Figure 9.27 (b)*.

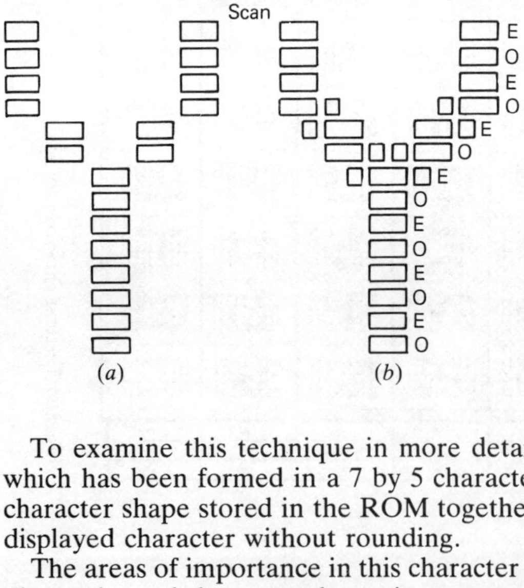

Figure 9.27 Character rounding

To examine this technique in more detail we will use an upper case letter 'K' which has been formed in a 7 by 5 character rectangle. *Figure 9.28 (a)* shows the character shape stored in the ROM together with *Figure 9.28 (b)* which shows the displayed character without rounding.

The areas of importance in this character are the upper and lower sections of the diagonals, and these are shown in greater detail in *Figures 9.29 (a)* and *9.29 (b)*. Assume the receiver is scanning line *n* on both the odd and the even fields. On the odd field the ROM dot pattern on line *n* is compared with the next line to be scanned ($n + 1$). This comparison will show that line $n + 1$ is displaced to the left of line n, causing the display circuits to add a half dot in front of the dot pattern for line n. This process is known as pre-rounding.

On the even field line, n is compared with the previous line scanned ($n - 1$). This comparison will show that $n - 1$ is displaced to the right of line n and consequently a half dot is added after the dot pattern for line n. This process is known as post-rounding. When both scans are interlaced as is shown in the centre of *Figure 9.29* the effects of character rounding can be seen.

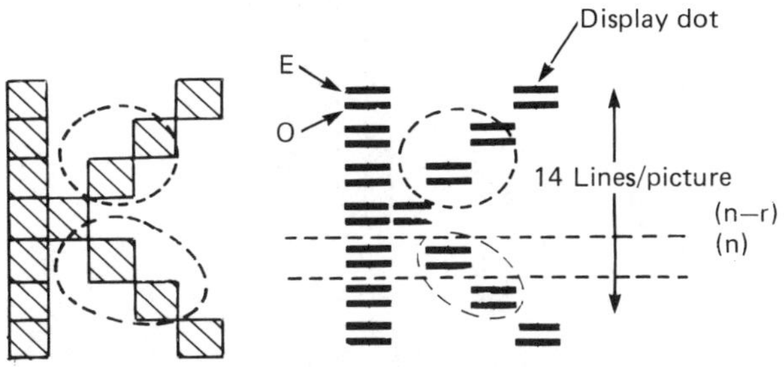

(*a*) Stored in ROM (*b*) Display without rounding

Figure 9.28

The same process takes place on the lower diagonal. On the odd field, line n is compared with line $n + 1$ showing that line $n + 1$ is displaced to the right of line n. This will result in a half dot post rounding on line n. On the even field line n is compared with line $n-1$ indication a displacement to the left. Half dot pre-rounding is therefore carried out on line n. Again the result of the two interlaced fields is shown in the centre of the diagram.

If the comparison process does not show a dot displacement then character rounding does not take place and the unmodified dot pattern is applied to the decoder output circuits.

A character rounding circuit requires the following criteria:

1. Two rows of information must be available simultaneously.
2. If the receiver is scanning an odd field then the next line of data ($n + 1$) must be available for comparison.
3. If the receiver is scanning an even field then the previous line ($n - 1$) must be available for comparison.

This criteria has led to the development of two types of character rounding circuits, the first of which is shown in *Figure 9.30*. Here two identical ROMs are used. A display ROM generates the full dot pattern of the character as previously described. A reference ROM receives the same character address as the display ROM, but the row address has been modified. On an odd field the row address to the reference ROM will be one ahead of the row address to the display ROM, resulting in an output of line $n + 1$. On the even field the row address to the reference ROM will be one behind the row address to the display ROM resulting in an output of line $n-1$. A combinational logic circuit compares the display data with the reference data and generates the half dot pattern for character diagonals. The half dot and full dot patterns are mixed to produce the character video output.

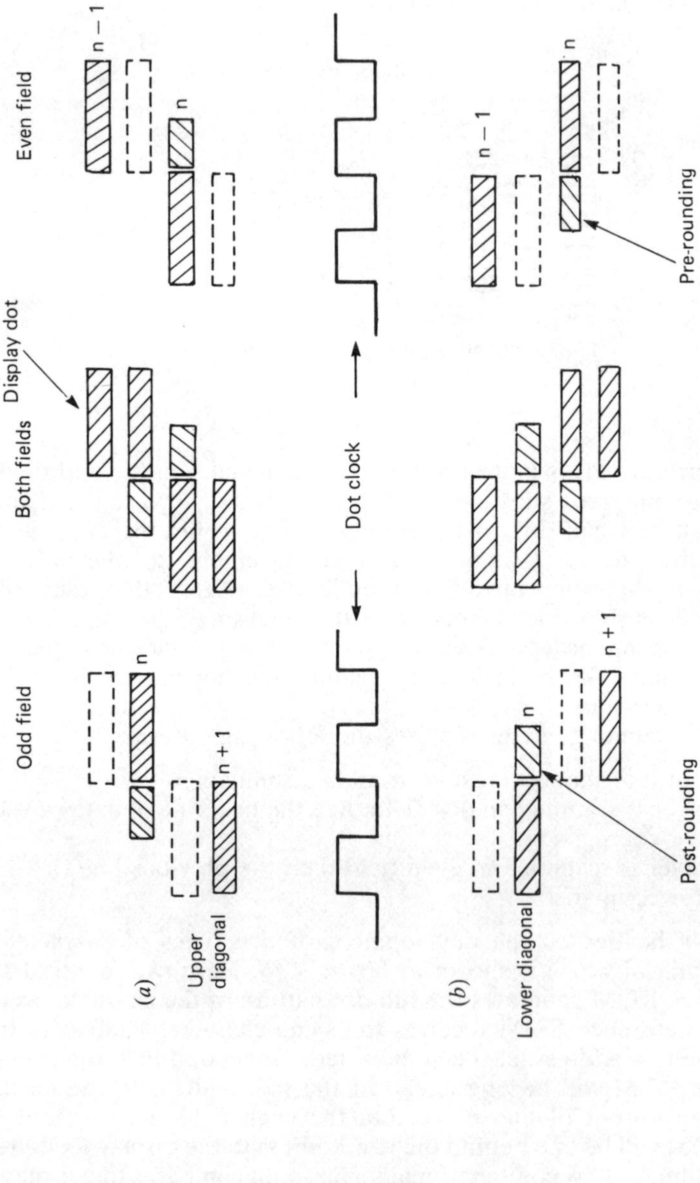

Figure 9.29 Detail of character rounding

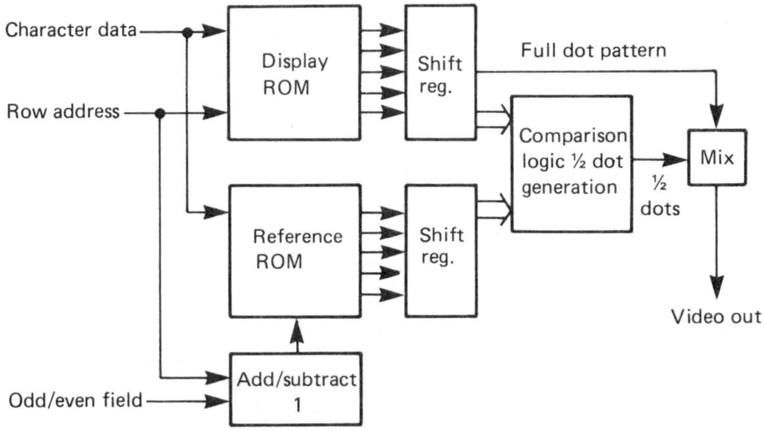

Figure 9.30 Character rounding circuit

This technique requires a display and a reference ROM, which adds to the cost and component count of the Teletext decoder. If the display circuits are to be housed in a single LSI chip then this will also produce a significant increase in the chip area. This problem can be solved by using a single ROM as shown in *Figure 9.31*.

A single ROM can only be used providing it has a very fast access time allowing two memory accesses to be made during the time it takes to display a single character. The first memory access to the display ROM will be for the reference line, either line $n+1$ or $n-1$. Data that is read from the ROM is latched into a

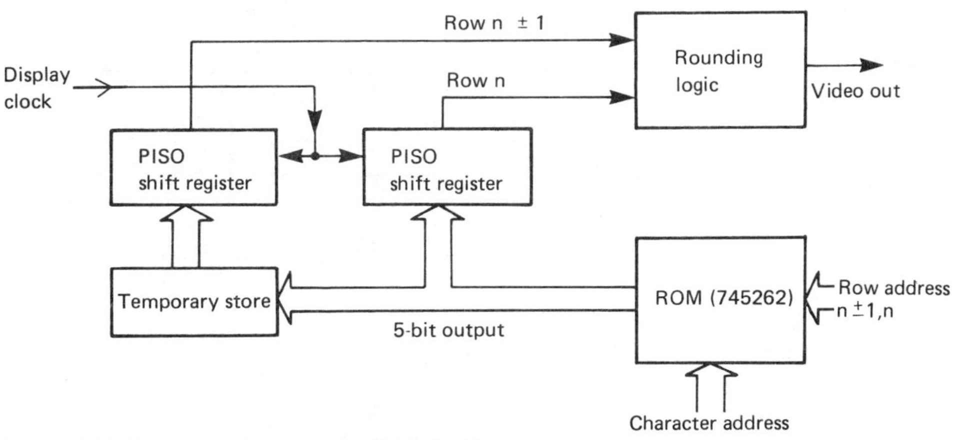

Figure 9.31 Alternative character rounding circuit

temporary store ready for transfer to a parallel in serial out shift register. The second access to the ROM will be for the display line (row n). Data read from this memory location is latched into a second shift register. After these two memory access row n and row $n+1$ will be housed in the two shift registers ready to be clocked out in serial form to produce the display. Comparison of these two dot patterns can now take place by the character rounding logic on a dot by dot basis. Half dot periods are added to the video waveform as required.

This second technique is favoured by manufacturers of LSI display chips because a single ROM makes significant savings in the chip area used, and allows character rounding to take place with the minimum increase in cost to the decoder.

9.15 Graphics

A simple graphics facility exists in the Teletext specification and this is used to enhance pages of text, and produce simple maps and diagrams. As the graphics format is very simple it is possible to produce a graphics generator using only a few logic gates. Before describing how graphics circuitry is implemented in the decoder it is necessary to examine the coding system used, and this should be done with reference to *Figure 8.3* of the Teletext character codes contained in chapter 8.

Control characters are shown in columns 0 and 1, and they determine the display mode of the decoder. These control characters are identified by bits 6 and 7 both set to a logic '0'. Some characters determine if graphics or alphanumerics are to be displayed, and others provide special facilities such as boxed display, normal or double height, graphics hold and release etc. At the beginning of a new line of text the decoder will automatically assume a certain number of prerequisites, and these are indicated by the number 2 in columns 0 and 1. The decoder therefore assumes that a new line of text will be displayed in white alphanumerics, with a steady display at normal height. Graphics characters will be contiguous on a black background in the release mode. If the Teletext decoder is to display graphics on a new line it must receive a control character to inform the decoder of the graphics format. *Figure 9.32* shows a typical sequence of control characters when changing from alphanumerics to graphics.

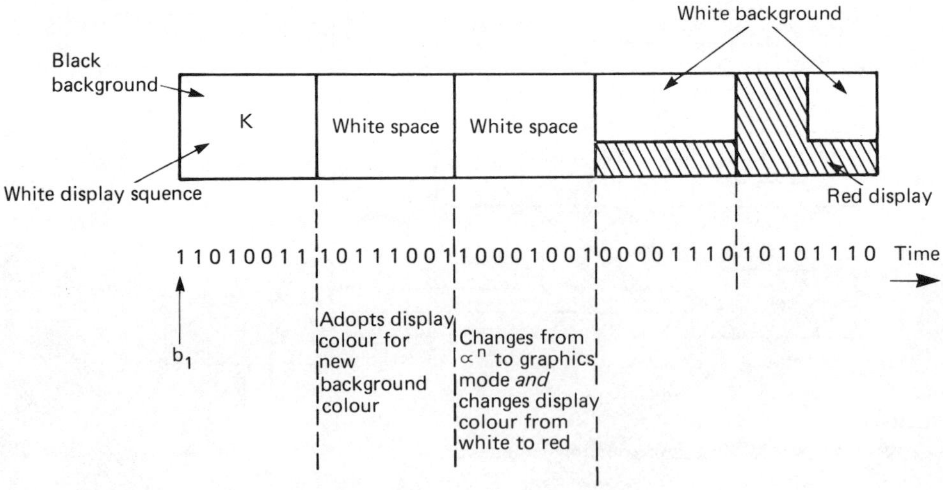

Figure 9.32 Typical sequence of control characters

The first byte of data is the character code for the upper case letter 'K' including the odd parity bit. We will assume that the background colour is black and the character is white. The second byte is the control character 'New Background'. Note that bits 6 and 7 are both at logic '0'. This control character causes the decoder to adopt the previous display colour as the new background colour. Consequently all alphanumerics or graphics following this control character will be displayed on a white background. The control character itself is displayed as a character rectangle containing the new background colour.

The third byte of data is for the control character 'Graphics Red'. This causes the display mode to change from alphanumerics to graphics, and also to change the display colour from white to red. In this case the control character will be displayed as a blank character rectangle of the current background colour which is white.

The transmission of bytes 2 and 3 in this sequence has changed the display mode in two stages from white alphanumerics on a black background, to red graphics on a white background. The character codes which follow the third byte of data will create red graphics shapes on a white background, and will continue in this mode until the arrival of further control characters, or until the end of the Teletext line. The next line of text will commence with the inbuilt default mode labelled 2 in *Figure 8.3* (chapter 8).

Graphics shapes can be displayed within the character rectangle in two basic formats. These are known as:

1. Contiguous Graphics (meaning touching).
2. Separated graphics (meaning not touching).

Both of these graphics formats are shown in *Figure 9.33*. As a general rule, the contiguous graphics mode is used for large areas of graphic information where as separated graphics is used for finer detail. Control characters are used to switch between contiguous and separated graphics modes. See *Figure 8.3*, column 1 rows 9 and 10.

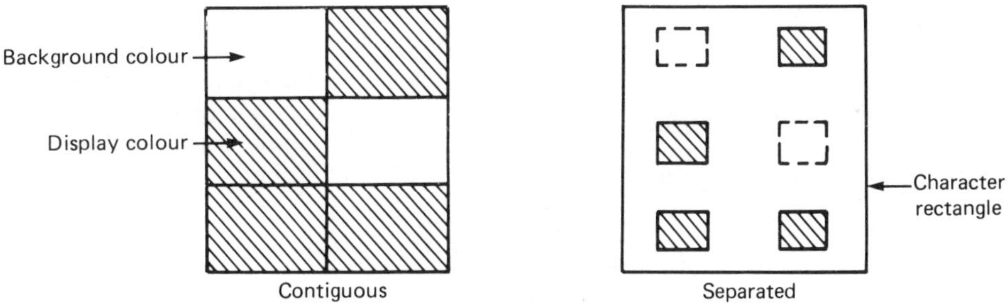

Figure 9.33 Contiguous and separated graphics formats

Graphics can also be transmitted using either the hold or release modes. In the default mode release graphics is selected by the decoder. This means that when a control character is received by the decoder the space on the screen that the control character occupies will be displayed as the selected background colour. In the hold mode the character space occupied by a control character will be displayed as the last graphics character received with bit 6 set to a logic '1'. This enables abrupt changes to be made between graphics shapes without the need to display a background space between graphic colours.

We have seen that the character rectangle used for Teletext is normally 8-display clock pulses wide by 10 lines high, and the graphic shapes occupy the whole of the rectangle. Two examples of character rectangles are shown in *Figure 9.34*. Both consist of 6 sections, each being illuminated on the CRT by setting its respective bit number to a logic '1'. Bit 6 is not included in the character rectangle format because it is used to indicate the difference between an upper case alphanumeric and a graphic symbol.

A complete set of the graphic shapes that are available is shown in *Figure 8.3* in columns 2a, 3a, 6a, and 7a. If we assume that the character rectangle is 8-display clock periods wide then the division between the two vertical halves of the rectangle will take place between display clock pulses 4 and 5. As there are 10 scanning lines per character rectangle then 1 of the blocks must be 1 line higher

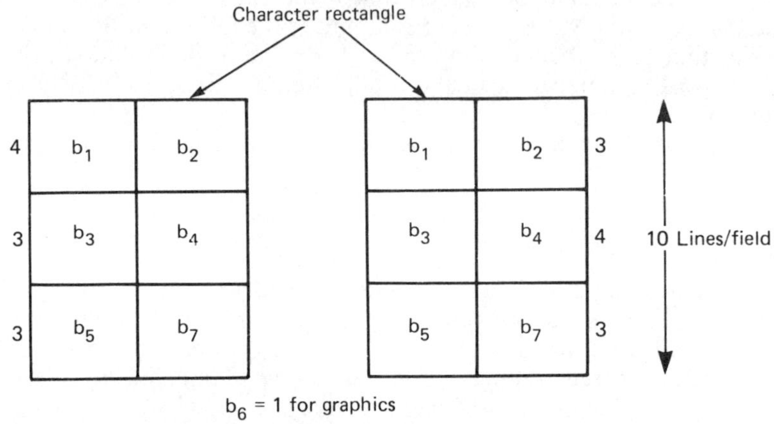

Figure 9.34 Two examples of character rectangles

than the other 2, as 10 is not exactly divisible by 3. In practice it makes little difference which of the three rectangles has four lines, but manufacturers prefer to spread the vertical division on a ratio of 4:3:3 or 3:4:3 lines per block.

An added subtlety to the graphics mode is the facility known as 'blast-through alphanumerics'. *Figure 8.3* shows that a graphics character can be identified when:

1. The Teletext decoder is in the graphics mode.
2. Bit 6 of the character code is set to a logic '1'.

Examination of columns 4 and 5 of *Figure 8.3* shows that these two columns contain the complete alphabet in upper case together with a few other characters. If the decoder is in the graphics mode bit 6 will normally be set to a logic '1', but if bit 6 falls to a logic '0' the corresponding alphanumeric character in column 4 or 5 will be 'blasted through' the graphics background colour. This enables graphic and alphanumeric characters to appear adjacent to each other without the need for control characters to switch from one mode to the other. There will therefore be no intervening space between the graphic and alphanumeric characters. An example of this facility is included in the engineering test page on *Figure 8.2*.

A practical graphics circuit, which includes the facility for blast-through alphanumerics, is shown in Figure 9.35. Also included with the diagram is the format of the character rectangle showing bits 1 and 4 set to a logic '1'. The circuit requires the following inputs:

1. Waveforms X1 and X2 are complementary square waves derived from the display clock divider and will complete one cycle per character rectangle width.

2. Waveforms Y1, Y2 and Y3 are square waves derived from the display ROM line divider and have a repetition rate of 10 lines.
3. Bits 1 to 7 are derived from the output of the decoder RAM.
4. The graphics/alphanumerics control input at the top of the diagram is derived from the control code detector (described later).
5. The serial alphanumeric characters are derived from the ROM output shift register.

NAND gates 1 to 6 have open collector outputs, all sharing a common load resistance R to the +5 V supply. When graphic block B1 is being scanned timing waveforms XI and Y1 are both at logic '1'. NAND gate 1 will therefore have three logic '1s' at its input causing its output to fall to a logic '0'. During display block B4, waveforms X2 and Y2 are both at logic '1' and consequently all three inputs to NAND gate 4 will be set to a logic '1'. The output from gate 4 will therefore fall to a logic '0' only when block B4 is being scanned.

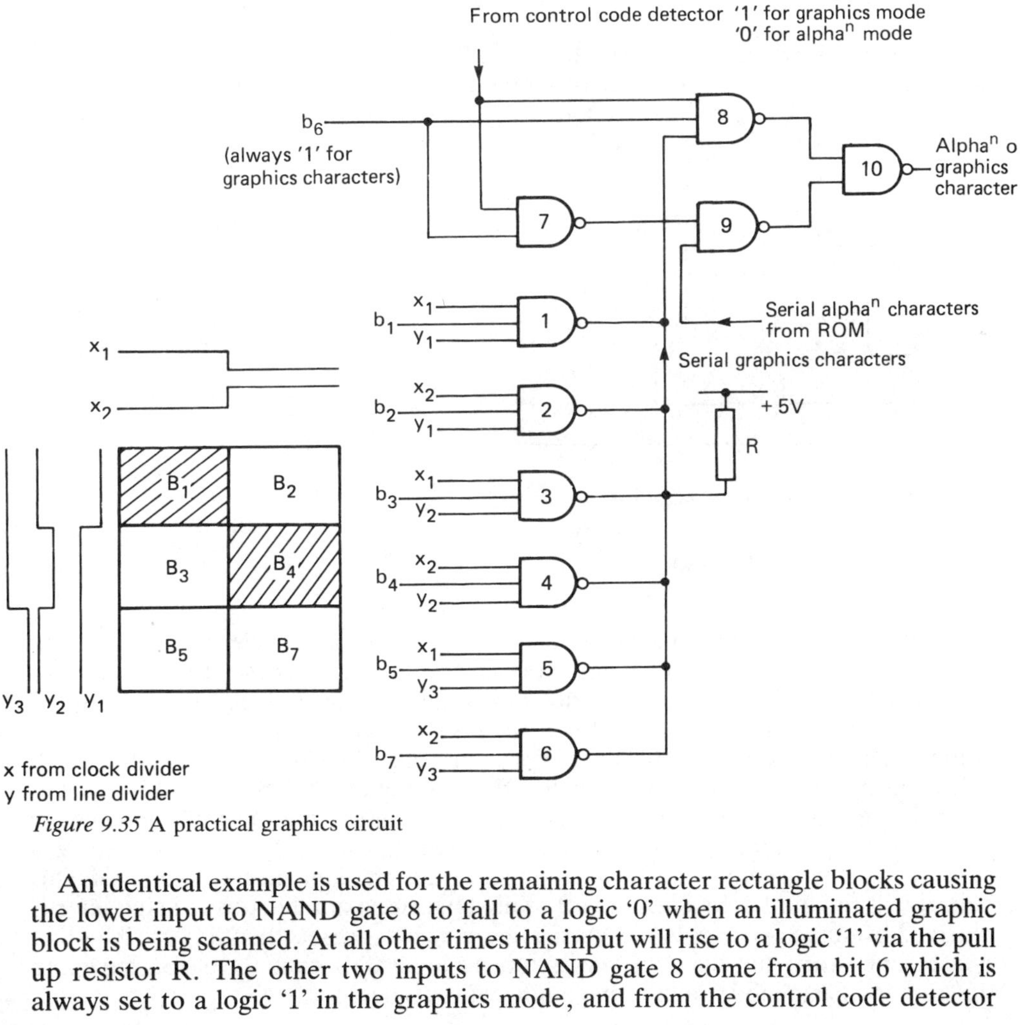

Figure 9.35 A practical graphics circuit

An identical example is used for the remaining character rectangle blocks causing the lower input to NAND gate 8 to fall to a logic '0' when an illuminated graphic block is being scanned. At all other times this input will rise to a logic '1' via the pull up resistor R. The other two inputs to NAND gate 8 come from bit 6 which is always set to a logic '1' in the graphics mode, and from the control code detector which is always a logic '1' in the graphics mode. Consequently when graphics are being transmitted the output from NAND gate 8 will rise to a logic '1' when a portion of the graphic rectangle is illuminated, and fall to a logic '0' when the graphic block is not illuminated.

NAND gate 7 will produce a logic '1' at its output when the decoder is in the alphanumerics mode, and a logic '0' in the graphics mode. Consequently gate 9 will produce an output which will always be a logic '1' when the decoder is in graphics. Should bit 6 fall to a logic '0' for blast through alphanumerics, or the decoder switch back to the alphanumerics mode, NAND gate 7 will produce a logic '1' at its output causing gate 9 to allow through the serial characters from the ROM. NAND gate 10 provides the final switching between the graphics from gate 8 and the alphanumerics from gate 9. The output from the circuit will therefore be the final video waveform which is then further processed to provide the addition of colour.

Control code detector

A simple circuit which can be used to detect the control codes and determine if the decoder is to work in the graphics or alphanumerics mode is shown in *Figure 9.36*. The first task of this circuit is to detect the presence of a control character which informs the decoder of a change between alphanumerics and graphics.

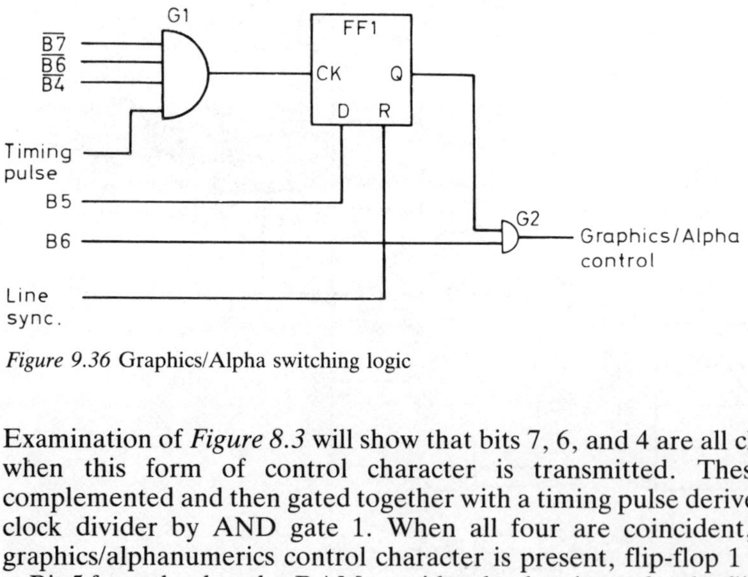

Figure 9.36 Graphics/Alpha switching logic

Examination of *Figure 8.3* will show that bits 7, 6, and 4 are all cleared to a logic '0' when this form of control character is transmitted. These three bits are complemented and then gated together with a timing pulse derived from the display clock divider by AND gate 1. When all four are coincident, indicating that a graphics/alphanumerics control character is present, flip-flop 1 is clocked.

Bit 5 from the decoder RAM provides the data input for the flip-flop. If it is set at logic '1' the graphics mode is required, but if it cleared to a logic '0' the decoder must select alphanumerics. The Q output of the flip-flop will therefore assume a logic '1' for the graphics mode and a logic '0' for the alphanumerics mode. This output will remain latched until another alphanumeric graphic control character is received from the decoder RAM, or if a line sync pulse resets the flip-flop. This line sync reset ensures that the Q output assumes a logic '0', setting the decoder back to the alphanumeric mode for the beginning of each new line.

9.16 Colour switching

The video waveform generated by the circuits discussed so far has been in monochrome only and, of course, Teletext is capable of producing a colour display. Circuitry is therefore required to detect the requested colour contained in the control characters and use this to gate RGB outputs from the decoder. Such a circuit is shown in *Figure 9.37*.

AND gate 4 detects the presence of an alphanumeric/graphic control character in exactly the same way as described in the previous circuit diagram. The output of this AND gate is used to clock 3 D-type flip-flops when an alphanumeric/graphic control character is present at the output of the decoder RAM.

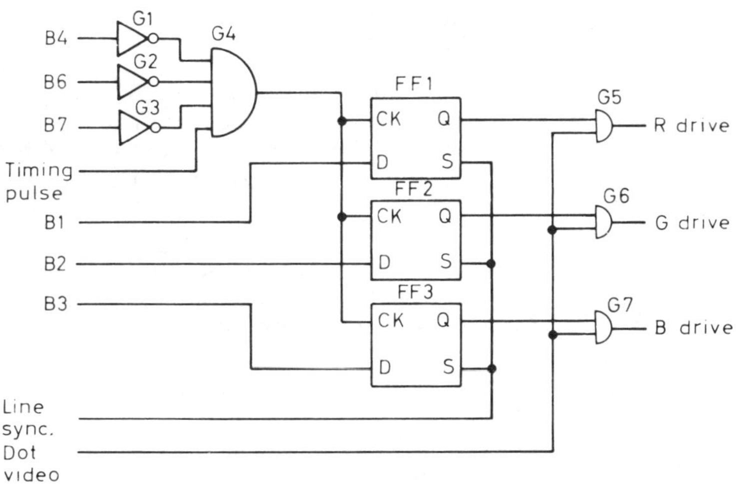

Figure 9.37 Colour switching circuit

Bits 1, 2, 3 from the RAM form the D inputs of these flip-flops. Examination of the binary codes for the control characters in *Figure 8.3* shows that bit 1 is set whenever the display colour contains red, bit 2 is set when the display colour contains green and bit 3 is set when the display colour contains blue. Bits 1, 2, 3 of the control characters therefore indicate the RGB drives for the CRT respectively. Each Q output from the flip-flops operates a 2 input AND gate switch. The second input to these AND gates is the video waveform containing the mixed alphanumerics and graphics. Whenever the Q output of a flip-flop is set to a logic '1' its respective AND gate will allow the video waveform to pass through and produce the appropriate drive for the CRT. Should the Q output clear to a logic '0' then the appropriate CRT drive will be removed. Receiver line sync pulses set each flip-flop at the beginning of a new TV line, producing a white on black display. This is the default mode.

Many of the circuits used by a full specification decoder have been omitted because of space restrictions and the necessity to limit this chapter to a 'digestible' size. These include boxed and flashing display, double height characters, background colour control, time coded pages etc. The aim of the descriptions just given is to provide the reader with a fundamental understanding of the most important sections of the decoder and include practical circuit diagrams where appropriate.

It was mentioned earlier in this chapter that all Teletext decoders currently available use custom made LSI integrated circuits which incorporate many sections of the decoder in a single IC. The circuit diagrams included in this chapter so far will be of little practical use because they will only be met if the reader has the misfortune to fault find a discrete component decoder. Not an easy task! !

The remainder of this chapter will introduce the techniques employed in some of the LSI decoders currently available, showing how the size and complexity of the decoder has been reduced using modern LSI techniques.

9.17 The Texas XM11 Teletext decoder

Texas Instruments Ltd were the first IC manufacturers to commercially produce a set of custom-made integrated circuits for use in a Teletext decoder. The decoder was named the TIFAX XM11 and contained fifteen integrated circuits on a double sided PCB 160 by 100 mm. Out of the fifteen ICs seven of these were used for the page memory.

This decoder represented the first significant advance in technology over earlier designs which used 74 series TTL devices. The first receiver manufacturer to make use of this decoder was Rank Radio International. Their receiver was designated the AC 6333, and required a considerable amount of interfacing between the TIFAX XM11 module and the rest of the receiver electronics. This interfacing was needed because Rank incorporated ultrasonic remote control, and the decoder was designed to be used with a directly wired key pad. Although the XM11 module was very small the complete interface panel was 29 by 20 cm.

The XM11 was introduced before the Teletext specification was reappraised in 1976 and therefore did not include such facilities as double height, background colour, separated graphics and graphics hold, etc. It used a combination of linear and low power Schotky TTL techniques and a simplified block diagram of the decoder module is shown in *Figure 9.38*. The video input is applied to TX 005 which is a linear IC containing all the input processing circuits. These can be broadly divided into three areas:

1. Synchronizing.
2. Data Slicing.
3. Clock Oscillator.

The incoming video signal is split two ways, to a data slicer and a sync separator. Line flyback pulses from the receivers line output stage are also required for the synchronizing circuits which produces and even field, conventional field, and line sync outputs.

The other video path feeds an adaptive data slicer which provides a TTL output for the clock synchronizing circuits and the serial to parallel converter. TX005 also contains the 6.9375 MHz data clock oscillator which is synchronized using the divide-by-444 technique explained earlier in this chapter.

A line frequency waveform from TX005 is fed to the clock sync circuits in X905. Here it is compared with the divided clock oscillator output to produce the first control voltage. A second control voltage is produced by comparing the data slicer

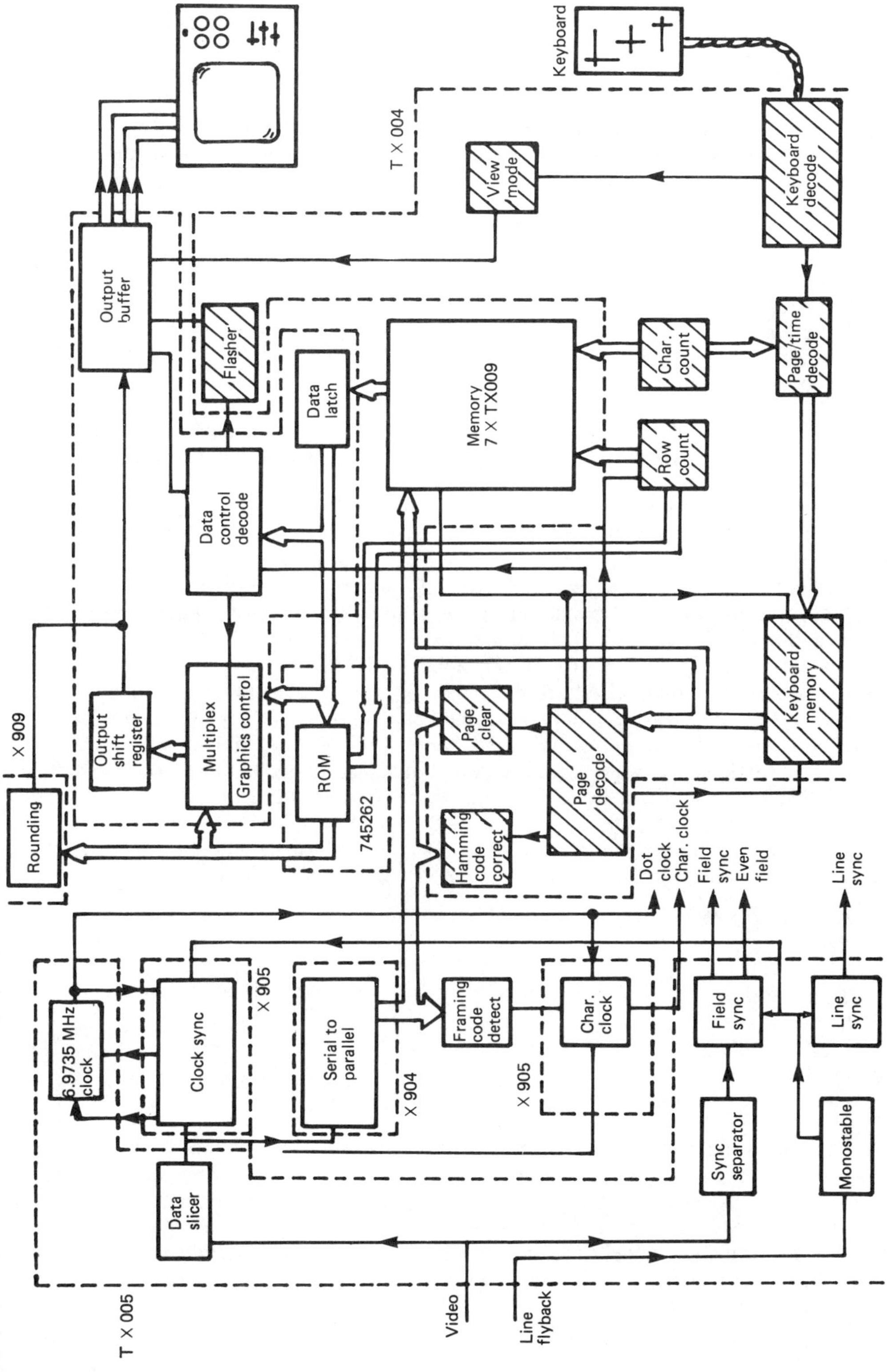

Figure 9.38 Block diagram of Texas XMII decoder

output with the clock oscillator frequency, and both error voltages are fed to the oscillator to produce a stable bit rate clock.

X904 contains a serial to parallel converter which produces the latched data bus used by a variety of circuits throughout the decoder.

The main control logic is contained in a 40 pin LSI chip TX004, and the sections included in this IC are shown shaded in the block diagram. Practically all the decoder control logic is contained in this single integrated circuit and these include:

Hamming Code Correction.
Parity Checking.
Page Clear.
Page Decoding.
Time Decoding.
Memory Addressing.
Keyboard Decoding.

As mentioned earlier the keyboard for this decoder is primarily designed to be a unit which is wired to the receiver, consequently the decoder will produce strobe pulses for the keyboard.

Seven integrated circuits form the page memory, which receives address and control signals from TX004 and input data from X904. Data which has been read from this memory is applied directly to another custom-made LSI X908. This IC handles all the display functions apart from character generation and rounding. The first stage of processing in this chip is carried out by a data latch which captures the data read from the memory. This latched data is applied to a 745262 Teletext ROM that is addressed by the control IC TX004. The ROM has a very fast access time and is read twice for each character, once for the display line and once for the reference line. Display line data is applied to a multiplexer, while reference line data is fed to a character rounding IC X909. Full dot and half dot patterns are generated by these two stages, and the combined signals are fed to an output buffer which also contains the video switching circuitry to create a flash and boxed text displays, and RGB outputs.

The direct wired keyboard consists of a 4 by 4 matrix of switches and an earth. The matrix contains a total of 17 keys having the following funtctions:

1. Numerals 0–9.
2. Page.
3. Time.
4. Text Only.
5. Superimpose.
6. Picture Only.
7. Update.
8. Display Concealed Text.

A positive-going video waveform with negative sync pulses between 2–3 V is required on the video input pin together with a negative line flyback pulse. A single 5 V supply rated at 1 A is the only power requirement of the decoder.

This Teletext decoder has been chosen as a first example of a commercial LSI approach because it closely resembles the block schematic outline in *Figure 9.1*, and the reader will be able to make direct comparisons between these two approaches.

9.18 The Mullard Teletext decoder

The first Mullard LSI Teletext decoder was introduced after the revised Teletext specification in 1976. It used a series of custom ICs which not only provided Viewdata and Teletext reception, but also interfaced directly with their remote control and digital tuning integrated circuit sets. Each integrated circuit is designed in such a way that it can be used with a variety of other ICs, to produce Teletext and Viewdata decoders having varying degrees of sophistication. Mullard manufacture the complete range of devices under the trade name of Multitext (RTM)[4], and in this section we will examine just those integrated circuits used in the teletext decoder. Other circuits used within the Multitext system are described in chapters 10, 11 and 12.

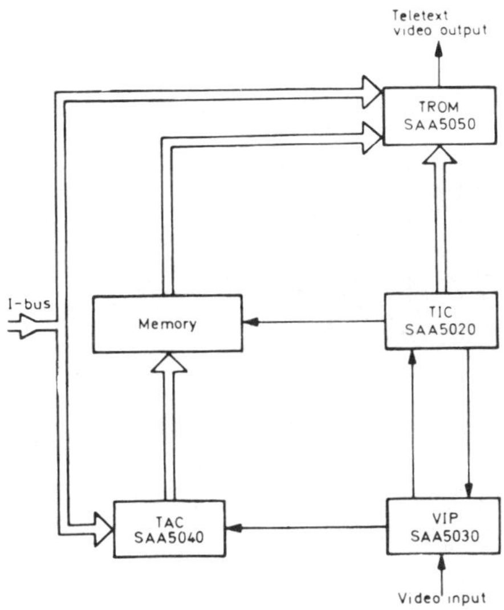

Figure 9.39 Block diagram of the Mullard Teletext decoder

The block diagram of the Mullard decoder is shown in *Figure 9.39*. It is immediately apparent that the use of LSI NMOS technology has reduced the number of devices used within the decoder significantly. The whole decoder is based around four integrated circuits, and these are:

1. Video Input Processor (VIP) SAA5030.
2. Teletext Acquisition and Control (TAC) SAA5040.
3. Timing Chain (TIC) SAA5020.
4. Teletext Read Only Memory (TROM) SAA5050.

These four custom-made chips operate in conjunction with two RAM chips (2614) each one containing 1 K by 4 bits of store. Four address decoding chips. (74LS02, 74LS83A, 74LS161), are also used. Three digital ICs (TIC, TAC, TROM), are NMOS devices while the VIP is a linear bipolar device. The Mullard Teletext decoder is designed to be used exclusively with either infra-red or ultrasonic remote control systems. In their design of the Multitext system, Mullard have developed a communications highway between the remote control receiver and the custom LSI

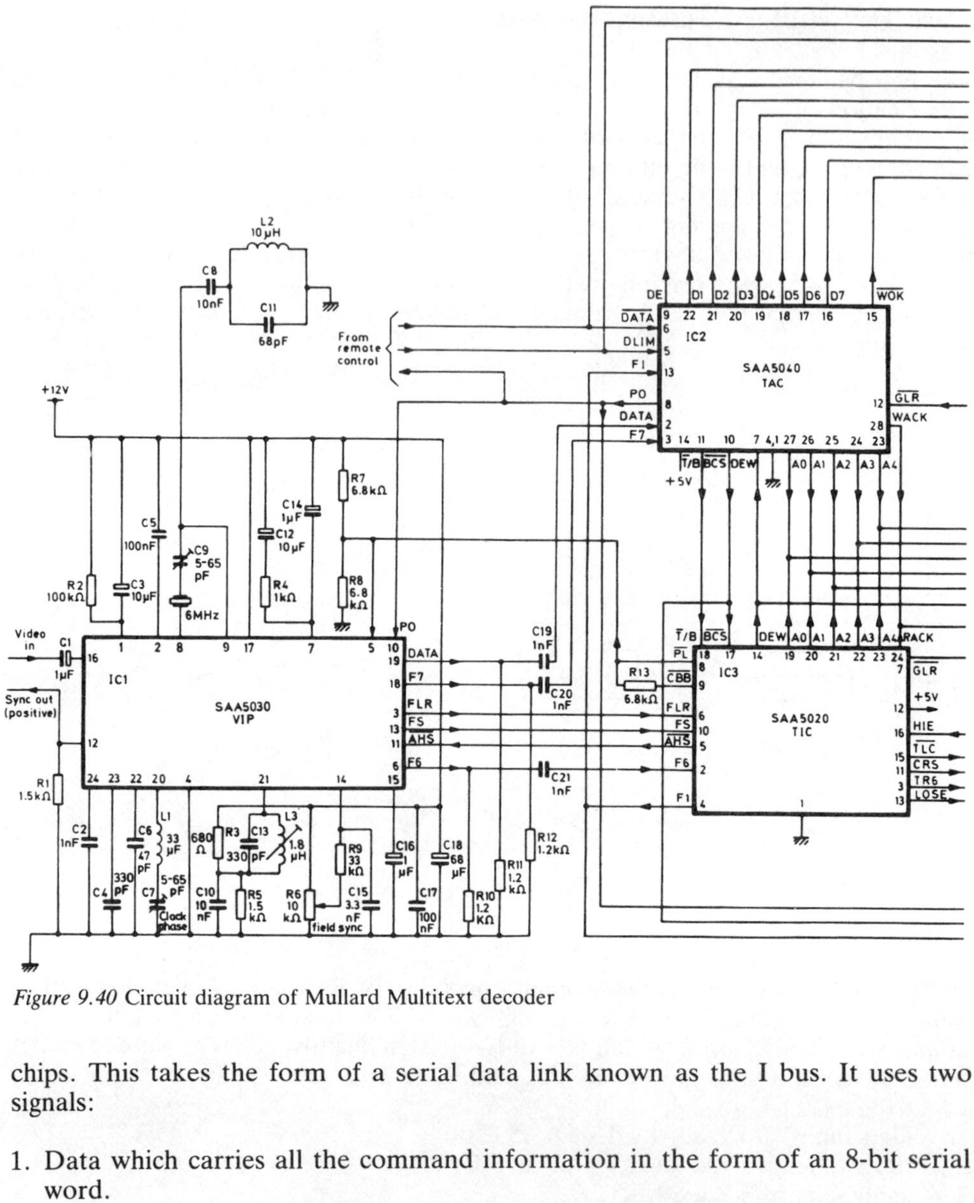

Figure 9.40 Circuit diagram of Mullard Multitext decoder

chips. This takes the form of a serial data link known as the I bus. It uses two signals:

1. Data which carries all the command information in the form of an 8-bit serial word.
2. The clock enable line DLM which carries a series of 14 clock pulses at a nominal 62.5 kHz rate whenever a command has been received.

These two signals can be applied to a variety of the Multitext integrated circuits to provide user control over the display modes. Conveying user command information in this way over two serial data lines considerably reduces the complexity of IC interconnections with the remote control system. It does however complicate the internal electronics of those ICs using the I bus as the serial data has to be decoded before the remote control command can be acted upon.

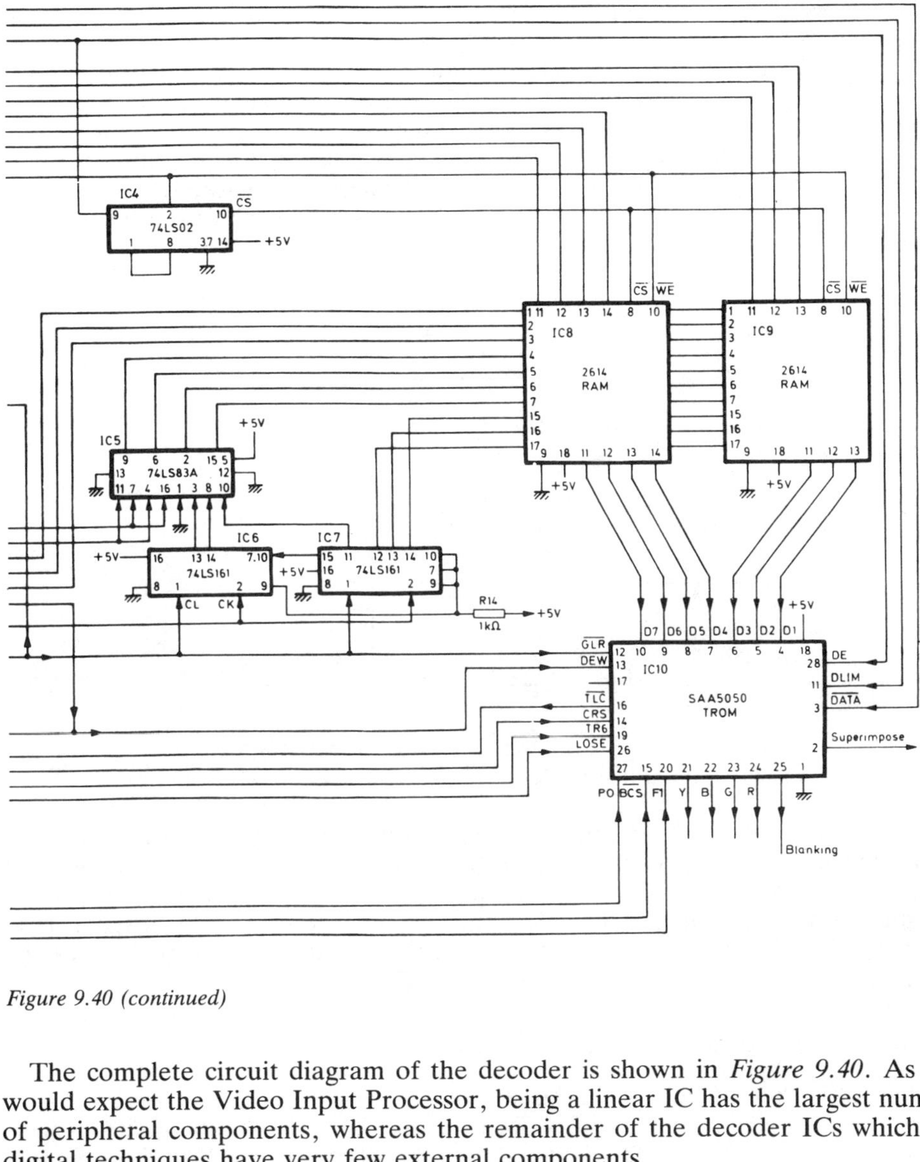

Figure 9.40 (continued)

The complete circuit diagram of the decoder is shown in *Figure 9.40*. As one would expect the Video Input Processor, being a linear IC has the largest number of peripheral components, whereas the remainder of the decoder ICs which use digital techniques have very few external components.

Video input processor (VIP) SAA5030

Referring to *Figure 9.41* it can be seen that the Video Input Processor is divided into two major sections—the circuits responsible for data retrieval and those responsible for the display clock generation. The incoming video input is split into two paths to provide signal feeds for both of these sections. Data retrieval consists of an adaptive data slicer that provides a serial data output for the TAC IC, and also a synchronizing input to the data clock oscillator. Mullard have chosen to use a relatively simple clock regenerator which relies on a parallel LC circuit for its stability. This oscillator is synchronized by the clock run in and data transitions throughout the Teletext line as previously described in this chapter. The clock output from the generator is again fed to the timing acquisition and control chip.

A display clock oscillator running at 6 MHz is also included in this integrated circuit. Under normal signal conditions this oscillator is flywheel synchronized by a phase detector which compares off air line sync pulses with sync pulses derived from the timing chain IC. This will ensure that the dislay is accurately synchronized to the off air signal, and provide a stable display when text is boxed or mixed with the picture.

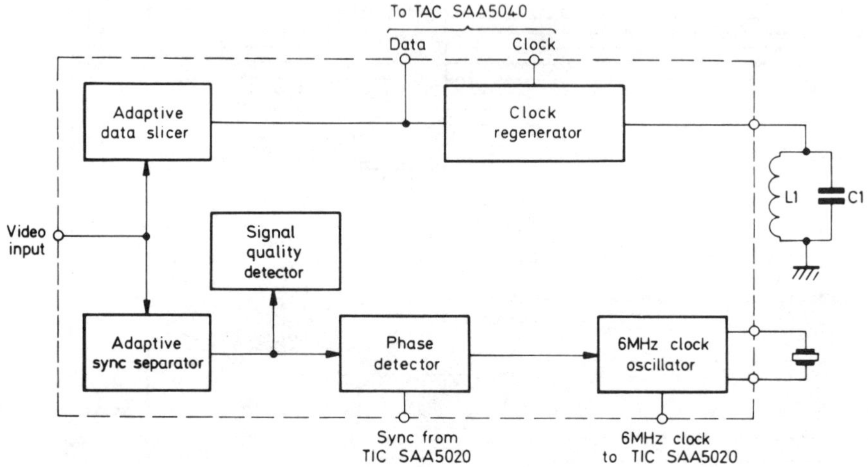

Figure 9.41 Video input processor (SAA 5030)

Should the quality of the off air signal degenerate to such an extent that accurate synchronization is not possible a signal quality detector comes into operation. It disables the data output to the TAC IC and allows the display clock oscillator to free run. This facility enables a stable display to be produced in the absence of an off air signal so that any time coded pages of text stored in the decoder RAM can be viewed after normal transmissions have gone off the air.

Teletext data acquisition and control (SAA5040)

The block diagram of the TAC IC is shown in *Figure 9.42*. There are four major inputs. These are the clock and data which come from the video input processor previously described, and Data and DLIM signals from the SAA5010 remote control receiver. These four inputs provide the TAC IC with Teletext data, together with its synchronized clock signal, and digitally coded serial instructions from the remote control receiver.

The incoming serial Teletext is transferred to the 8-bit parallel data bus by an input serial to parallel converter. 8 bits of this bus are fed directly to the Hamming and parity checking circuits while 7 bits (BO to B6) are fed to the decoder RAM.

Unlike the previous Teletext decoders, the Mullard system addresses the memory by using a 5-bit row address in the normal way, and a write address clock (WACK) signal which causes the row address counters (IC6 and IC7) to increment to the next memory address after each character has been stored. This is carried out by the bit counter which produces a new clock pulse every time a complete byte of data has been received and latched on to the data bus.

A comparator in the decoder detects a match between the magazine and page number transmitted with that selected by the user. It then allows the write control circuit to generate a signal designated as Write O.K. (WOK), which tells the decoder RAM when valid data can be written into the memory.

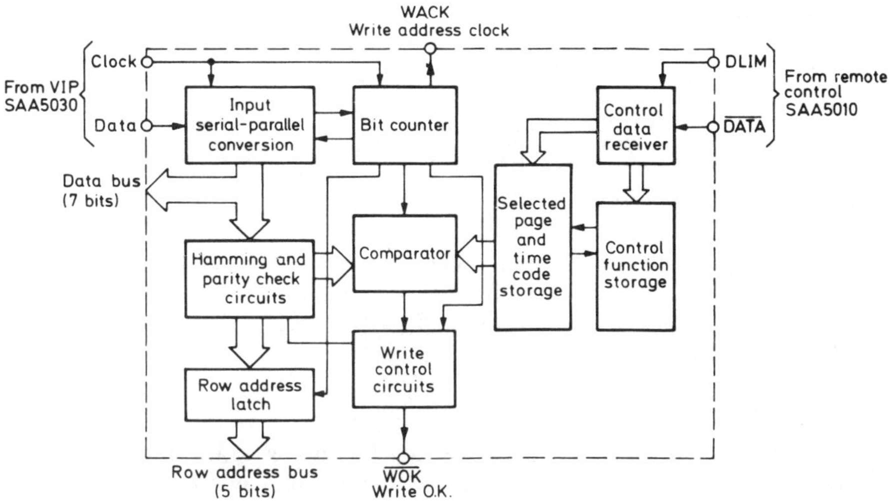

Figure 9.42 Timing acquisition and control (SAA 5040)

The remainder of this integrated circuit is dedicated to receiving the serial digital signals from the I bus, decoding them, and then storing the received instructions that are relevant to the IC in a small memory. It should be noted that all the integrated circuits connected to the I bus will receive all remote control instructions. Therefore if the user transmits an instruction to change channel then the serial digital signal generated by the remote control receiver will be fed to all ICs on the I bus. Obviously this form of instruction will be ignored by the TAC IC and only those instructions relevant to Teletext functions will be stored and acted upon.

Timing chain (SAA5020)

A single integrated circuit, IC3 (see *Figure 9.43*) provides all the timing facilities for the whole of the display electronics. It receives a 6 MHz input from the VIP chip which is then subsequently divided by 6, 64, and 625 to produce timing signals for the internal decoder. These timing signals are used by the TROM IC to control the generation of graphics, parallel to serial conversion, character rounding and the ROM row address. The decoder also produces a read address clock (RACK), for the decoder RAM during the display period. A divide-by-10 counter receiving timing pulses from the decoder has its output divided by 24 to produce a 5-bit row address. This is used by the decoder RAM in the display mode.

The decoder display circuits have been designed to provide synchronized memory read addresses, even in the absence of an off air signal. It is therefore

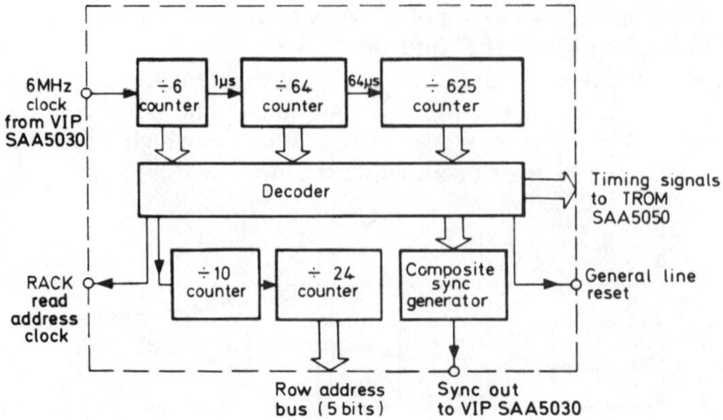

Figure 9.43 Timing chain IC TIC (SAA 5020)

necessary to generate an after hours sync signal for the receiver's timebases. Consequently the TIC IC contains a composite sync generator which will produce mixed line a field sync pulses that can be used by the receiver's timebase circuits when the off air transmission is lost.

Decoder RAM

The decoder RAM (*Figure 9.40*) consists of 2–4 bit by 1 K static random access memories (IC8 and IC9). Each RAM can be considered as four 32 by 32 bit matrices which are addressed by a 10-bit binary code on the address pins 1–7 and 15–17. IC5 provides the 11- to 10-bit address code conversion to select one of the 960 locations in the RAM. Pins 11, 12, 13, 14 on each IC share data input and output facilities, and IC4 is included to avoid data conflicts from occurring when the memory is activated. IC8 stores data bits 7, 6, 5, 4, while IC9 stores bits 3, 2 and 1. I K bit of storage in IC9 is unused. The three data outputs from IC9 and the four outputs from IC8 combine to provide a 7-bit data input for the ROM.

Teletext read only memory (SAA5050)

The major inputs for the TROM IC (*Figure 9.44*) are the 7 bits of data from the decoder RAM and the timing signals from the TIC IC (SAA5020). The incoming 7-bit data bus from the decoder RAM is split into two directions to provide signal feeds for the ROM and the control character detection and storage circuits. A line frequency input from the TIC IC is divided by 10 to produce the horizontal row address for the ROM, while the other two provide timing signals for graphics generation, parallel to serial conversion, and character rounding. A separate graphics generator which receives data from the RAM provides full graphics facilities including all the latest additions to the Teletext specification. Graphics and serial text are combined in the display control circuits under the direction of the character detection and storage stage. This display control provides RGB and monochrome text outputs together with line blanking.

Signals from the I bus are also fed to the TROM integrated circuit so that display mode commands such as Superimpose, Text, Picture, Reveal, etc can be received and acted upon. The I bus decoders have not been shown in *Figure 9.44* in order to keep the diagram as simple as possible.

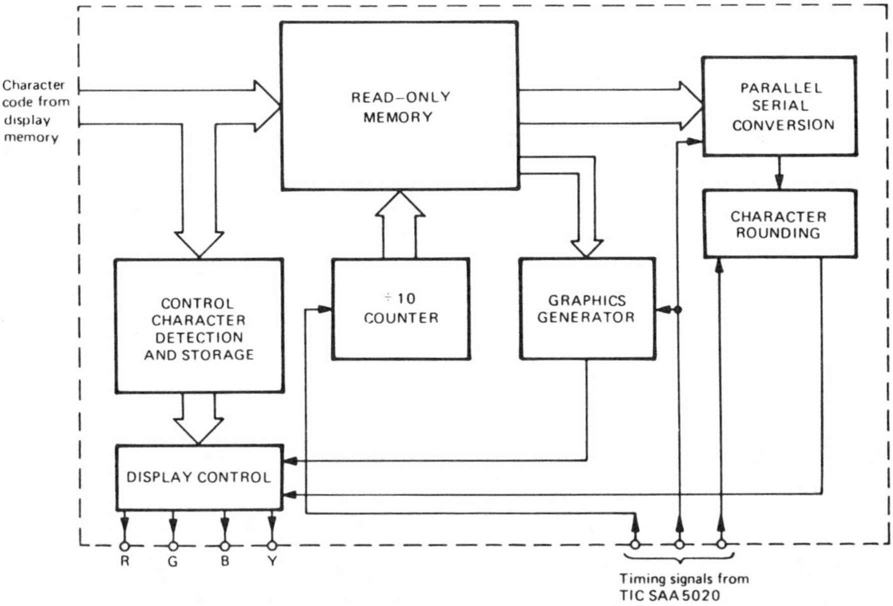

Figure 9.44 Block diagram of Teletext Read Only Memory (SAA 5050)

This set of ten integrated circuits provides a full specification Teletext decoder complete with remote control interface on a very small double-sided printed circuit board using the minimum of components. The alignment procedure is relatively straight-forward with the whole decoder having four pre-set controls. All the dedicated LSI chips described here are also used with the Mullard Viewdata/ Teletext decoder which only requires the addition of a further two LSI chips together with a Line Terminating Unit to interface the decoder with the telephone line. This decoder is fully described in chapter 10.

References

1. *Wireless World Teletext Decoder*
2. *'Television' Teletext Decoder*
3. *Multitext: Technical Information 72*, Mullard Ltd, Mullard House, Torrington Place, London WC1E 7HD

Chapter 10

Viewdata system and decoders

Viewdata is the term given to an interactive alphanumeric/graphics data service, transmitted and received via a telephone line. It was the original name for the information system developed by the Post Office Research Department in the early 1970s, but the UK Patents Office ruled that the word Viewdata was a generic term and could not be registered, consequently the name Prestel (Registered Trade Mark), was chosen for the British Telecom Viewdata service.

Throughout this chapter the term Viewdata will be used, but the system specifications mentioned are based on the British Telecom Prestel Terminal Specification[1]. When a TV receiver is connected to the Viewdata computer it becomes a computer terminal, consequently the receiver is called 'the terminal' throughout the chapter.

Unlike the Teletext system offered by the BBC and the IBA the British Telecom Viewdata system is two way and interactive. This means that requests entered via the user's keypad not only control functions within the terminal but some commands such as page numbers etc, are transmitted via the telephone line to the Viewdata computer. On receipt of the required page number the computer will access its data base to locate the requested page and then transmit the page back to the user via the telephone line. This fully interactive system allows the user to respond to instructions issued by the computer and interrogate the data base at will.

Data transmission rates used in the British Viewdata system are considerably slower than those used in Teletext, these being:

1200 baud between the computer and the terminal.
75 baud between the terminal and the computer.

This transmission speed does not compare very favourably with the 6.9 Mega-baud rate used with Teletext. Despite this considerable disadvantage Viewdata does have a much faster average access time to data, because the computer response time to a request is virtually instantaneous. With the Teletext system the user has to wait for the requested page to be transmitted at its allocated time with all the other pages being broadcast. This can lead to access times of several tens of seconds in the case of pages having a low priority.

Another advantage that Viewdata has over Teletext is the physical size of its data base. Teletext transmissions are limited to a maximum of 800 pages, discounting sub-pages. The Viewdata data base contains many hundreds of thousands of pages of information, and most of these are available to the user.

Some pages of the data base are maintained by the Post Office, such as indexes, user accounts, etc, but the vast majority are maintained by independent commercial organizations known as Information Providers (IPs). These information providers rent the data base capacity they require on the computer, in order to produce their own information services. Companies can store their own information in the data base, and modify or retrieve it anywhere in the country via a Viewdata terminal. They are of course charged for this facility, and this cost may be recovered by supplying the same information to all the users of the Viewdata system. When a user accesses a page of data belonging to an information provider the charge for the page is automatically debited from the users account, and credited to the IPs account. The cost of each page is determined by the IP and can range from 0 to 50p. Information Providers can protect their data from being accessed by unauthorized users by including a password to be used with each page. This password must be correctly entered by the user before access to a page, or series of pages is granted.

Another facility provided by most Viewdata systems is known as the closed user group. Here a number of mutually interested companies, or parties, may rent from the Post Office a group of pages which may only be accessed by that group of users. This enables information to be exchanged between the users of the group via the data base. Again this information can be protected by a password, or series of passwords, to ensure that unauthorized users cannot interrogate or corrupt the information.

This interactive system between the data base and the computer provides a whole host of facilities, many of which have not yet been fully exploited. Many of the larger retail companies provide pages of data which advertise goods and commodities etc, and offer the user a purchasing service via the Viewdata system. The user may scan through a series of pages free of charge in order to select the item to be purchased. When the item has been found the user enters his account number, the method of payment, and the order is made. The user does not have to enter his name or address because all this information is stored on the data base. At regular intervals during the day the information provider can retrieve from the data base a list of the orders placed, together with the names and addresses of the buyers.

This interactive system can be further expanded to include programme learning packages which would enable the data base to ask students questions which have multiple-choice answers displayed on the screen. Each student would enter the number of the multiple-choice answer which is thought to be correct, before the computer provides them with the next question. This facility can be invaluable in distance learning applications such as those used by the Open University or the Technician Education Council where students study at home for qualifications offered by these organizations. Students could also be supplied with information relating to their course of study using the same techniques. Facilities are also being developed so that Viewdata can be used with the major banks and finance houses. Customers will be able to access information from their own accounts, pay bills, and make simple transactions.

10.1 Basic principles of transmission

Viewdata uses asynchronous techniques which were fully described in chapter 6 ‘Data transmission’. The format illustrated in section 6.2 is based on Viewdata practice, and consequently the following description will be a quick revision of the techniques employed. The reader should refer to section 6.2 on asynchronous transmission for a more detailed explanation.

Communication between the data base and the user, and between the user and the data base, takes place using asynchronous NRZ transmission over a telephone line. The 7-bit symbol code used is compatible with the Teletext character set, apart from the inclusion of an eighth even parity bit. Each symbol code is preceded by a logic '0' start bit and superceded by a logic '1' stop bit. When data is not being passed along the telephone line a continuous logic '1' level is transmitted by both the terminal and the computer. The transmission format of an upper case letter 'M' is shown in *Figure 10.1*.

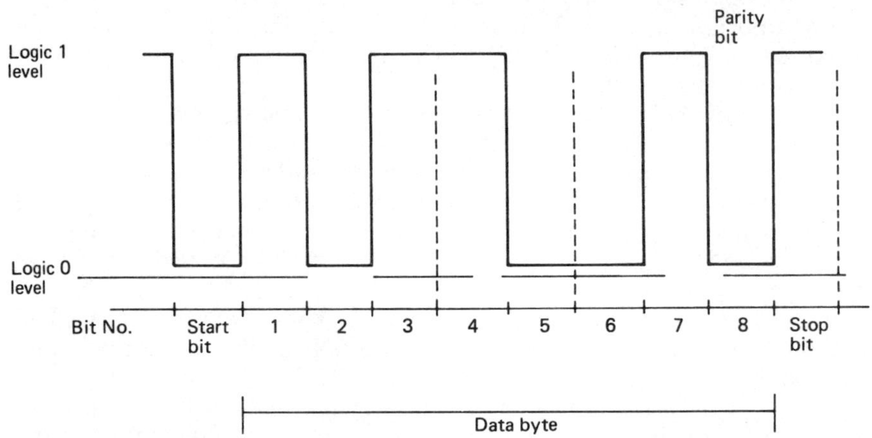

Figure 10.1 Example of asynchronous transmission

A typical telephone system has a limited bandwith of about 3 kHz. If the digital signal of *Figure 10.1* was subjected to this bandwidth restriction severe distortion would result due to the attenuation of the high frequencies, and because of noise spikes on the line. Additional difficulties also arise because it is possible for the telephone exchange to interpret the digital signal as dialling pulses, causing the system to malfunction.

These problems are overcome by using a standard PO data transmission technique known as Frequency Shift Keying (FSK). Instead of transmitting binary '0s' and '1s' along the line, each logic level is converted into a discrete frequency which falls within the 3 kHz bandwidth of the PO line. These frequencies are as follows:

FROM THE COMPUTER TO THE TERMINAL
Logic '1' = 1300 Hz.
Logic '0' = 2100 Hz.
FROM TERMINAL TO THE COMPUTER
Logic '1' = 390 Hz.
Logic '0' = 450 Hz.

When data is not being transmitted, the computer outputs a continuous tone of 1300 Hz (logic '1'), and the terminal outputs a continuous tone of 390 Hz (logic '1'). This protocol is very important as it indicates that the transmission link is open.

Should either tone fail it would indicate that the line has been broken and would result in both the computer and the terminal automatically 'ringing off' preventing the user from incurring unnecessary computer charges. The symbol code and start/stop bits of *Figure 10.1* are shown in *Figure 10.2* as an FSK encoded signal that would be transmitted from the computer. Data transmitted from the terminal to the computer uses an identical format to that shown in *Figure 10.2*, apart form the fact that a logic '1' = 390 Hz and a logic '0' = 450 Hz.

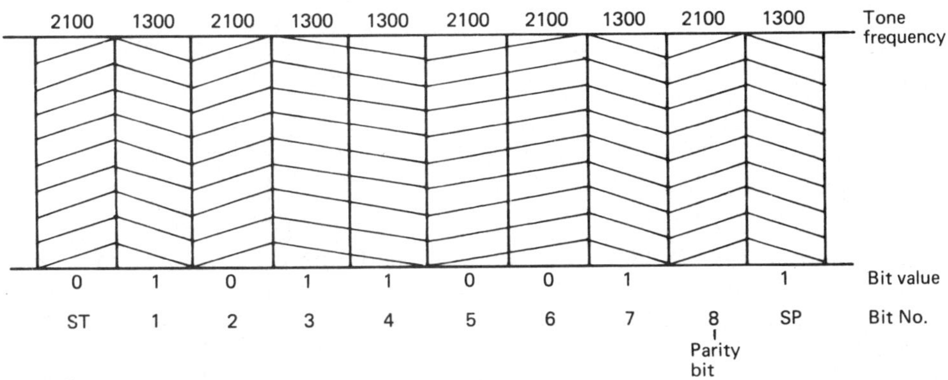

Figure 10.2 FSK encoded signal

Information sent to the computer via the user's keypad is not displayed on the CRT immediately. A technique known as 'echo' is employed, where transmitted instructions are echoed back to the user and then displayed. This technique has the following advantages:

1. If an error occurs due to a transmission failure or to corruption of data on the line then the user will be made aware of the error and can re-enter the information.
2. The system will show up any errors brought about by computer malfunction.

As a large proportion of the pages are charged to the user's account it is important that he is aware of the page number the computer has received and accepted. The echo facility is the easiest way of achieving this.

10.2 The Viewdata computer network

It is important that the reader has an appreciation of the computer system used to provide the Viewdata service, because the Viewdata terminal is a computer peripheral using the telephone line as a communications link between the two units. As the service expands television engineers will increasingly come in contact with the Viewdata service, not only in the role of installing and maintaining receivers which have Viewdata decoders, but also in the installation of editing terminals used by the Information Providers.

This description is based on the British Telecom Prestel System which uses a network of GEC 4082 computers and high speed data links. It can be seen from *Figure 10.3*, that each user is connected to an Information Retrieval Centre (IRC)

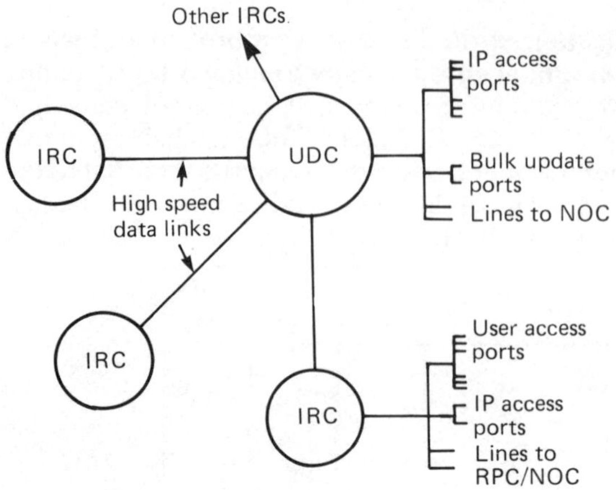

Figure 10.3 The Prestel system

via a Modem (Modulator/Demodulator) and I/O port. The location of the Information Reception Centres within the telephone network is determined by the anticipated demand for the system. If a user is not in an area which has an IRC then access to the nearest centre is provided by a data multiplexer (Dataplex), and Modems which connect the user via a high speed private circuit. This configuration is shown in *Figure 10.4*.

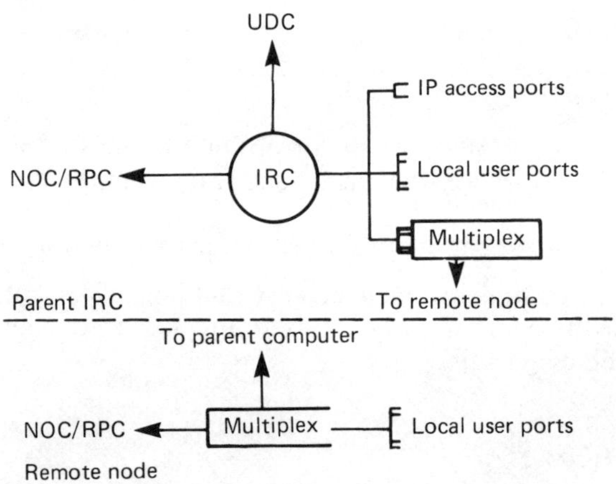

Figure 10.4 Prestel configuration

In addition to the user ports connected to the IRC there are:

1. Ports for the Information Providers.
2. Lines connected to the Regional Prestel Centre (RPC), which provide customer services, sales, etc.
3. Lines to the National Operations Centre (NOC), which is responsible for system management of all Information Retrieval Centres and the complete Prestel network.

Each IRC is connected via an inter-computer link, using synchronous transmission at 4800 baud, to an Update Centre (UDC). The function of the UDC is to give the Information Providers access to the data base so that they can enter and amend the frames of information they rent from the system. ('Frames' is the Viewdata term for pages). The UDC holds the 'master copy' of all the frames stored in the data base. The frames are distributed via the inter-computer links to the IRCs where they are held on magnetic discs. Identical copies of the data base are currently held at each computer centre.

As the system expands, additional Update Centres will be brought on line to improve efficiency. Each UDC will be able to communicate with all other UDCs within the system, while also supporting its own network of IRC computers. A typical simplified schematic is shown in *Figure 10.5*. Each IRC or UDC computer

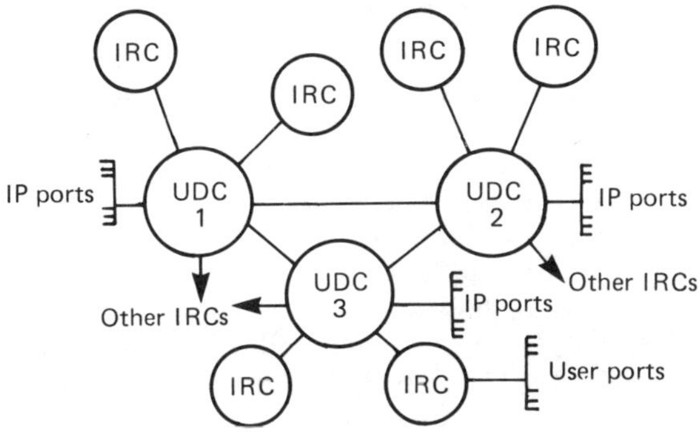

Figure 10.5 Simplified schematic for UDC

stores the frames of information on hard magnetic discs. These discs have very fast access times, typically a few milliseconds. It is therefore possible for the computer to find and retrieve any frame of information in a fraction of a second, but this information will be transmitted over a period of several seconds to the user as the limited bandwidth of the telephone line restricts the transmission speed between the IRC and the user.

10.3 Viewdata and the user

Most of the major differences between Viewdata and Teletext become apparent to the user immediately. Although control of the system is still governed by a hand held remote control unit, operation of the complete system differs considerably from Teletext.

To access the Viewdata system the terminal must be connected to the telephone line by means of a Post Office jack plug, and be registered with the Viewdata service. When Viewdata is requested at the terminal the following sequence of events takes place.

1. The telephone line is sized by the Viewdata terminal and waits for a dialling tone to be returned from the exchange. This dialling tone is made audible via the terminal loudspeaker.
2. The terminal will then automatically dial the number for the local Viewdata Informations Retrieval Centre, and this process can be monitored via the terminal loudspeaker.

3. When the call has been connected, the local IRC answers by sending a 1300 Hz tone, (logic '1'), back to the terminal.
4. On receipt of the 1300 Hz tone from the IRC computer the terminal will return a 390 Hz, (logic '1'), informing the computer that a successful link has been made.
5. On receipt of the return 390 Hz tone from the terminal the computer transmits at 1200 baud an enquiry (ENQ) code which requests the terminal's identity.
6. The terminal responds to the enquiry code by sending to the IRC an identification number of up to sixteen digits. This information is stored in a ROM within the Viewdata terminal.
7. On receipt of the identification number the computer validates the received code and checks that it belongs to a registered user. Programmes within the IRC computer are then executed to update the user's account as frames which incur charges are used.
8. Once the IRC has validated the user's identification number it will return frame 0 causing the terminal to display an introductory 'Welcome' frame and an index.
9. Frames of Viewdata can then be accessed, and these use the same character format as Teletext. A cursor indicates to the user where the next character is to be displayed. This is a facility that is not provided by Teletext.
10. As the user gains access to pages of information which incur a charge the IRC computer will debit the user's account accordingly.
11. When the user has finished with the system the terminal will disconnect itself from the telephone line by removing the 390 Hz tone. This causes the computer at the IRC to release the I/O port and complete programme execution for that user. The last frame of information accessed by the user will remain in the terminal RAM until:
    (a) A page of Teletext is accessed.
    (b) A new page of Viewdata is accessed.
    (c) Power is removed from the terminal.

To gain access to the desired frame of information stored in the data base a tree structure is used, and this is outlined in *Figure 10.6*. Assume the user wanted to book two tickets for a London show. Once the terminal has been connected to the local IRC the introductory field 0 will be displayed, giving a series of options. Assume one of these was entertainment and, hence, the user would key in the

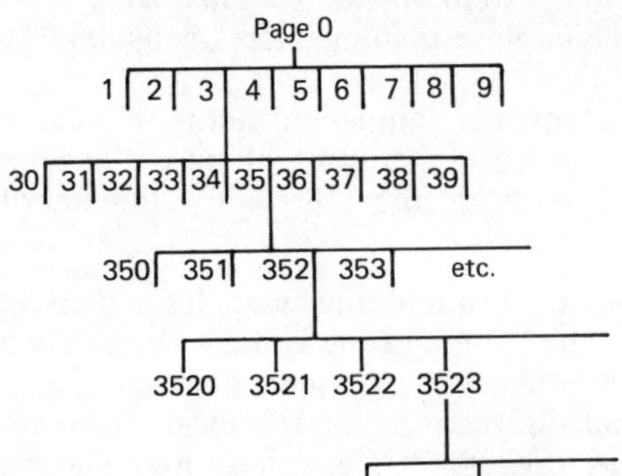

Figure 10.6 Typical tree structure used for Viewdata page selection

appropriate number. The computer would then respond by returning another field of information to the terminal listing the major types of entertainment, e.g. cinema, theatre, opera, classical music, rock concerts, etc.

In our example the user would select theatre, by entering the appropriate number causing the IRC to return a list of regions where shows are taking place, e.g. London, the South-east, the South, the South-west, the Midlands, etc. Once again the user would select the required area causing the computer to return a list of shows available in one or more fields. At this stage a particular show would be selected and the IRC would return detailed information on that particular show, such as seating, prices, times etc. The user could then select a day, seat and time for the requested show by answering simple questions which appear on the screen. Finally payment could be made by entering a credit card number or an account number.

The cost of this exercise to the user would be:

1. The standard charge for the telephone line between the user's terminal and the nearest IRC. This is normally charged at local rates, but in some circumstances where a local IRC is not available the user may incur STD charges.
2. The charge per page incurred by the user while going through the tree structure to reach the final page or pages required. In the example outlined above no charge would be made because the Information Provider would bear the cost completely. If charges were made, for example in obtaining up-to-date information on stocks and shares, etc, then the cost of each frame is displayed in the top right hand corner of the CRT. A user can enquire as to the current state of his or her account at any time.

One of the major disadvantages of a Viewdata system for the general public is the cost incurred while accessing the data base. A considerable amount of information supplied by Viewdata is also available on the Teletext systems and cost conscious users opt for the cheapest source. Consequently the major users of Viewdata at the present time are the larger businesses and corporations which require up-to-date and accurate information. For example, stocks and shares, exchange rates, bookings for air travel, etc.

10.4 The Viewdata character code

Figure 10.7 shows the complete set of Viewdata transmission codes without the even parity bit. Character and graphic codes are identical to those used by Teletext, therefore maintaining compatibility between the two systems and allowing a single display circuit to generate the video signals for both systems.

However, there are some major differences in the coding of the control characters. This has been necessary because standard data communication systems use the codes in the first two vertical columns, (0 and 1), for a number of control functions. It is important that the Viewdata system maintains compatibility with other data communication systems which use the telephone network, and for this reason, it has been necessary to change the binary coding of the control characters. Control characters contained in vertical columns 0 to 1 are primarily reserved for cursor control (up, down, left, right etc.), so that the computer in the IRC can position the cursor anywhere in the screen under programme control.

In the Teletext system all the control characters are contained in columns 0 and 1, and in order to allow Viewdata to use these control characters they have been

Bits $b_4 b_3 b_2 b_1$	b_7 / b_6 / b_5 → Col / Row	0	1	2	2a	3	3a	3b	4	4b	5	5b	6	6a	7	7a
	b_7	0	0	0	0	0	0	0	1	1	1	1	1	1	1	1
	b_6	0	0	1	1	1	1	1	0	0	0	0	1	1	1	1
	b_5	0	1	0	0	1	1	1	0	0	1	1	0	0	1	1
0 0 0 0	0	NUL				0			@		P		–		p	
0 0 0 1	1		Cursor on	!		1		Set verify mode (1)	A	Alphanumeric Red	Q	Mosaics Red	a		q	
0 0 1 0	2			"		2		Set verify mode (2)	B	Alphanumeric Green	R	Mosaics Green	b		r	
0 0 1 1	3			£		3		Skip block	C	Alphanumeric Yellow	S	Mosaics Yellow	c		s	
0 1 0 0	4		Cursor off	$		4		Set programme mode	D	Alphanumeric Blue	T	Mosaics Blue	d		t	
0 1 0 1	5	ENQ		%		5		Tape pause on playback	E	Alphanumeric Magenta	U	Mosaics Magenta	e		u	
0 1 1 0	6			&		6		Tape start	F	Alphanumeric Cyan	V	Mosaics Cyan	f		v	
0 1 1 1	7			'		7		Tape stop	G	Alphanumeric White	W	Mosaics White	g		w	
1 0 0 0	8	Cursor left	Delete line*	(		8			H	Flash	X	Conceal Display	h		x	
1 0 0 1	9	Cursor right		)		9			I	Steady	Y	Continuous Mosaics	i		y	
1 0 1 0	10	Cursor down		*		:			J	End edit*	Z	Separated Mosaics	j		z	
1 0 1 1	11	Cursor up	ESC	+		;			K	Start edit*	←		k		¼	
1 1 0 0	12	Cursor home & clear screen	Single shift 2	,		<			L	Normal Height	½	Black Background	l		‖	
1 1 0 1	13	Cursor return	Single shift 3	–		=			M	Double Height	→	New Background	m		¾	
1 1 1 0	14	Shift out	Cursor home	.		>			N		↑	Hold Mosaics	n		÷	
1 1 1 1	15	Shift in		/		?			O		#	Release Mosaics	o		■	

NOTE:

Columns 0 and 1 form the C0 control character set.

Columns 4b and 5b form the C1 set of display attribute control codes.

Columns 2, 3, 4, 5, 6 and 7 form the G0 character set.

Columns 2a, 3a, 4, 5, 6a and 7a form the mosaic character set. The shaded area represents foreground colour.

Column 3b is a non-standardised set of control characters, used from computer to terminal only.

*From terminal to computer only.

Figure 10.7 Viewdata transmission codes (*Reproduced by courtesy of 'Prestel', British Telecoms Viewdata Services*)

shifted to columns 4b and 5b. Control character positioning within these columns is identical to that used in Teletext, and the only difference between the two versions is that bit 7 of the Viewdata character is inverted.

A decoder capable of processing both Teletext and Viewdata characters will have to identify these differences in character coding and take the appropriate action. This is achieved by using a system known as shift coding. When any of the Viewdata control characters of columns 4b and 5b are to be transmitted they are preceded by an escape (ESC) control character which has a binary code of 0011011. When the decoder receives the ESC character it will simply invert bit 7 of the next byte of data received before it is stored in the decoder RAM.

Inversion of bit 7 will convert the Viewdata control character back in to a Teletext control character, and when the decoder RAM is read during the display period the control character will appear as though it came from a Teletext transmission. Characters in columns 0 and 1 of *Figure 10.7* are processed separately by additional circuitry when the decoder is operating in the Viewdata mode. This circuitry is required because of a second major difference between the Teletext and the Viewdata systems.

It has been seen in chapter 8 that every row of data transmitted contains a Row and Magazines Address Group which is used to generate a memory address for the beginning of the received row of text. In the Viewdata system the Magazine and Row Address group are not transmitted, but in its place a series of cursor control characters are used to address the page store. The control characters APB, APF, APD, APU, APR and APH allow the cursor to be positioned anywhere within the field of text, and the decoder identifies the memory address in to which data will be written by 'keeping track' of the cursor location.

On transmission of a clear screen (CS) command the decoder returns the cursor to the top left-hand corner (known as the home position) and erases the decoder RAM. From this point, the cursor position is automatically shifted one place to the right on receipt of each character until the end of the row is reached. At the end of the row the character address will return to zero and the row address is incremented causing the next transmitted character to appear at the beginning of the next horizontal row. The cursor commands that have been mentioned allow the cursor position to be moved anywhere within the frame without writing characters of text into the decoder RAM, and also allows the cursor to be turned on and off. This control system is very versatile and provides the following facilities:

1. Blank spaces at the end of a line of text do not have to be transmitted, but can be replaced by an APR control character which returns the cursor to the beginning of the next horizontal row. This will increase the speed of page transmission therefore reducing access time to the data base.
2. Lines of text can be 'over-written' by moving the cursor to the desired location and then transmitting new data.
3. Cursor control commands do not have to be preceded with the ESC command as is the case with other control characters, therefore cursor control is relatively fast.

User commands

A remote control unit designed for use with a Viewdata system will have two additional keys marked '*' and '#'. These keys are used when entering page requests using the format '* page number #'. For example, page 600 would be entered as '*600#'. It was explained earlier that a tree system is used to allow the user to progress to the page required, but if the required page number is known it can be accessed directly. The '*' and '#' commands are also used for correcting errors. For example, if the user accidentally selects the wrong number the error can be corrected by entering '*#' which causes the previously used page to be displayed.

It has also been mentioned that an 'echo' system is used, so that the user receives the requested page number from the IRC and not from the hand-held unit. Should the data become corrupt during the transmission process, either between the user and the IRC, or from the IRC to the user, then the wrong digit or page number will be displayed. If this situation arises then entering '*#' will return the user to the previous page and the number can be re-entered. Should data corruption take place during the transmission of the page, causing errors to appear on the screen then '**' will cause the computer to transmit the current page a second time.

10.5 The Viewdata terminal

Viewdata terminals differ in many respects to a Teletext decoder despite the fact that a large proportion of the electronics is common to both systems, *see Figure 10.8*. For our purposes the decoder can be split into two major areas:

1. The interface electronics required to connect the decoder to the telephone line. This section consists of the Line Terminating Unit (LTU) and a MODEM.

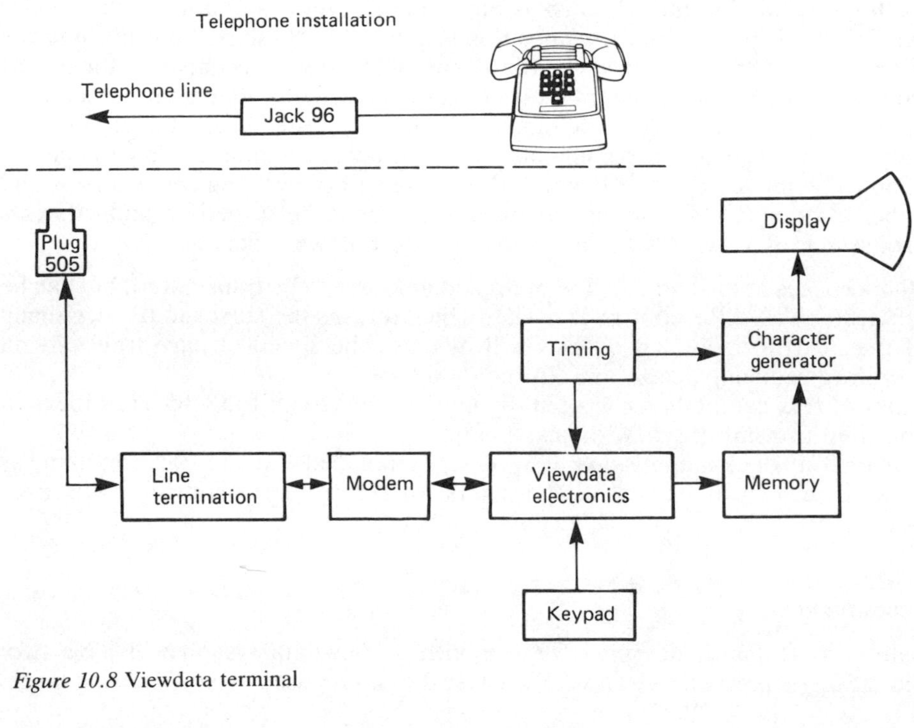

Figure 10.8 Viewdata terminal

2. The decoder electronics which consists of the user's keypad, timing electronics, RAM, and character generator. All of these have been described in various chapters throughtout this book. Additional Viewdata electronics is required however to:
   (a) Decode the asynchronous signal received from the modem.
   (b) Generate memory address signals so that character data may be stored in the decoder RAM.
   (c) Generate an asynchronous signal for some of the commands from the keypad.
   (d) Generate control signals such as autodialling so that the terminal can be connected to the IRC computer.

In addition to the facilities in (2) above various options may also be included such as:

1. To receive and process data from an alphanumeric keyboard for message writing and page generation. These facilities are required by the information providers.
2. To store Viewdata pages on domestic cassette tape.
3. Provide additional address and control information for optional RAM, so that more than one page may be stored in memory.
4. Provide an RS232 serial signal for a printer so that Viewdata information may be transferred to paper.
5. Provide a serial interface with a micro or mini computer so that received Viewdata pages may form part of a private data base.

It can be seen from the above description that a full specification Viewdata decoder is far more complex than the passive Teletext decoder. Fortunately this increase in complexity is not necessarily reflected in the hardware, because the transmission speed between terminal and computer, and *vice-versa*, is very slow. This allows microprocessor techniques to be employed in managing the control of data, therefore greatly simplifying the electronics and keeping the cost of the entire system down to an economical level. Some decoder manufacturers have developed their own LSI chips which interface directly with microcomputer ICs resulting in very compact but complex decoders.

Much of the decoder electronics is now reduced to a microcomputer plus dedicated peripheral ICs, but the interface between the decoder electronics and the telephone line still tends to use an array of discrete components. This is largely due to:

1. The interface is processing analogue signals in the form of FSK data. These signals require filtering, and therefore circuitry is difficult to reduce in size or complexity.
2. Protection circuits are required to safeguard the telephone line from the hazards of a television receiver using live chassis techniques. These safeguard components have high voltage ratings and are therefore totally unsuitable for inclusion in integrated circuits.
3. Automatic dialling of the local IRC is carried out by pulsing relays to generate the required dial pulses. The relays themselves are naturally bulky devices and require relatively high drive currents from discrete semiconductor circuits.

This natural division between large scale ICs, processing digital signals in the Viewdata decoder, and discrete component analogue circuits used in the line terminating unit, results in the electronics being split into two separate halves. The

MODEM tends to be in 'no man's land' because it itself is an interface device. It converts the received FSK signal from the computer into a digital signal, and converts the digital signal from the user's keypad into an FSK signal before it is fed to the computer. Some manufacturers such as Mullard include the MODEM in the decoder electronics, while others such as General Instrument keep the MODEM separate from the decoder.

In the following descriptions of the techniques used in Viewdata decoders, only those circuits used exclusively by Viewdata will be described, as the remainder of the decoder electronics has been adequately covered elsewhere.

10.6 The Line Terminating Unit (LTU)

The Line Terminating Unit (LTU) interfaces the decoder electronics with the telephone line. It contains all the analogue circuitry responsible for autodialling together with line drivers and filters for the FSK signal, see *Figure 10.9*. Any LTU designed to work with a Viewdata system must have the following sections:

1. Safety and isolation components to protect the telephone line from the hazards of a life chassis receiver.
2. Dialling relays and their drive circuits which are used to pulse the telephone line with the correct dialling code for the local IRC.
3. A receive channel bandpass filter which will allow the received signals at 1300 Hz and 2100 Hz from the computer to pass and reject all others.
4. A carrier detection filter which will allow 1300 Hz to pass and reject all other frequencies. It indicates to the terminal that the IRC computer is 'on line'.
5. A signal level detector which will sample the received signal from the IRC computer and determine if it is of an acceptable level for use by the terminal.
6. A return channel filter which allows frequencies of 390 to 450 Hz to pass and reject all others. It provides the correct FSK signal for the return path to the computer.
7. An audio AGC amplifier which provides a gain controlled audio signal for the terminal audio amplifier. This allows the user to monitor various functions of the terminal.

A brief description of each block will now be given before dealing with individual stages in circuit detail.

The telephone line enters the LTU at two points on the left-hand side of the block diagram. The upper two points simply loop through a normally closed relay contact, allowing the telephone receiver to gain access to the line. When Viewdata is requested at the terminal this relay contact will open, disconnecting the telephone receiver from the line and connecting the Viewdata terminal.

The second pair of entry points to the LTU are connected directly to the protection components previously mentioned. Between the protection circuits and the isolation transformer are a series of relay contacts known as:

1. Seize Line (SZLN).
2. Impulsing (IMP).
3. Dial Off Normal (DON).

All of these relay contacts, and the protection circuits, are on the 'safe side' of the isolation transformer. Everything to the right-hand side of the transformer is known as the 'dangerous side', and these circuits are connected and powered by the

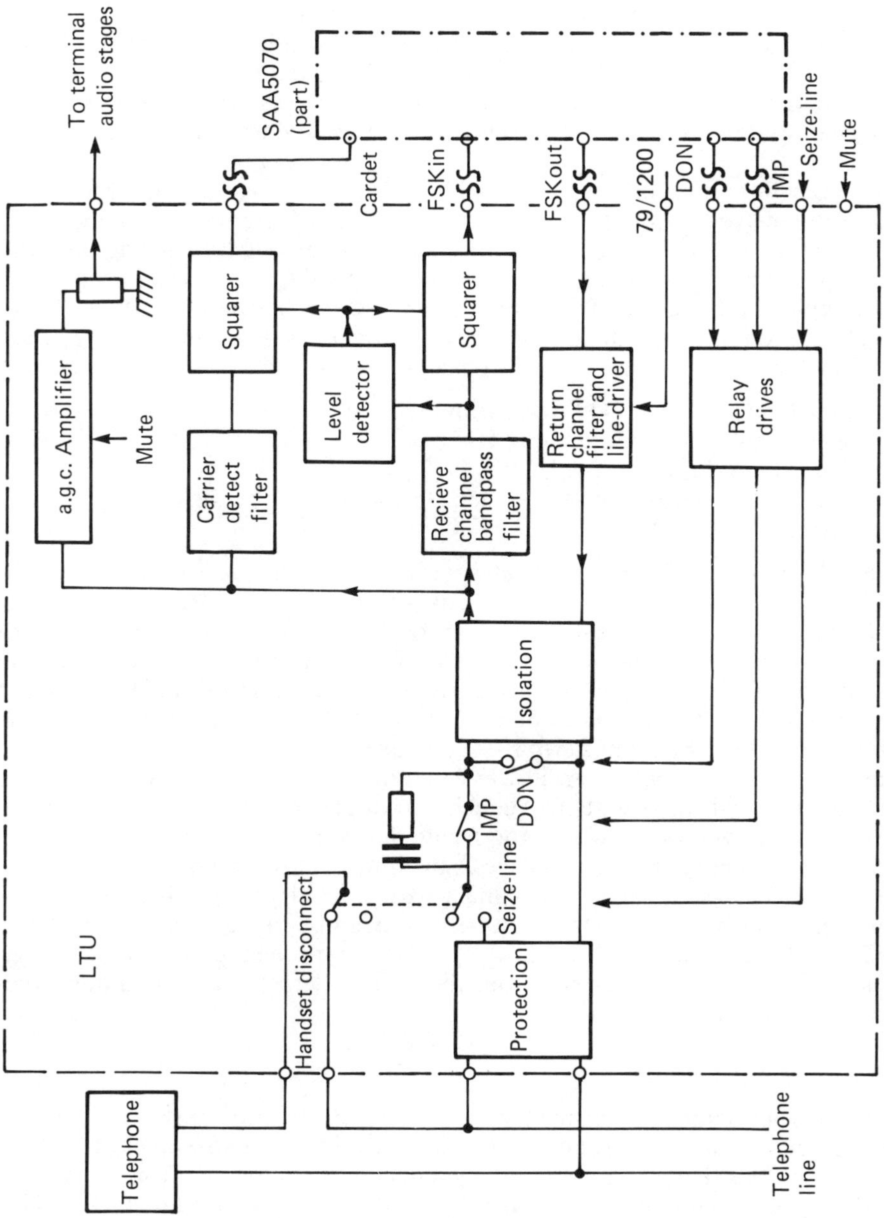

Figure 10.9 Line Terminating Unit (LTU)

terminal electronics. The Seize Line, Impulsing and Dial Off Normal relay contacts are powered from a series of relay drivers. These drivers receive their appropriate signals from the control circuits of the terminal, and their operation is completely automatic.

FSK signals sent to the IRC computer are filtered and amplified by a return channel filter and line driver. This circuit also has facilities for changing the baud rate filters between 75 and 1200 baud. Seventy-five baud is used for the Viewdata system and 1200 baud is an option for private communication systems, such as business computers using the telephone line for terminal to computer communication.

The upper output from the isolation block is fed to a variety of circuits which process the received FSK signal. At the top of the block diagram is an a.g.c. audio amplifier which provides audio monitoring of the telephone line during the call set up period. British Telecom require that signals on the telephone line should be made audible via the terminal loudspeaker while the call is being made to the local IRC. This requirement is mandatory for all supervisory tones in the range of 0 to −30 dBM. These tones are:

1. Number Unobtainable (400 Hz continuous).
2. Engaged Tone (400 Hz interrupted).
3. Pay Tone (400 Hz interrupted).
4. Ringing Tone (400 Hz or 400 + 450 Hz modulated by 50, 25 or 17 Hz and interrupted).

This amplifier must also ensure that speech and other signals below −57 dBM produced as a result of system crosstalk or interference are prevented from reaching the receiver loudspeaker. This action protects the privacy of the network.

The second output from the isolation transformer is fed to a carrier detect filter and squaring circuit to produce a Carrier Detect signal (CARDET). This signal will be generated when the 1300 Hz. Logic '1' received from the local IRC is equal to or greater than −53 dBM at the terminating resistance of the line (600 Ω).

The third and final output from the isolation stage is fed to a receive channel band pass filter, which accepts frequencies from 1.3–2.1 kHz. This action ensures that the decoder will only process the frequency components of the received FSK signal. Level detection also takes place at the output of the band pass filter because British Telecom specify that any signals with a level greater than −39 dBM at 1300 Hz or −41 dBM at 2100 Hz must be processed by the terminal. If the level of the received signal falls below these two predetermined points then the level detector will remove the outputs from the carrier detect and FSK input circuits, causing the terminal to shut down.

LTU protection circuits

Every terminal manufacturer must satisfy the safety requirements outlined in the British Telecom Prestel Terminal Specification[1]. These requirements call for the protection of the telephone line and its associated equipment from a variety of conditions that could arise from faults in the Viewdata terminal. Briefly the requirements laid down in the specification cover the following points:

1. A barrier transformer must be included so that the only connection between the terminal electronics and the telephone line is electromagnetic. This transformer must conform to BS 3535: 1962.

2. Isolation between the terminal and the telephone line must be capable of withstanding a minimum of 2.85 kV for 60 seconds.
3. The telephone line must be protected against transverse voltages by means of:
   (a) A 20 mm fast blow 250 mA fuse which meets the BS 4265 standard.
   (b) A gas discharge tube having a maximum striking voltage of 115 V and is capable of carrying the fusing current of the series line fuse.

A typical circuit which conforms to these requirements, and includes the necessary relay contacts for dialling, etc, is shown in *Figure 10.10*. All circuitry to the right of the barrier line is said to be on the safe side of the line terminating unit, and these components are connected directly to the telephone line. All components to the left of the barrier line are said to be on the dangerous side and form part of the terminal electronics. All relay energizing coils are shown on the dangerous side. The critical safety components are:

1. The isolation transformer.
2. The series line fuse.
3. The gas discharge tube.

When the terminal is not being used, the relay contacts are as shown in the diagram. The PO jack plug (505) has contacts 2 and 3 connected across the twin wire telephone line, and RLD with its closed contacts is wired in series with the conventional telephone receiver (*Figure 10.9*). Under these conditions the telephone receiver is connected to the telephone exchange via the contacts of RLD and the Viewdata terminal is disconnected because RLB contacts are open.

When the Viewdata terminal accesses, the telephone line RLD energizes, disconnecting the telephone receiver from the line. RLB also energizes, connecting the isolation transformer safe side across the telephone line on pins 2 and 3 of plug 505. Note that RLA is not energized at this point and its contacts are closed. Connection of the safe side transformer impedance across the telephone line causes the exchange to respond by returning a dialling tone, which will be heard at the terminal loudspeaker.

When the telephone line has been 'seized' circuits in the terminal initiate an automatic dialling process, which will dial the correct number for the local IRC computer. The number is dialled by closing RLC, which is the Dial Off Normal relay, and then by pulsing RLA contacts open a predetermined number of times depending on the digit being dialled. For example if the number 5 was being dialled the following sequence of events would take place:

1. The DON relay RLC would close placing a very low impedance across the telephone line.
2. The IMP relay RLA would be pulsed open and closed 5 times.
3. The DON relay RLC would then open.
4. The process would be repeated for the next number to be dialled, changing the number of impulses in step 2.

On completion of the autodialling sequence, the Viewdata terminal will wait for the 1300 Hz tone to be returned from the computer. On receipt of this tone a number of digits identifying the terminal to the computer are transmitted. Throughout the length of time that the terminal uses the telephone line the relays

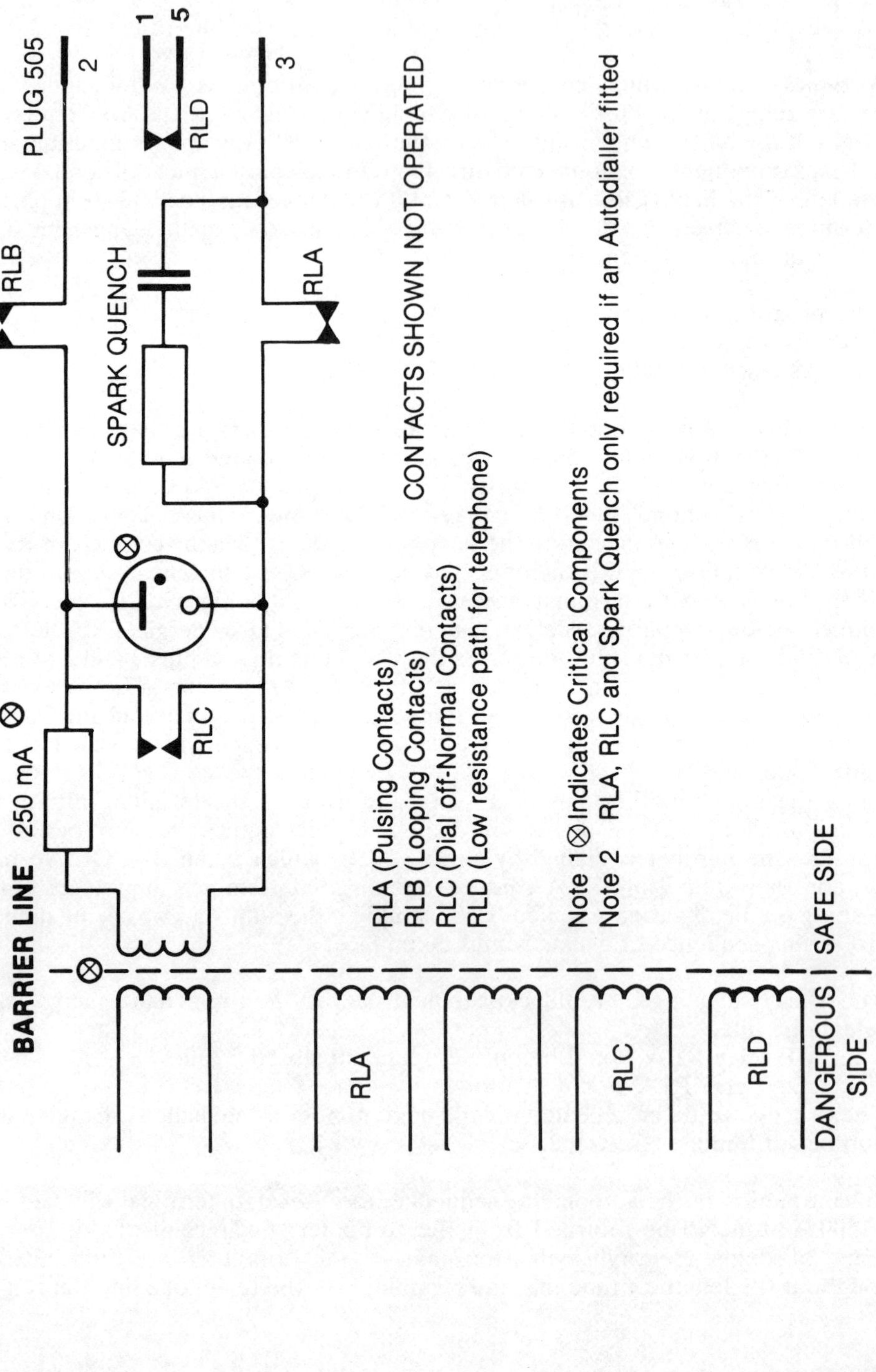

Figure 10.10 LTU protection circuit

assume the following conditions:

1. RLB (seize line) will be energized.
2. RLD (handset disconnect) will be energized.
3. RLA (impulsing) will be de-energized.
4. RLC (dial off normal) will be de-energized.

When the terminal is switched off RLB de-energizes, opening the seize line contacts to disconnect the transformer safe side from the telephone line. RLD, the handset disconnect relay, will also de-energize. Its contacts close and re-connect the telephone receiver to the line.

Receive channel band-pass filter

The purpose of the receive channel band-pass filter is to process the FSK signal sent by the IRC computer prior to demodulation at the modem. This filter must have the following specification:

1. A pass band from 1.3 kHz to 2.3 kHz which will allow the transmitted FSK signal to pass through.
2. Uniform group delay over the pass band.
3. Rejection of 390 and 450 Hz signals to remove the return FSK components present at the filter input.
4. A high frequency roll off to attenuate noise.
5. An adequate dynamic range which will handle the expected changes in signal levels.

A filter with this specification is complex in design, and this complexity is reflected in its circuit diagram which is shown in *Figure 10.11*. This circuit diagram (Mullard) is divided into six separate stages.

The first of these stages consists of the operational amplifier IC1 and its associated components. It is a relatively simple low-pass filter which attenuates unwanted high frequency noise but amplifies the lower frequency inputs signals. The output from IC1 is coupled to the second stage of the band-pass filter.

This consists of IC2, IC3, IC4 and their peripheral components. These integrated circuits form a band stop filter which has a centre frequency of 420 Hz. The return FSK signal from the terminal to the computer uses frequencies of 390 Hz for logic '1' and 450 Hz for logic '0'. These are either side of the centre band stop frequency of 420 Hz, and careful design of this stage has ensured the rejection characteristic is wide and deep enough to attenuate the return FSK signal by the required amount.

Stage three, consisting of IC5 and its associated components, is a second band stop filter at a higher frequency and is required to maintain a fixed group delay over the pass band of the filter.

IC6, together with its peripheral components forms a conventional second order band pass filter. Its output is fed to the final two stages which are low pass filters, that provide additional HF noise attenuation.

Carrier detect filter

Earlier in this chapter the frequencies of the supervisory tones were outlined. All of these have harmonics which lie within the 1.3 to 2.1 kHz bandwidth of the receive channel pass-band filter, and can pass through the filter on to the electronics of the decoder. As soon as the LTU accesses the telephone line at the beginning of a

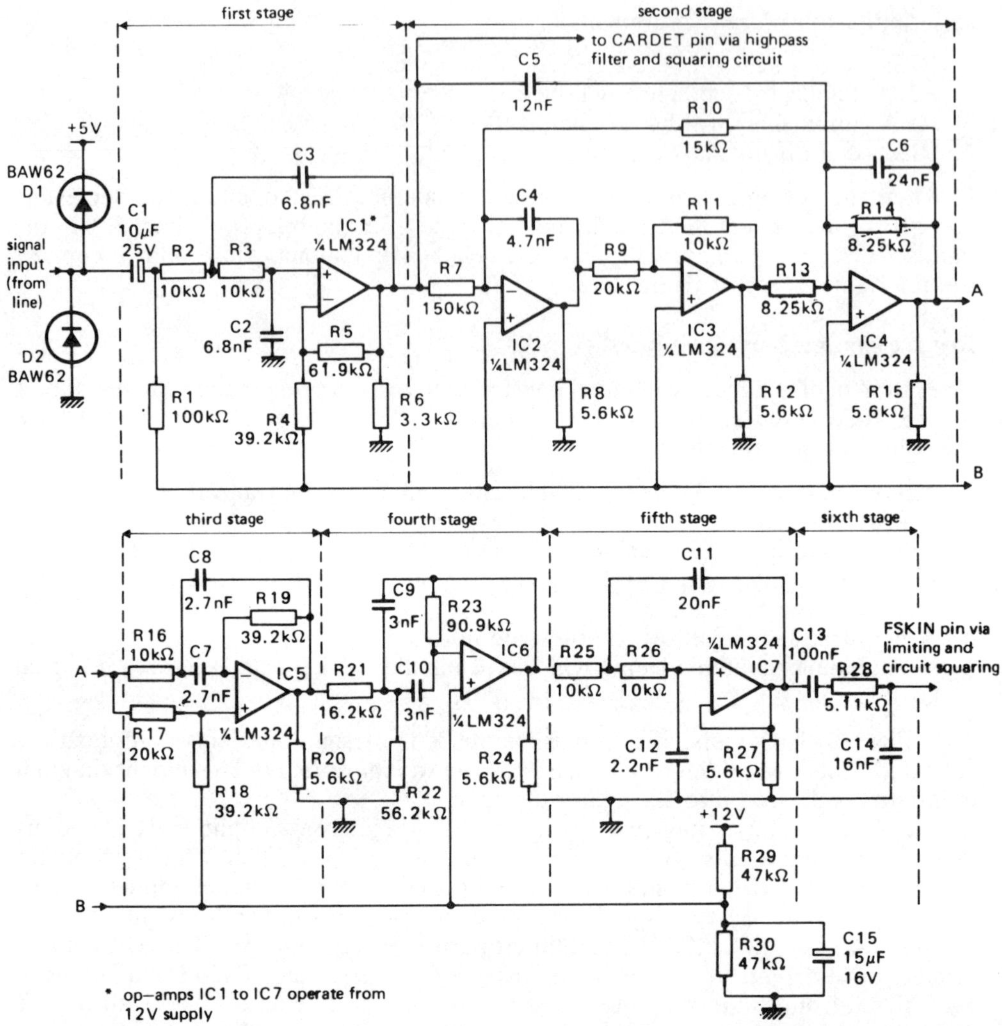

Figure 10.11 Circuit diagram for band-pass filter

Viewdata call one of these tones will be present on the line, and it must be stopped from reaching the decoder electronics if terminal malfunction is to be prevented. To achieve this, the FSK input to the decoder is disabled during initial dialling operations, and when a carrier signal from the IRC computer has been detected for approximately two seconds the FSK input is then enabled allowing the received information to be processed.

The carrier detect filter consists of IC1 and its associated components from *Figure 10.11*, together with IC1 and IC2 from *Figure 10.12*. The first stage of *Figure 10.12* is a high-pass filter, which receives its input signal from the first low-pass section of *Figure 10.11*. The overall response of these two stages in series produces a peak at the received carrier frequency of 1.3 kHz. IC2 squares up the filtered signal from IC1 before passing it to the carrier detect input of the terminal decoder.

The CARDET output from IC2 can be disabled by applying a negative control voltage from the level detector to the cathode of D1.

Also included in the circuit diagram of *Figure 10.12* is the limiting and squaring circuit for the FSK input signal. IC3 and its associated clipping diodes D2 and D3 form the limiter, and its output produces signal feeds for two stages.

1. To the limiting circuit formed by IC4, which produces the received FSK output from the Line Terminating Unit.
2. To the level detector which monitors the amplitude of the received FSK signal to determine if it is of an acceptable level.

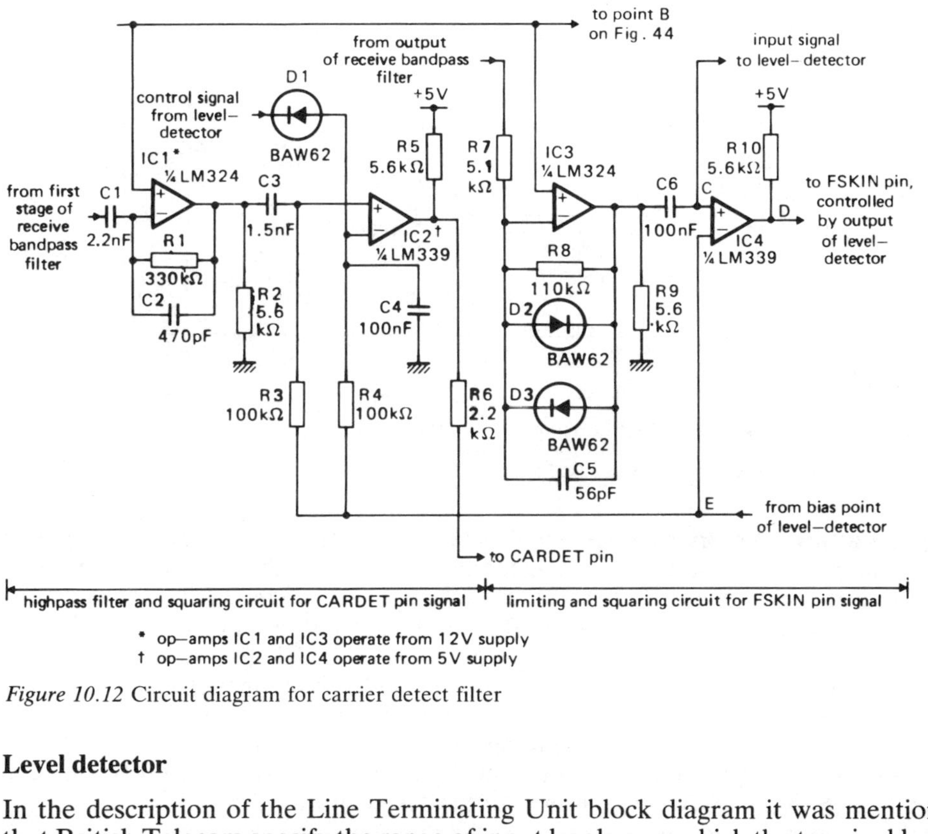

* op–amps IC1 and IC3 operate from 12V supply
† op–amps IC2 and IC4 operate from 5V supply

Figure 10.12 Circuit diagram for carrier detect filter

Level detector

In the description of the Line Terminating Unit block diagram it was mentioned that British Telecom specify the range of input levels over which the terminal has to function correctly. Should the input signal level fall below −53 dBM into 600 Ω the terminal must inhibit the incoming signal. This facility is carried out by the level detection circuit shown in *Figure 10.13*.

The input to this level detector is taken from the limiting stage (IC3), of the band-pass filter. Assume that no input is present. In the absence of a signal on the non inverting input of IC1 an offset voltage will be developed between the two inputs of the differential amplifier. This will cause the open collector output of the IC to go high and allow C4 to charge via R8 and R7 to the positive supply potential. This causes the output of the second differential amplifier IC2 to go low. TR1 is therefore cut off, causing TR2 to saturate and place a very low impedance across the squared up and filtered FSK signal from point D of *Figure 10.12*. This low impedance shunts the FSK signal to chassis removing it from the input of the decoder.

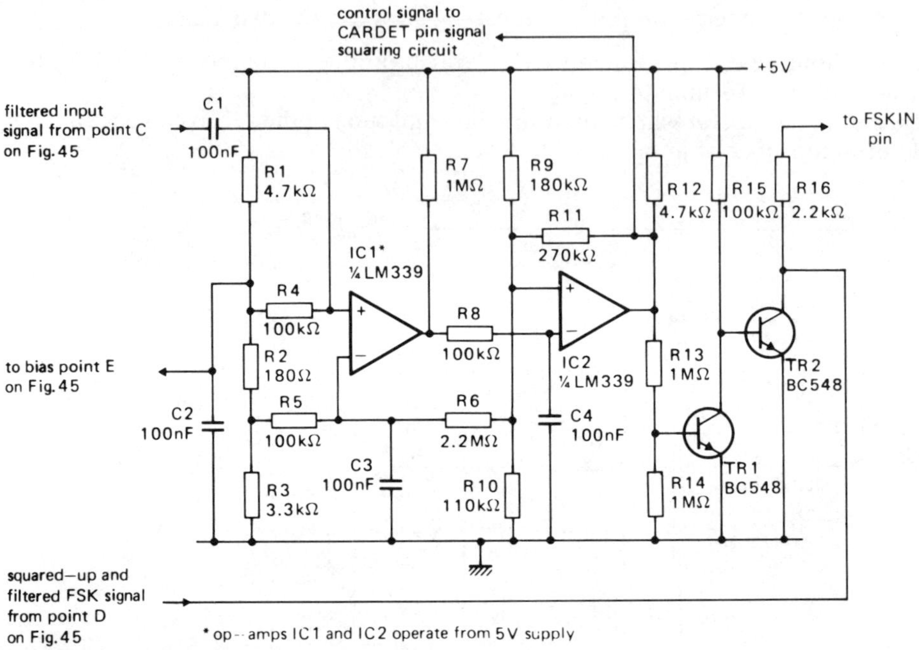

Figure 10.13 Level detection circuit

When a filtered carrier detect signal is present on the input to IC1 its negative half cycles will cause the non inverting input to drop below the inverting input. This will drive the output of IC1 low for the negative half cycles of the input signal. During this time C4 will discharge via R8 and the low output impedance of IC1. As the time constant C4 × (R8 + the output impedance of IC1) is very much less than the time constants C4 × (R8 + R7), C4 will gradually discharge until a point is reached where the inverting input of IC2 will fall below its non inverting input. The output of IC2 will therefore go high very shortly after a continuous input to the circuit is present. This action will saturate TR1 which in turn will cut off TR2 and enable the FSK and CARDET inputs to the terminal decoder.

Return channel filter and line driver

Before the FSK output of the terminal decoder can be fed to the telephone line, it has to be filtered to ensure that its frequency components lie within the pass band specified by British Telecom. It must also be impedance matched to the 600 Ω telephone line. A suitable circuit is shown in *Figure 10.14*. The isolation transformer T1 has to present a 600 Ω termination to the telephone line, and in this example, the termination resistance is produced by R1 and the impedance of T1 dangerous side winding. British Telecom specify that the output level from the terminal into this 600 Ω impedance should not exceed −9 dBM, and in this circuit

(Mullard) a level of −13 dBM has been chosen. This is the same as the transmit level of the IRC computer. To achieve an output level of −13 dBM the FSK input coupled via C6 has to be attenuated, and this is carried out by the input biasing resistors to the operational amplifier IC1.

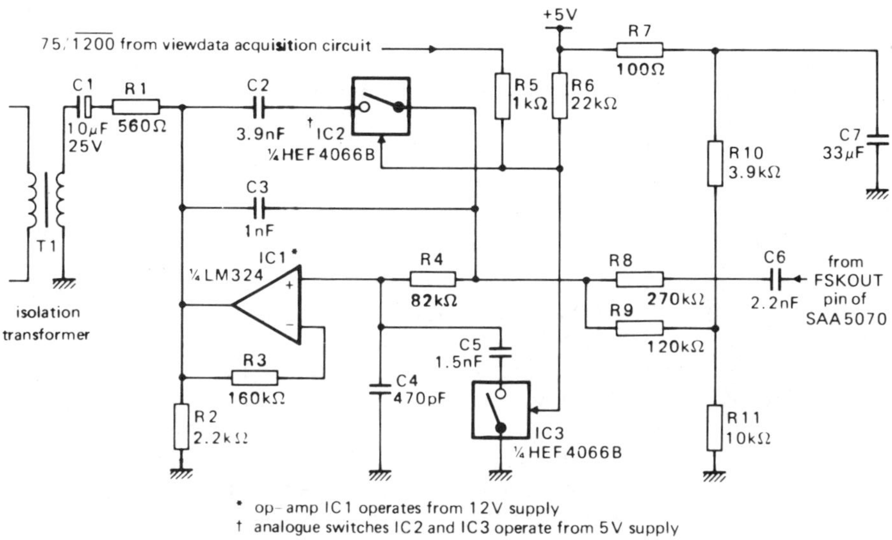

Figure 10.14 Circuit for return channel filter and line drive

In this example of the Return Channel Filter and Line Driver two transmission baud rates can be used:

1. 75 baud for the normal Viewdata system.
2. 1200 baud for communication between non-Viewdata computers.

Filtering has to be switched accordingly, and this is carried out by the analogue switches IC2 and IC3. When communication is required at 75 baud both of these switches are closed insetring C5 on the input of IC1 and C2 in the feed back path of IC1.

10.7 Viewdata electronics

Practically all Viewdata decoders use a microcomputer-based system to process data received from the modem, and write this data into the decoder RAM. It was mentioned earlier that the memory, character generation, and timing circuits use identical techniques to those previously described in the chapter on Teletext. Consequently the Viewdata electronics carries out the following tasks:

1. It receives data in serial form from the modem and converts this in to parallel data ready for its storage in the RAM.
2. It detects errors and control characters in the received data.
3. It generates control and address signals which enable character storage in the RAM.

4. It stores the terminal identity number which is remotely programmed by the Prestel Registration Centre (PRC), and will transmit this number at the request of the IRC computer.
5. It stores a series of telephone numbers for up to four different Viewdata services.
6. It initiates autodialling of the pre-stored numbers to enable the terminal to gain access to the IRC computer.
7. It controls various functions within the modem such as seize line etc.
8. It monitors the users key pad for commands and outputs these commands in serial form to the modem.
9. It provides a variety of ancillary functions such as:
   (a) Tape storage of data.
   (b) Accessing multi field memories.
   (c) Printing of field information.

All of these control and data manipulation processes can be carried out using a single chip microcomputer, and this technology is used in all decoders currently in production.

A considerable amount of interfacing is required however in order to provide a microcomputer IC with all the data and services it requires to control the operation of the decoder. This interfacing may be carried out by a number of digital integrated circuits specifically constructed for individual tasks, or it may be condensed into a dedicated LSI which carries out all the interfacing requirements between the decoder peripherals and the microcomputer chip.

It is virtually impossible to give the reader a detailed explanation of the complete range of microcomputer-controlled Viewdata decoders. The following sections are designed to give an understanding of the many possible techniques that are becoming available to receiver manufacturers. The advent of the microcomputer is making the production of Viewdata and Teletext decoders much more cost effective, and results in fewer components, lower power dissipation, simpler construction and lower costs. Unfortunately this advance in technology is making fault diagnosis and repair an increasingly complex process, and as technology progresses this situation will become more evident.

10.8 The Mullard LUCY Viewdata Decoder

A dedicated LSI interfacing system for a microcomputer-based Viewdata decoder has been designed by Mullard, who have dramatically reduced the electronics required down to three integrated circuits. These being:

1. A single chip microcomputer (Intel 8048/8049).
2. A dedicated interface integrated circuit (SAA 5070).
3. An electrically-alterable Read Only Memory (ER 1400).

These three integrated circuits operate in conjunction with a range of dedicated Teletext LSI chips which form the display electronics. These being:

4. Timing Chain (SAA 5020).
5. Teletext Read Only Memory (SAA 5050).
6. Display Random Access Memory (2 × 2114).
7. RAM Address Circuits (2 × 74LS193, 74LS83).

The ICs outlined in 4 to 7 above are identical in every respect to those previously described in chapter 9 'Teletext decoders'. They have been interfaced in this application with the microcomputer and the SAA 5070 to form a complete Viewdata decoder. A simplified block diagram of the system is shown in *Figure 10.15*.

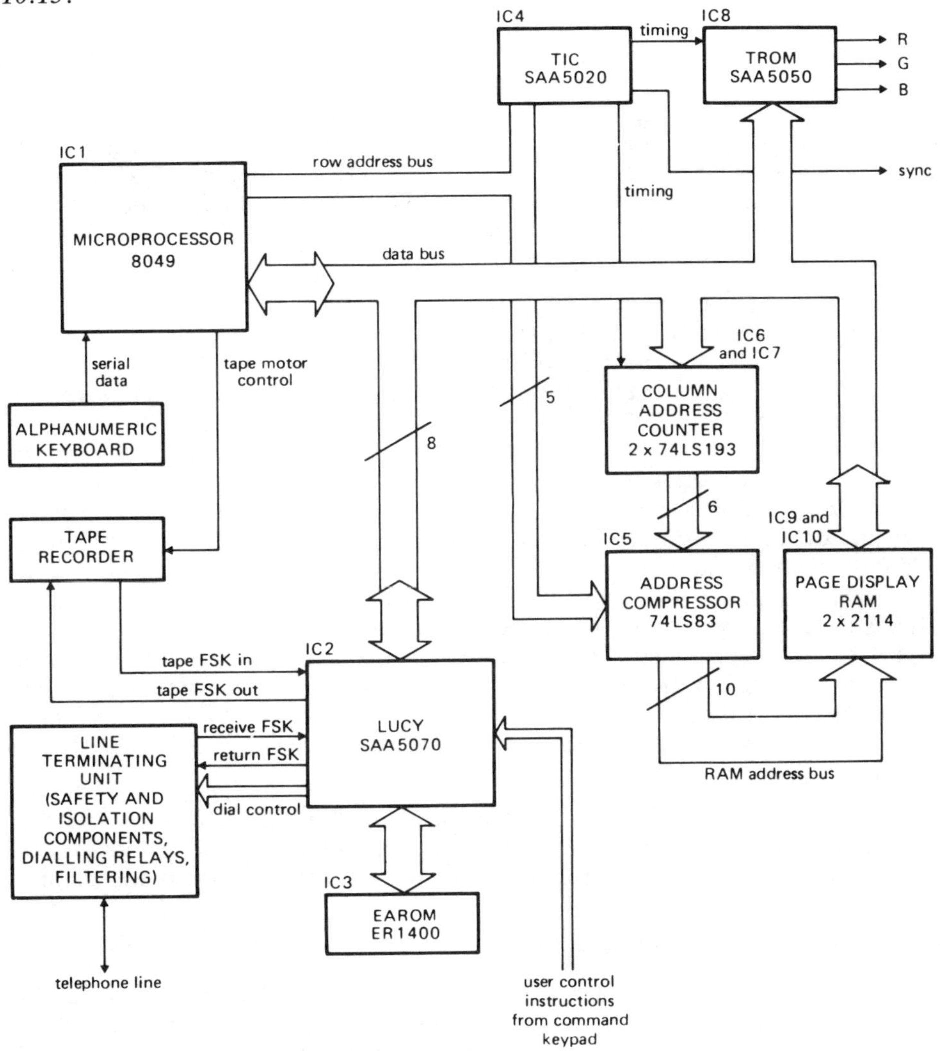

Figure 10.15 Block diagram for Mullard SAA 5070 system

To the right of centre of this diagram are the integrated circuits which form the display electronics, these are ICs 4, 5, 6, 7, 8, 9 and 10. To the left of the centre line is the Viewdata Electronics consisting of:

IC1 an 8049 microcomputer.
IC2 an SAA 5070 interface chip designated LUCY (derived from the term 'Line Coupling Unit').
IC3 an ER 1040 Electrically alterable Read Only Memory.

I/O signals which are interfaced by the LUCY IC are:

1. User instructions from either a remote or local key pad.
2. Receive and return FSK signals from the LTU.
3. Auto dial signals to the LTU.
4. Tape FSK input and output signals for an audio cassette recorder.

The LUCY IC block diagram is shown in *Figure 10.16*[2]. It contains a number of registers which are up to eight bits wide. Each register is used to interface a peripheral device to an internal data bus which can in turn be accessed by the microcomputer under programme control. In addition to the interfacing registers, there are two status registers, a mode register and a command register. These determine the current status of the integrated circuit, dictate the mode of operation, or initiate a specific function.

The integrated circuit contains the following electronics:

1. An interface between any one of its inputs or outputs and the decoder data bus.
2. Autodialling circuits for the Impulsing and Dial Off Normal relays which meet both British Telecom and European requirements.
3. 60 and 1.5 second time out circuits which are used in the connection protocol to the IRC computer.
4. A 1200 baud demodulator and asynchronous receiver which converts the received FSK signal from the modem into an 8-bit parallel byte of data.
5. A 75/1200 baud FSK modulator and asynchronous transmitter which converts parallel data into a serial FSK signal and outputs this signal to the modem.
6. A tape recorder modem using FSK techniques to a modified 'Cansas City' standard together with a tape asynchronous receiver/transmitter. This enables fields of text to be stored on magnetic cassette tape.
7. An I bus receiver/transmitter which enables the integrated circuit to interface with the remote control receiver, local keypad, and other integrated circuits which use the I bus.
8. Two general purpose I/O ports which provides access to the EAROM, Line Seize relay, etc.

In order to make the LUCY IC as flexible as possible receiver manufacturers are able to make use of the FSK modulator/demodulators within the IC, or they can use their own modems, and bypass the internal modulator/demodulator circuits.

LUCY microprocessor interface

An 8-bit data bus, D0 to D7, is connected to the LUCY IC on pins 26 to 33. This data bus is connected to all the major components within the Viewdata decoder. It also provides a communication highway between the LUCY IC and the microcomputer, and between the microcomputer and the display circuits. It is not only used to transport data between the decoder ICs but also memory addresses which are generated by the microcomputer.

It has become common practice to multiplex the data and address information so that the number of pins on the microprocessor chip can be reduced. Consequently, during certain periods of time, the data bus will be used to transport address information, and at other times it will be used to transport data. A signal known as Address Latch Enable (ALE) is generated by the microcomputer so that circuits using the data bus will know whether the bus contains data or address information.

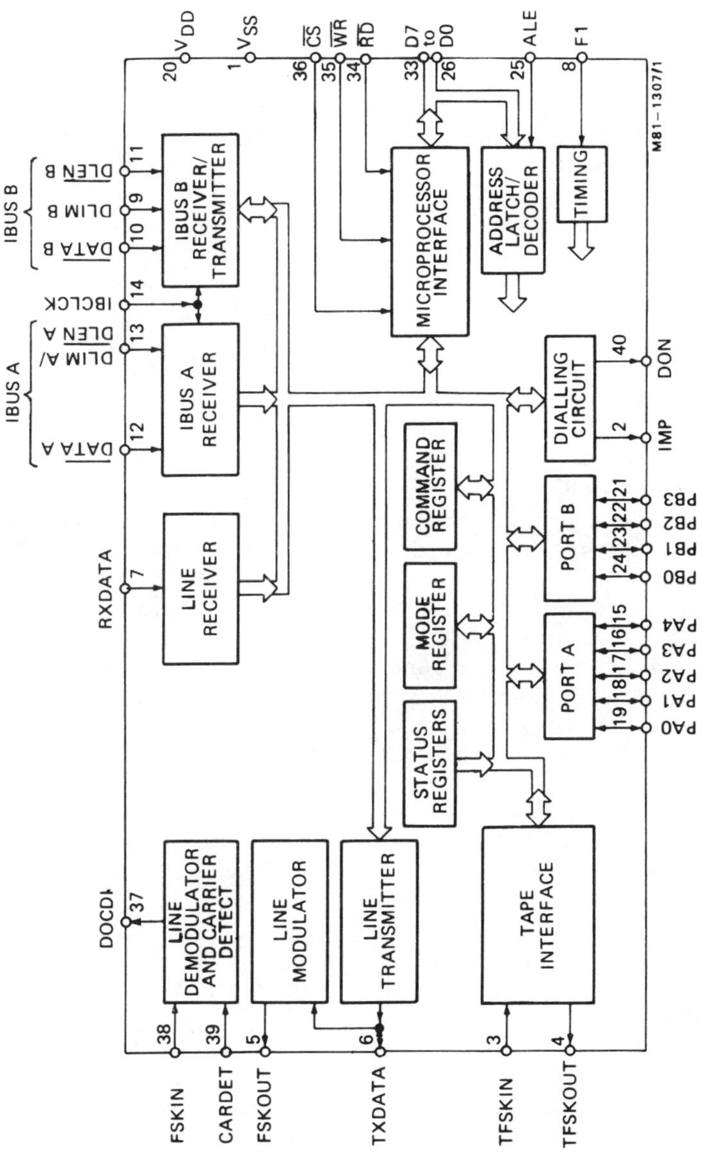

Figure 10.16 Block diagram for the LUCY IC

The LUCY IC must therefore be able to use the ALE signal to 'capture' memory addresses placed on the data bus, and then decode them so that individual registers in the LUCY IC can be written to or read from.

When the microcomputer communicates with the LUCY IC it does so by treating it as a small block of memory. For example. If the microcomputer was to read the contents of one of the status registers in the LUCY IC the following sequence of events would take place under the control of the microcomputer:

1. The address of the status register required would be placed on the data bus.
2. When this address is stable and present on the input pins of the LUCY IC the Address Latch Enable signal, generated by the microcomputer, would cause the LUCY IC to store the address in its address latch.
3. The LUCY IC will then decode this address and select the requested register.
4. The memory read control line from the microcomputer would then fall to a logic '0', causing the addressed status register within the LUCY IC to place its data on the internal data bus. The interface electronics in the IC would also be enabled so that data is transferred to the decoder data bus.
5. Once the data from the addressed status register is stable on the data bus the microcomputer will write this data into one of its internal registers.

The sequence of events required to write data from the decoder data bus into an internal register in the LUCY IC takes place using a similar process. In this instance a memory write control signal from the microcomputer is generated causing the data to be transferred from the decoder data bus into the addressed LUCY IC register.

It is imperative that other integrated circuits that share the data bus are inoperative while the microcomputer is communicating with the LUCY IC. To achieve this the microprocessor will generate a special control signal on one of its output ports known as Enable Microprocessor Access (EMA). This signal has several functions, one of which is to cause the outputs of all other circuits which use the data bus to go tri-state, thus releasing the data bus for use by the microcomputer.

Dialling circuits

The dialling circuit generates two signals, known as Impulsing and Dial Off Normal, which are used to dial the telephone number of the local IRC computer. A series of telephone numbers can be used, and these are stored in an Electrically Alterable Read Only Memory which the LUCY IC addresses via the five bit parallel Port A, (PA0 to PA4).

A programme stored in the microcomputer ROM causes Port A to address and read the contents of the EAROM and then transfer the coded data, via the internal data bus of the LUCY IC, to the dialling circuit. The dialling circuit then energizes the DON relay and impulses the IMP output the desired number of times for each digit in the number. Impulsing takes place at the rate of 10 pulses per second, and the impulsing ratio can be altered to cover both UK and European requirements. An internal timing circuit which produces either a 1.5 or 60 second time out facility can be optionally used to release the telephone line if a connection is not made within 60 seconds of dialling the last digit.

LUCY parallel I/O ports

Two parallel I/O ports are contained in the LUCY IC, port A having five bits and port B having four. In the Mullard Viewdata decoder port A is used exclusively to interface the EAROM with the system, while port B is a general purpose port used for handling miscellaneous I/O signals such as Audio Mute and the Seize Line relay. Receiver manufacturers are not limited to these applications, and can use the two ports for any input or output processing.

LUCY line demodulator

An FSK input to the LUCY IC on pin 38 is provided by the band-pass filter and squaring circuits previously described in the section on the Line Terminating Unit. The FSK input is demodulated using a pulse counting method, similar to that used for demodulating the FM video signal in domestic video cassette recorders. Each transition in the waveform from a logic '1' to a '0', or from a logic '0' to a '1', is used to generate a negative pulse of fixed duration. When 1300 Hz, (logic '1'), is being received, the pulse repetition rate generated by this circuit will be 2600 pulses per second. When the 2100 Hz, (logic '0'), is being received then 4200 pulses per second will be produced.

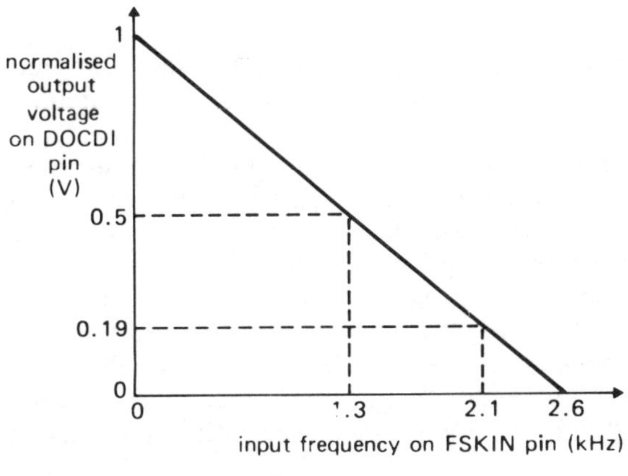

Figure 10.17 Pulse counter dc value

The average d.c. value of the output from the pulse counter will therefore decrease linearly with frequency as shown in *Figure 10.17*. At 2100 Hz, an average d.c. voltage of 0.19 V is produced, while at 1300 Hz this rises to 0.5 V. The output from the pulse counter leaves on pin 37 and is applied to a post detection filter which passes the d.c. and low frequency components which correspond to the data but rejects the high frequency harmonics. Pulse edges leaving this filter have very slow rise and fall times, and these are sliced by a comparator before the resulting 'squared up' data waveform, (RXDATA), is fed back to the LUCY IC on pin 7.

Receiver manufacturers have the option of using modems of their own design, as an alternative to the Mullard circuit. If this is the case then the receive data modem output would be applied directly to pin 7, and the line demodulator in the LUCY IC would not be used.

LUCY line receiver

Data from the FSK post detection filter, which has been sliced and cleaned up, is fed back into the LUCY IC on pin 7. A line receiver de-serializes this 10-bit data word, and then checks and removes the start and stop bits. The remaining 8 bits are loaded into a holding register which can be read by the microcomputer. Parity detection within the line receiver can be set for odd, even, or no parity at all.

LUCY line transmitter

Any byte of data stored within the Viewdata decoder can be transmitted in serial form to the IRC computer. Data from any register or memory location is loaded into the line transmitter by the microcomputer, where parity, start and stop bits are added. The data is then clocked out to the TXDATA pin (6) at either 75 or 1200 baud.

If an external modem is used then the output from the line transmitter will be taken directly from pin 6. Alternatively the TXDATA is fed to a line modulator within the LUCY IC and leaves in modulated form on the FSK output pin (5). From this point the FSK output signal is applied directly to the line driver and filter circuits previously described in section 10.6.

LUCY tape recorder interface

The LUCY IC provides an interface with a domestic cassette tape recorder that allows Teletext or Viewdata pages to be recorded on tape. These pages can then be retrieved by replaying the tape to the decoder.

Cassette tape suffers from similar bandwidth limitations to the telephone line, and FSK techniques are again used to overcome these. This tape interface is a modified form of the 'Cansas City' standard which was originally devised for recording and replaying programmes for domestic 'home' computers. A logic '1' is represented by a single cycle of 1300 Hz and a logic '0' by two cycles of 2600 Hz. The baud rate is 1300 baud on both the record and playback modes, and the data format is identical to that used by Viewdata. The tape baud rate is deliberately chosen to be higher than that used by Viewdata so that the microcomputer has time to process the received characters and output them to the cassette recorder without experiencing timing difficulties.

The tape receiver section is divided into two separate stages. The first of these detects the logic '1' carrier frequency off tape and demodulates the incoming FSK signal. The tape receiver section de-serializes the incoming data bit stream and stores it in a temporary register, ready for the microcomputer to read. When a byte of data from tape is ready to be processed, a flag is set in one of the status registers to inform the microcomputer that data is available. The microcomputer can then read the contents of the tape receiver register and process the data accordingly before the next byte of data has been received from the tape recorder.

A tape transmitter and FSK modulator provide the interface between the internal data bus of the LUCY IC and the tape FSK output at pin 4. The parallel byte of data to be recorded is latched into a register within the tape transmitter where parity, start and stop bits are added. The complete word of 10 bits is then clocked out in serial form at 1300 baud to the tape modulator which produces the FSK signal for the tape recorder.

LUCY I bus receivers/transmitters

The LUCY IC receives command information from the I bus using the same techniques as other integrated circuits in the Mullard Multitext chip set. I bus communication within the LUCY IC is divided into two areas:

1. I bus (A) receiver. This is primarily designed to receive instructions from either the remote control decoder, or a local key pad mounted on the receiver front panel.
2. I bus (B) receiver/transmitter. This is designed to provide communication between other integrated circuits requiring I bus signals, e.g. between the Teletext read only memory and the LUCY IC.

I Bus receiver A can be set to receive 2- or 3-wire signals. In the three-wire mode, a 62.5 kHz square wave is generated within the LUCY IC to provide the required I bus clock, (IBCLCK), output on pin 14. Receiver/transmitter B is a two wire system which is able to generate all the necessary signals for two or three wire I bus transmissions.

These two stages, in conjunction with the microcomputer, provide a very powerful communication system because coded instructions received on I bus A can be code converted under the control of a microcomputer programme and re-transmitted on I bus B. In future years other integrated circuits may use an I bus system which is not directly compatible with existing circuits, but if this is the case then the microcomputer can carry out the necessary code conversion and the LUCY IC will provide the interface between old and new chip sets.

The description above has outlined the basic techniques of a Viewdata decoder using a single chip microcomputer and a dedicated interface IC. This example can be further expanded to include the reception of Teletext, and a simplified block schematic of an extended specification Viewdata decoder is shown in *Figure 10.18*. If this diagram is compared with *Figure 10.16* on the standard Viewdata decoder several factors become apparent:

1. The display electronics consisting of ICs 4 to 10 of *Figure 10.15* have been condensed into a single block, and a printer interface has been added.
2. The LUCY IC still provides most of the interfacing required between the peripheral components and the microcomputer based system.
3. Additional peripherals have been added in the form of an alphanumeric keyboard, and other sub systems. These are interfaced via the two I/O ports within the LUCY IC.

Teletext reception is pocessed by the Teletext Data Acquisition and Control block towards the top left-hand side. This is basically the two-integrated-circuit chip set previously described in chapter 8, namely a Video Input Processor (VIP), SAA 5030 and the Teletext Data Acquisition and Control (TAC) SAA 5040. These two integrated circuits are controlled by the LUCY IC via an I bus interface, and place the requested Teletext data on the decoder data bus. Once data is on this bus it may be placed either in the display RAM and subsequently displayed on the CRT, or it may be recorded on tape, printed, or fed to another computer system, either directly or via the telephone line.

LUCY microcomputer

Mullard have used the Intel 8048/8049 single-chip microcomputer ICs. These are 8-bit processors and will therefore interface directly with the Teletext and Viewdata

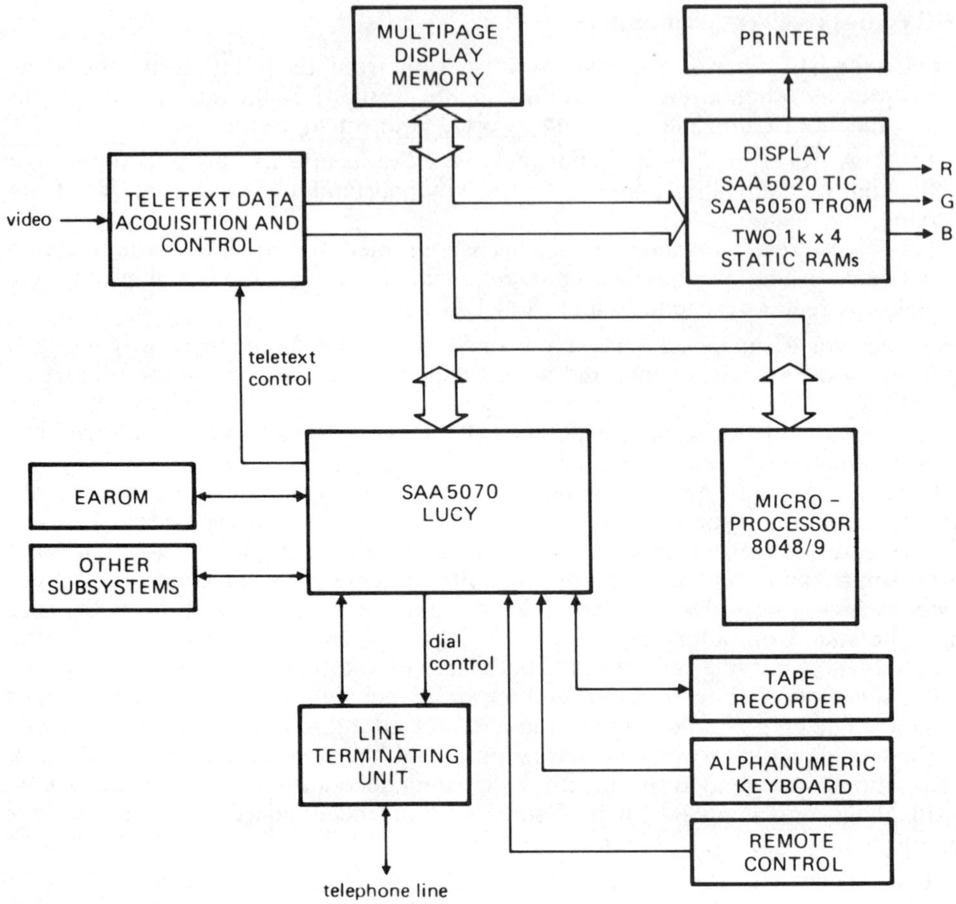

Figure 10.18 Block schematic of Teletext/Viewdata recorder

systems. This range of microcomputer chips follow 'standard' 8-bit microprocessor technology, and a simplified block diagram of the architecture is shown in *Figure 10.19*. The heart of the microcomputer is shown in the lower left-hand corner and consists of the control and timing section, Arithmetic Logic Unit, Flag Register, Accumulator and Decimal Adjust Logic. All of these sections follow the principles outlined for the 8080/8085 microprocessors in chapter 7.

It is common practice with a single chip microcomputer to limit the number of internal registers which can be used in conjunction with the ALU and memory. The reasons for this practice are two-fold:

1. The reduction in the number of internal registers will simplify the instruction set and make programming easier.
2. As the microcomputer has its own internal RAM data can be transferred from the ALU to the resident RAM at very high speed making the need for a range of dedicated registers unnecessary.

Seven 8-bit registers are allocated memory addresses and form part of the resident 64 byte random access memory. The remainder of this memory is used for data storage and a small stack.

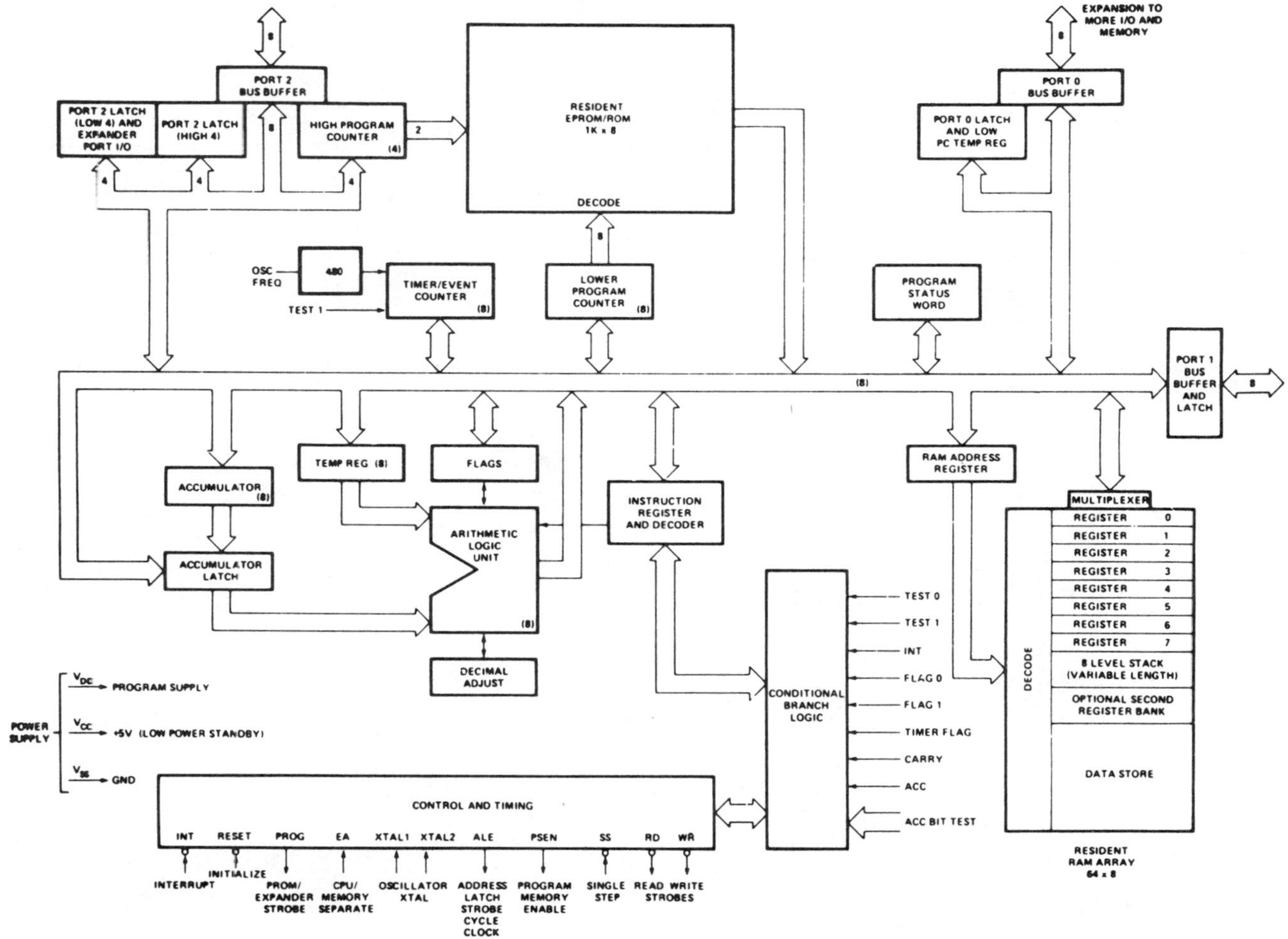

Figure 19.19 Intel 8048 CPU block diagram (*Courtesy of Intel Corporation*)

A separate 8-bit by 1 K ROM is contained within the IC, and has its own address decoders which latch the memory address from the internal multiplexed address/data bus. If communication with additional memory is required then the two ports 0 and 2 provide latched addresses for the additional external memory while port 1 provides the latched data from the internal 8-bit data bus.

In most microcomputer based Teletext and Viewdata systems the resident ROM is adequate for the immediate needs of these integrated circuits. As technology advances it is quite possible that the single-chip microcomputer will be used in conjunction with additional Read Only or Random Access Memory in more complex decoders. At the same time integrated circuit technology will advance, creating larger internal ROMs and RAMs. At the present time it is difficult to speculate if future decoders will use existing single chip microcomputers with additional external memory, or whether technology will provide microcomputers with sufficient internal storage to meet future technology trends.

References

1. *The British Telecom Prestel Terminal Specification Edition 1,* 1980, Prestel Headquarters, Telephone House, Temple Avenue, London EC4Y 0HL
2. *Technical Publication M 81–0013. Microprocessor Peripheral for Viewdata.* Mullard Ltd, Mullard House, Torrington Place, London WC1E 7HD

Chapter 11

Remote control systems

Remote control over colour receiver functions first became available in the early 1970s. The facilities provided by these early systems were primitive when compared with those currently available, partly because complex remote control was not required at this early stage, and secondly because early systems were constructed using discrete components, which significantly increased the cost of the receiver.

The first remote control systems to appear used ultrasonic communication between the receiver and transmitter, and all the processing of this signal was carried out using analogue circuitry. Theoretically, therefore, the earlier type of remote control system lies outside the scope of this book because we are primarily concerned with digital applications in TV receivers. However, it is important for the reader to understand how remote control systems have evolved since their introduction. For this reason one of the simpler analogue ultrasonic remote control systems will be covered to give the reader an appreciation of the system fundamentals, and to show how rapidly this aspect of television receiver design has advanced.

The system described below was employed in a range of Rank Radio International colour receivers, and provided control over channel selection and audio mute. Channel selection was achieved by quickly depressing the single button on the ultrasonic transmitter. This caused the receiver to step up to the next position on the tuner panel. Six preselected channels were available, and once the stepping sequence reached channel six a further press of the transmitter button would return the receiver to channel one.

Sound muting was accomplished by transmitting a longer signal of approximately two seconds, and the sound was restored by a further short signal. The user was not able to change channels while the sound was muted.

11.1 The simple ultrasonic transmitter

The transmitter is a discrete component device powered by a single 9 V battery, and a circuit diagram is shown in *Figure 11.1*. All component numbers are prefixed by eleven. Components VT1, R1, C2, and T1 form a Hartley oscillator which generates a 40 kHz ultrasonic sine wave when S1 is depressed. T1 is a step-up transformer producing a 300 V peak-to-peak sine wave on its secondary winding, and this is coupled by C6 to the ultrasonic transducer LS1. This transducer requires a d.c. polarizing voltage which is produced by the doubler circuit C4, D1, D2 and C5. The d.c. restorer section of the doubler is formed by D1 and C4, and the half-wave rectifier and smoothing circuit by D2 and C5. A series resistance in the output of the doubler (R2), prevents the low impedance of C2 from shunting the ultrasonic signal across the transducer.

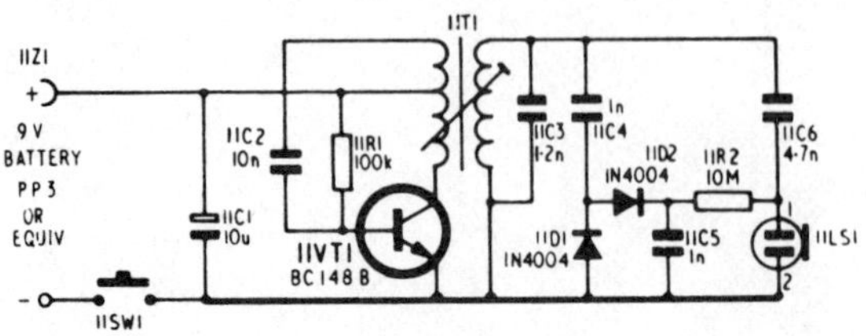

Figure 11.1 Circuit diagram for simple ultrasonic transmitter

This simple circuit is only capable of transmitting a fixed frequency ultrasonic signal. The length of time that this signal is transmitted determines the function of the remote control command.

11.2 The simple ultrasonic receiver

Figure 11.2 shows the circuit diagram of the remote control receiver. All component numbers are prefixed by ten. Ultrasonic sound radiated from the transmitter is converted back into a 40 kHz sine wave by the receiving transducer MIC1. This transducer is polarized by 188 V d.c. from the bias resistors R1 and R2. The received signal is coupled via C2 to a two stage amplifier formed by VT1 and VT2. Each of these transistors incorporates a high-pass filter in its base circuit to remove low frequency noise. An amplified and filtered signal therefore appears on the collector of VT2, and from here it is applied to the differential amplifier VT3 and VT4. Diodes D1 and D2 form a simple limiting circuit which clips the signal before it is applied to the amplifier input.

The output from the collector of VT4 is fed to a parallel tuned circuit L1 and C7, which oscillates at the received frequency of 40 kHz. This frequency is then half-wave rectified and smoothed by VT5 and its associated components. VT5 uses class C bias and only conducts on the positive half cycles of the voltage developed across the tuned circuit. During this conduction period the collector voltage of the transistor will fall towards 0 V.

A smoothing circuit in the collector of VT10 is formed by C8 and R15. It has a time constant of approximately 10 mS which reduces interference from spurious low frequency signals. The voltage on the collector of VT5 will therefore be a waveform which is normally 12 V and will fall to 0 V when the ultrasonic signal is being received. This switching waveform is coupled via D3 to the Schmitt trigger VT6 and VT7 to produce a clean switching waveform which can then be further processed by a pulse length discriminator.

The remainder of the circuit is responsible for detecting the length of the received ultrasonic signal. If a short pulse is received then channel change will be implemented, but if the pulse is long the sound will be muted. Should the sound already be muted then the next short pulse will restore the sound without changing channels. This is achieved as follows.

When a short pulse is received, VT7 will switch on and off again quickly, and during this time the electrolytic coupling capacitor C10 will not have sufficient time

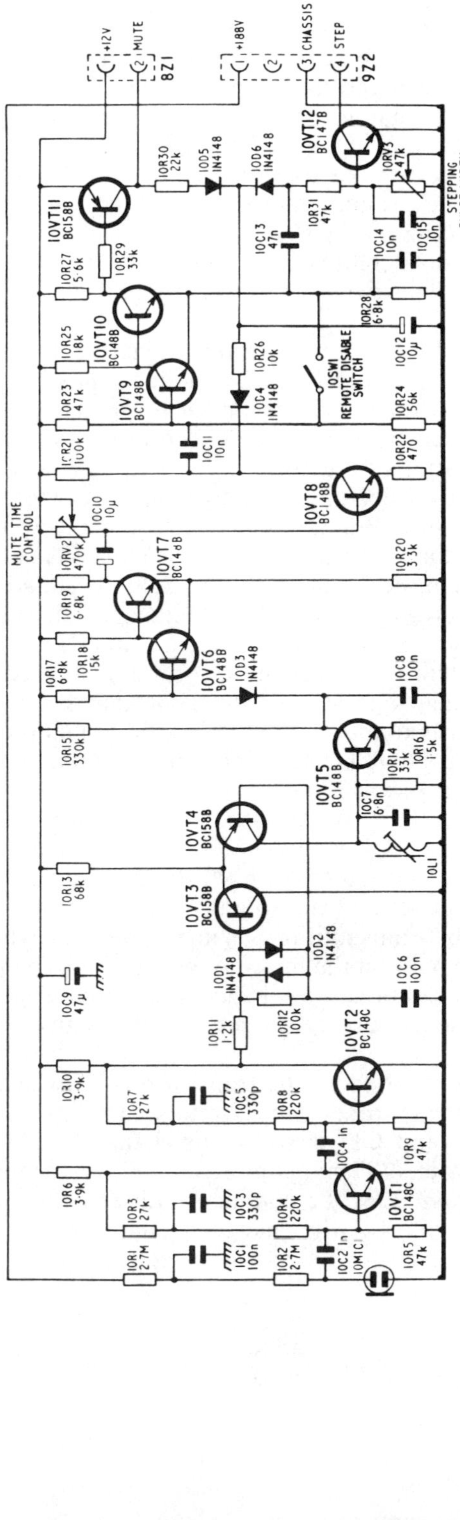

Figure 11.2 Circuit diagram for remote control receiver

to discharge through RV2. The short switching pulse is therefore coupled via C10 to the base of VT8 which in turn will switch off and on again. In the collector of VT8 is the coupling capacitor C11, which together with the input resistance of VT9 will differentiate the switching waveform. Consequently when a short pulse is transmitted a positive and then negative pulse will be present on the base of VT9.

If a long ultrasonic command is transmitted the electrolytic capacitor C10 will have sufficient time to discharge and then recharge in the opposite direction. This will cause the base voltage of VT8 to fall rapidly and then slowly rise, gradually turning the transistor back on. As the conduction of VT8 is now a slow process the collector voltage is not differentiated by C11 and consequently a negative pulse will not be produced on the base of VT9.

To summarize, if a short command is transmitted a positive and negative pulse will be produced on VT9 base but if a long command is transmitted only a positive pulse appears.

Transistors VT9 and VT10 form a Schmitt trigger bistable which is triggered by the differentiated pulses produced by VT8. When a positive pulse is produced by VT8 the bistable will latch in the condition VT9 conducting, VT10 cut off. When a negative pulse is received from VT8 the two transistors will change state so that VT9 is cut off and VT10 is conducting.

An output from VT10 collector is directly coupled to the base of VT11 which is the audio muting transistor. When a positive pulse is received on the base of VT9 the bistable will latch causing the base voltage of VT11 to rise to 12 V, ensuring that this *p-n-p* transistor is cut off. Under this condition the sound will be muted. When a negative pulse is received the bistable will change state and VT11 base voltage will fall below its emitter voltage. VT11 therefore conducts and its collector will rise towards the supply potential. Under these conditions the audio is restored.

During channel change operations, which produces a positive followed by a negative pulse as described above, the audio will momentarily mute and then be restored. This will ensure that a channel change operation is not accompanied by a large sound disturbance.

Permanent audio muting takes place when a long ultrasonic command is received. This results in a single positive pulse being generated by VT8, causing the bistable to latch in the condition VT9 on VT10 off. The output voltage from VT10 ensures that VT11 is cut off and the sound is permanently muted.

Sound is restored by transmitting a short command which produces both a positive and a negative pulse at the input to the bistable. The positive pulse will not change the bistable condition, and hence the audio will remain muted. The following negative pulse will change the bistable state causing VT11 to conduct and restore the audio.

Channel changing is initiated by the transistor VT12 which acts as a simple low impedance switch for the tuner step circuits. The time constant of its base components RV3 R31 and C13 is so arranged that VT12 will only conduct for approximately 1 mS when VT10 is conducting. This is achieved by sampling the voltage developed across R28, the common emitter load resistor for the Schmitt trigger bistable. This voltage will change as the state of the bistable changes because VT9 has a collector load of 18K Ω and VT10 collector is only 5K6 Ω. Consequently when:

VT9 conducts VR28 = 3.2 V approx.
VT10 conducts VR28 = 6.5 V approx.

During the change-over from VT9 to VT10 conducting the voltage across the emitter load will rise from approximately 3.2 to 6.5 V causing the timed 1 mS pulse to occur on the base of VT12. The emitter pulse amplitude required to switch VT12 on is about 6 V. The circuit formed by R26, R30, D4, D5, D6 and C12 ensures that the channel cannot be changed while the sound is muted, this prevents the next negative pulse which restores the sound from changing the channel.

The design of this remote control system by modern standards is rather crude, heavy on components and therefore high in cost, and only provides limited control over the receiver functions. These facilities were quite adequate for a simple colour receiver, but with the advent of Teletext, the complete concept of remote control had to undergo a serious 're-think' if the sophisticated facilities which Teletext offers were to be controlled from a simple hand held transmitter.

11.3 Simple Teletext remote control system

It was mentioned in chapter 9 that Rank Radio International were the first receiver manufacturers to produce a commercial Teletext receiver. This used the TIFAX XM11 Teletext Decoder which was interfaced with an ultrasonic remote control system. The system devised provided the following facilities:

1. Selection of any one of eight channels.
2. Control over volume, brightness and colour.
3. Picture modes of:
   off air picture.
   Teletext.
   picture and text mixed.
4. Sound mute.
5. Teletext page selection.
   Time code selection.
   Reveal.
   Clear.
6. Restore receiver analogue controls to their present position.

It is clearly not possible to use a system based on the previous description because of the large increase in the number of commands to be transmitted. This leaves two possible techniques:

1. To pulse code modulate a single carrier with digital information representing the function requested.
2. To transmit a continuous carrier which changes in frequency depending on the function requested.

Rank opted for the latter of these two techniques in the remote control system of their first Teletext receiver. The hand-held transmitter housed twenty-five push-buttons and was therefore considerably larger than one used in their earlier system. It was based on a CMOS integrated circuit (SAA 1024) which contained a variable divider chain which received a signal from a 4.433 MHz master oscillator. This divider chain produced a total of thirty discrete ultrasonic frequencies between 33.945 kHz, and 43.99 kHz, each command being separated by about 350 Hz. Only twenty-five of the thirty frequencies were used.

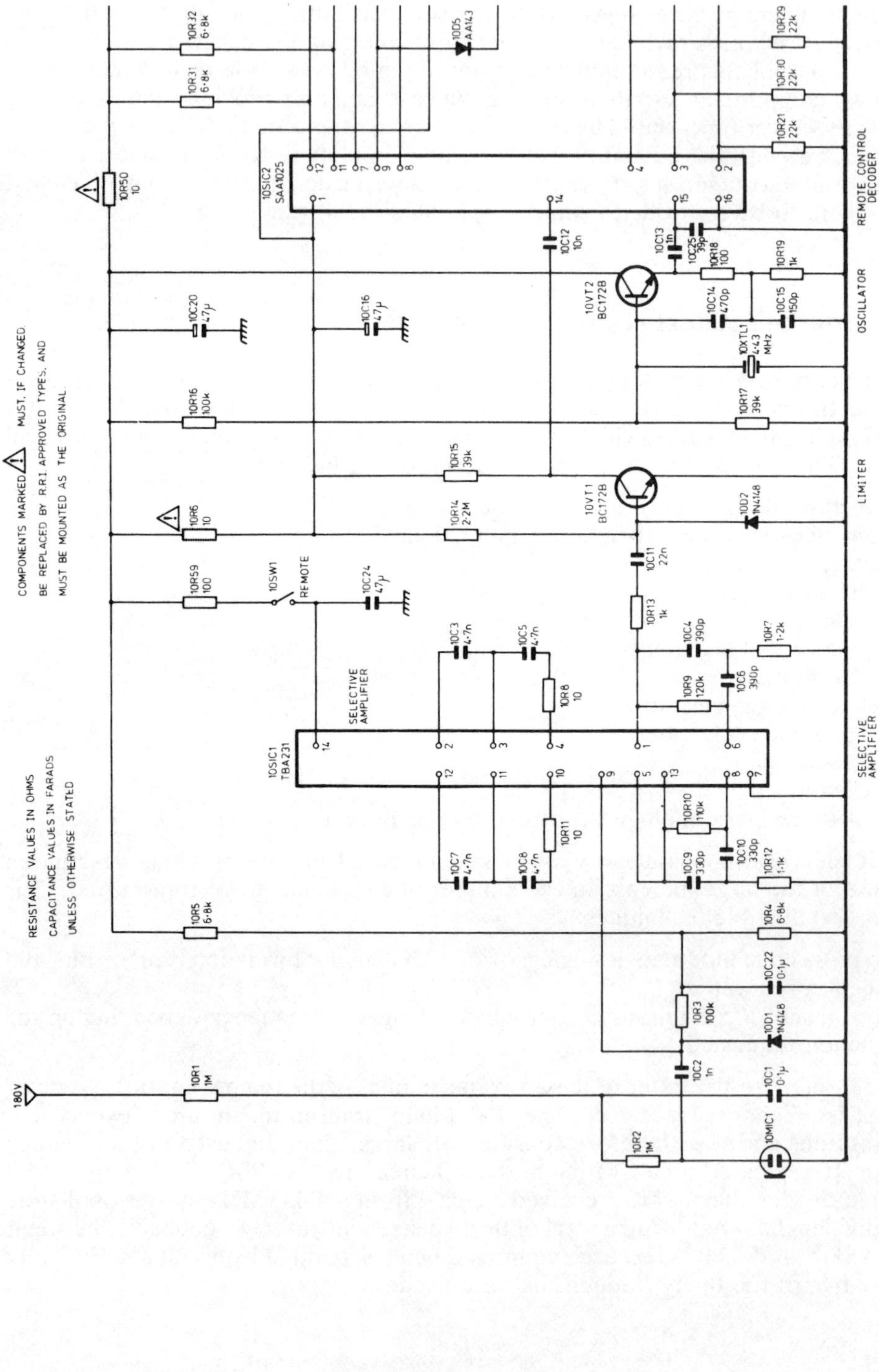

Figure 11.3 Circuit diagram for receiver

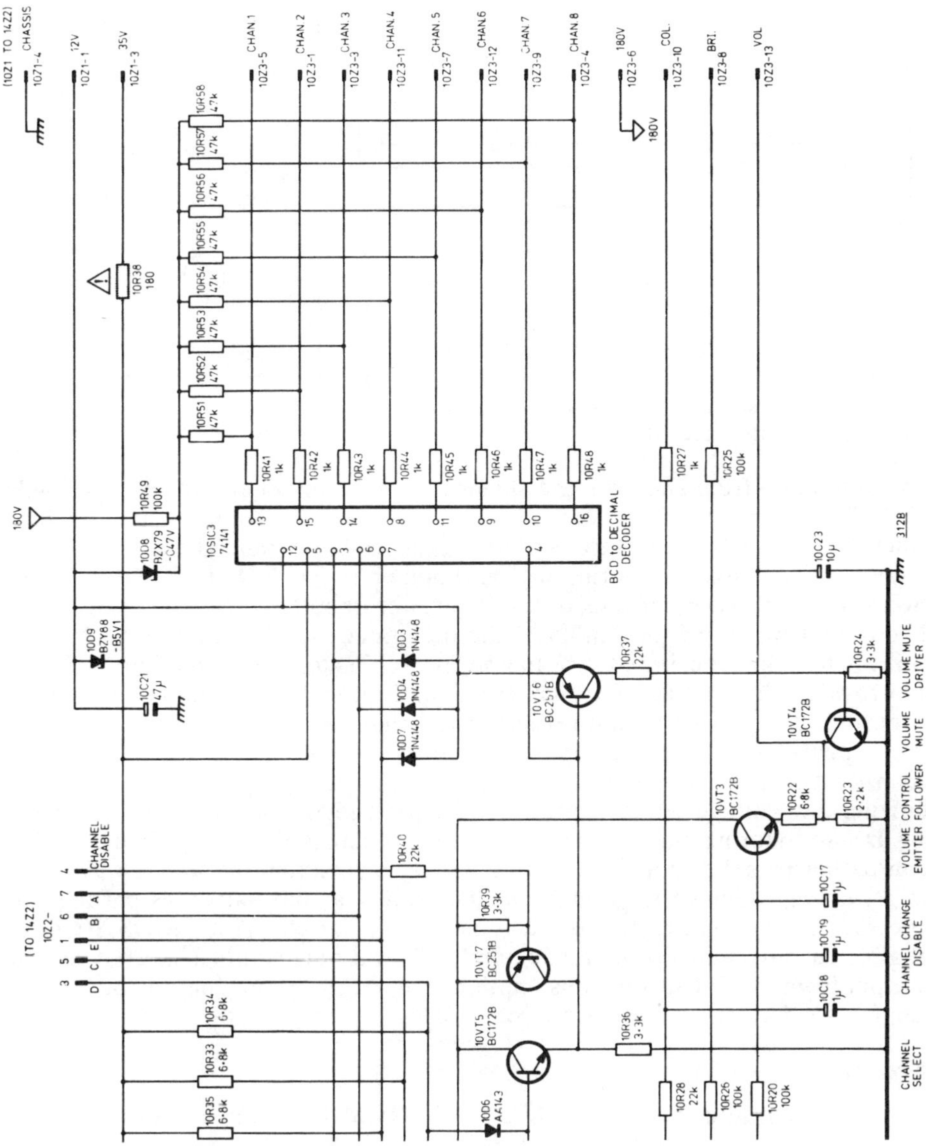

Figure 11.3 (continued)

The transmitter

The circuit diagram is shown in *Figure 11.4*. All component numbers are prefixed by eleven. Although the facilities offered are far more comprehensive than those offered by the previous system, the circuit is much simpler.

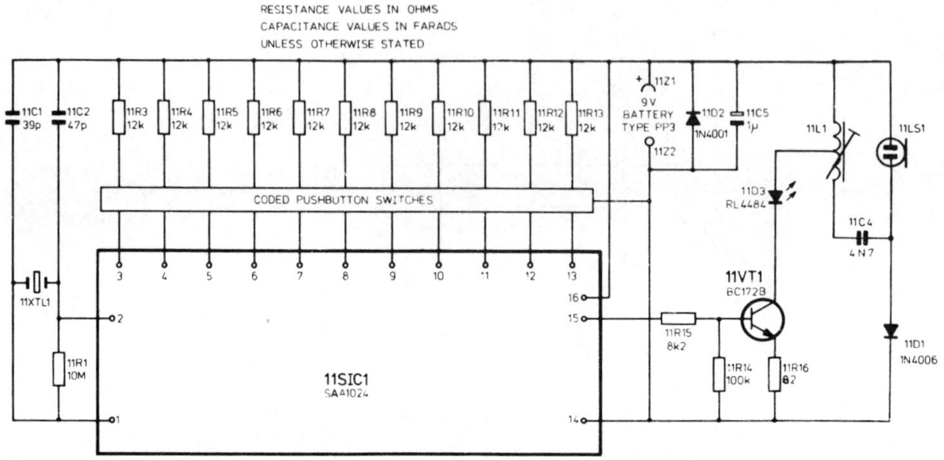

Figure 11.4 Transmitter circuit diagram

The reference frequency for the divider chain is provided by the 4.433 MHz crystal XTL1. The division ratio is programmed by connecting a pair of input contacts from between pins 3 to 13 to the negative supply rail. These 11 input pins are normally held high by a series of 12 K Ω pull up resistors R3 to R13, but when a button is pushed two input pins of the IC are returned to 0 V.

The output waveform on pin 15 of the integrated circuit saturates VT1 on the positive half cycles, causing current to flow in its collector circuit and illuminate D3. This provides the operator with a visual indication that a command is being transmitted. L1 is a step up auto-transformer which generates an ultrasonic signal of 120 V peak-to-peak for the capacitive transducer LS1.

As with the previous remote control transmitter this transducer requires a polarizing d.c. voltage which is produced by peak rectifying the transducer signal using D1 and storing the resultant d.c. voltage across C4. D2 provides a short circuit path across the supply should the battery be inserted the wrong way round, thus preventing damage to the CMOS IC. Note that the battery is permanently connected to the circuit, and there is no on/off switch. The CMOS IC draws negligible current in its dormant state and VT1 will only conduct when it receives an output from SIC1. Current consumption is therefore so low that the inclusion of a supply on/off switch is not necessary.

The receiver

The circuit diagram in *Figure 11.5*, can be subdivided into three major areas:

1. Frequency selective amplifier.
2. Remote control decoder.
3. BCD to decimal decoder.

All components are prefixed by ten.

The received ultrasonic frequency is converted back into an analogue sine wave by MICI. This transducer is polarized by 180 V d.c. from the resistance network R1 and R2.

IC1 is a TBA 231 which contains two operational amplifiers that have a high gain and a pass band of approximately 40 kHz. The incoming frequency from the transducer is therefore amplified and filtered before being fed to the limiting stage VT1.

With no input signal VT1 is held into saturation by the base bias resistors R6 and R14. When an ultrasonic command is received the shunt rectifying diode D2 clips the negative half cycles of the transistor input waveform ensuring that the base of the transistor is returned to 0 V, and VT1 is cut off.

On positive half cycles, the transistor is driven into saturation by a combination of the input waveform, and also the d.c. bias present on the base. The transistor is therefore either saturated or cut off resulting in a collector waveform which will swing between well defined maximum and minimum limits. This output is coupled via C12 to the remote control decoder IC SIC2 (SAA 1025).

The remote control decoder compares the incoming ultrasonic frequency on pin 14 with a reference frequency provided by a 4.433 MHz crystal oscillator formed by VT2, XTL1, and their associated components. When the incoming remote control frequency has been identified SIC2 will produce a pulse waveform on one or more of the output pins 7, 8, 9, 11, 12. These pulses take the form of a square wave

	Ultrasonic Frequency (KHz)	D	C	B	A	E						
MUTE	34·291	0	0	0	1	1						
COL +	34·638	0	0	1	0	0						
RESET	34·984	0	0	1	0	1						
COL–	35·330	0	0	1	1	0						
BRI +	36·023	0	1	0	0	0						
BRI–	36·716	0	1	0	1	0						
VOL +	37·409	0	1	1	0	0						
VOL–	38·101	0	1	1	1	0						
REVEAL	38·448	0	1	1	1	1						
CLEAR	38·794	1	0	0	0	0						
PICTURE	39·141	1	0	0	0	1						
MIXED	39·487	1	0	0	1	0						
TEXT	39·833	1	0	0	1	1						
PRESELECT	40·180	1	0	1	0	0						
PAGE	40·526	1	0	1	0	1						
9	40·872	1	0	1	1	0						
8	41·219	1	0	1	1	1						
7	41·565	1	1	0	0	0						
6	41·912	1	1	0	0	1						1
5	42·258	1	1	0	1	0					1	
4	42·604	1	1	0	1	1				1		
3	42·951	1	1	1	0	0			1			
2	43·297	1	1	1	0	1		1				
1	43·643	1	1	1	1	0	1					
0	43·990	1	1	1	1	1						
	10Z2 Pin No.	3	5	6	7	1						
	10IC2 Pin No.	8	9	11	12	7						
	10IC3 Pin No.	–	–	6	3	7	13	15	14	8	11	9
	10Z3 Pin No.						5	1	3	11	7	12
	'0' level is +17V '1' represented by 23mS negative pulses reaching approx. +11·5V.						Unselected level is approx. +59V '1' represented by 23mS negative pulses reaching approx. +12V.					

Figure 11.5 Truth table for pulsed outputs

having a mark to space ratio of 23 mS negative and 185 mS positive while a signal is being received. In the absence of a remote control command these output pins are returned to +17 V via the pull up resistors R31 to R34. *Figure 11.5* shows the truth table for these pulsed outputs together with the input frequencies to the remote control decoder chip. A '0' represents +17 V and a '1' represents the train of 23 mS negative pulses.

In addition to the five coded outputs, SIC2 also produces a further three outputs on pins 2, 3, 4. These are used to control the volume brightness and colour of the receiver. Each output consists of a 17 V peak-to-peak square wave having a frequency of approximately 9 kHz. An up/down command for any of these three control functions will alter the mark to space ratio of the 9 kHz square wave from 1:30 to 30:1 in 30 steps. Each output is integrated by a C,R network consisting of a series resistor R28, 26, 20 and a shunt capacitor C18, 19, 17. The voltage developed across these capacitors will therefore be variable from approximately +1 to +16 V, the lowest voltage corresponding to the minimum and the highest voltage corresponding to the maximum setting of each command.

The volume control voltage is processed by an emitter follower VT3 whose output is attenuated by R22 and R23 before it is fed to the manual control circuit. Transistor VT4 shunts this control voltage during channel change so that spurious noise is not produced at the loudspeaker when the receiver changes channel. The audio mute instruction causes the square wave on pin 2 of SIC2 to revert to a 1:30 mark to space ratio. This will reduce the audio control voltage to its minimum setting.

All five outputs on pins 7, 8, 9, 11, 12 of SIC2 are used during Teletext commands, and they are therefore coupled to the Teletext decoder interface panel via the connector Z2 on pins 7, 6, 1, 5, 3.

The truth table in *Figure 11.5* shows that when channel change instructions are transmitted (0 to 7) only outputs A, B, and E on pins 7, 11, 12 of SIC2 will change. These three signals are connected to a BCD to decimal decoder SIC3, its fourth input on pin 4 is held low during a channel change by the operation of VT5. This BCD to decimal decoder is a 74141 which is a standard TTL logic chip designed to operate from a +5 V supply. In its application in this circuit the 5 V supply pin (5) is returned to the 17 V supply, and its 0 V pin (12) is returned to the +12 V supply making the total potential across the supply pins 5 V. Its outputs on pins 8, 9, 10, 11, 13, 14, 15, 16 are pulled up to +59 V via R51 to R58.

When a channel is selected, the BCD inputs to this IC cause a single output to be pulled down to +12 V producing 47 V negative pulses which are 23 mS wide. These are then applied to the channel select circuit via the edge connector Z3. The three diodes D7, D4, and D3 clamp the BCD inputs to SIC3 so that they cannot fall significantly below +12 V. If these diodes were not included the outputs from SIC2 would fall to 0 V when an output occurred, and this would cause damage to SIC3.

SIC3 must be prevented from operating under two conditions:

1. When the received remote control command is not a channel change instruction.
2. When the numerical inputs to the receiver represent Teletext page or time coded data.

By referring to the truth table of the remote control decoder IC in *Figure 11.5* it can be seen that the outputs on pins 8 and 9 of SIC2 both contain negative pulses during the transmission of a channel change command. During the transmission of any other remote control command either pin 8 or pin 9 or both will rise to +17 V.

This action will cause D5 or D6 or both of the diodes to conduct, saturating the emitter follower VT5. VT5 emitter will now become positive, due to the current passing through R36. This positive voltage is applied to the 2^3 input of SIC3, which in turn will generate a decimal output that is above the range of the tuner selection circuits. This effectively inhibits a channel change operation.

A similar process takes place when a page or time select command is transmitted. The TIFAX XM11 interface circuits used in this receiver produce a negative pulse which lasts for four consecutive digits when a page instruction is transmitted (page plus three digits), or for five consecutive digits when a time select instruction is transmitted (time select plus four digits). This negative inhibit pulse is coupled from the decoder interface electronics to the base of VT7 via R40. During the presence of this pulse VT7 will saturate, and again return the 2^3 input on pin 4 of the BCD to decimal decoder to a positive potential.

Whenever a valid channel change instruction is received, pin 4 of SIC3 is taken low, producing a valid decimal output for the tuner circuits. This action will also cause VT6 to conduct, which in turn will saturate VT4. The volume control voltage is therefore shunted by VT4 to 0 V. This mutes the sound during channel changes, avoiding unwanted bursts of noise.

This circuit uses a combination of analogue and digital techniques to provide 'full specification' remote control over a Teletext colour receiver. Although its component count is still high by modern standards it represents a considerable improvement both in the facilities offered, and in circuit design, over the simpler system previously described.

Ultrasonic transmission does have some disadvantages, primarily its lack of sensitivity and susceptibility to spurious audio signals in the ultrasonic range. The high polarizing voltage required for the transducers in the receiver and transmitter also complicates circuit design.

Whichever ultrasonic remote control system is used the receiver will depend on either the very accurate timing measurement, such as that used in the first example, or accurate measurement of frequency such as that used in this example. All of these disadvantages can be overcome by a remote control system using pulse coded modulation on an infra-red beam of light.

11.4 Infra-red remote control system (Mullard)

Practically all receiver manufacturers have now opted for the infra-red system, because it overcomes the previously described disadvantages, is considerably more flexible, and uses fewer components. One of the potential disadvantages of an infra-red system is its susceptibility to other sources of infra-red light, but these problems are overcome by using pulse coded modulation. An example (Mullard) is shown in *Figure 11.6*.

When a key is pressed on the hand-held unit, a short psuedo random bit sequence is initially transmitted, followed by a start code, message code, start and message. The start code is seven bits of data (1101010), and is used to synchronize the receiver circuits using a similar technique used by the start bit in Viewdata. Each bit in the transmitted sequence occupies 1 mS, with a binary '0' being transmitted as two short pulses at the beginning of the 1 mS period, and a Binary '1' transmitted as two widely spaced pulses within the 1 mS period. Refer to *Figure 11.6*.

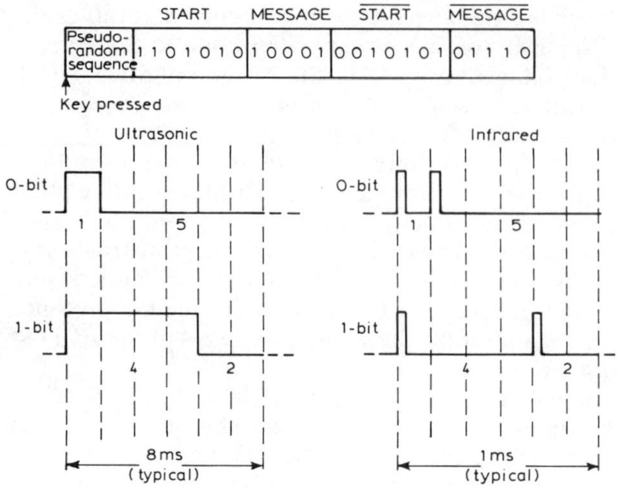

Figure 11.6 Infra-red remote control system

After the start code, a five-bit message code is transmitted which provides remote control for up to thirty-two commands. These coded commands have been carefully chosen so that they are compatible with conventional station selectors, binary input selectors, and the Mullard digital selection system (DICS). A complete table of all the possible binary codes together with their remote control commands is shown in *Figure 11.7*.

Once the message data is complete the whole sequence of the start and message information is repeated in complemented form. When a key in the hand-held unit has been pressed the complete sequence of data is transmitted automatically and the user does not have to keep the button depressed. It takes just over 24 mS to transmit all the information, and the receiver circuits will not respond until the complete sequence of data has been received and checked.

Two versions of this remote control system are available using either ultrasonic or infra-red techniques. The circuits described here are those used for the infra-red system.

The infra-red transmitter

The transmitter shown in *Figure 11.8* is powered by a 6 V battery and uses three infra-red LEDs to radiate the command. An SAA 5000 dedicated remote control IC is used to generate the correct bit pattern corresponding to the key depressed. The key pad is arranged in a matrix of 6 by 6 with four switch positions being unused. This matrix is scanned by placing an output signal on one of the vertical columns of the matrix. The IC then looks for the return of this signal on one of the six input lines between pins 4 and 9. If the scanning signal is not returned because a key in that column has not been closed then the IC shifts the scanning signal to the next vertical column and interrogates the input pins again.

When a switch is closed the scanning signal will be returned to the SAA 5000 on one of the input lines. For example if the key in the bottom right hand corner of the matrix was closed the scanning output on pin 10 would be returned on pin 4. Scanning is arranged so that only one vertical column is accessed at a time, so in this

No.	Code				b_7 / b_6	0 / 0	1 / 0	1 / 1
	b_5	b_4	b_3	b_2	b_1	Television mode	Teletext mode	Viewdata mode
1	0	0	0	0	0	Reset (on)	Reset (on)	Reset (on)
2	0	0	0	0	1	Mute	Mute	Mute
3	0	0	0	1	0	Standby	Standby	Standby
4	0	0	0	1	1	TV/on	TV/on	TV/on
5	0	0	1	0	0	Status	Status	Status
6	0	0	1	0	1	–	Hold	Ring off
7	0	0	1	1	0	1*	Reveal	Reveal
8	0	0	1	1	1	Time	Text cancel	Picture display
9	0	1	0	0	0	Analogue 1+	Analogue 1+	Analogue 1+
10	0	1	0	0	1	Analogue 1–	Analogue 1–	Analogue 1–
11	0	1	0	1	0	Analogue 2+	Analogue 2+	Analogue 2+
12	0	1	0	1	1	Analogue 2–	Analogue 2–	Analogue 2–
13	0	1	1	0	0	Analogue 3+	– (RA)	Tape rec. (RA)
14	0	1	1	0	1	Analogue 3–	– (RA)	Tape play (RA)
15	0	1	1	1	0	Analogue 4+	Timed page off (RA)	* (RA)
16	0	1	1	1	1	Analogue 4–	Timed page on (RA)	# (RA)
17	1	0	0	0	0	Station 1 (BBC1/on)	Number 1	Number 1
18	1	0	0	0	1	Station 2 (ITV/on)	Number 2	Number 2
19	1	0	0	1	0	Station 3 (BBC2/on)	Number 3	Number 3
20	1	0	0	1	1	Station 4 (BBC1/on)	Number 4	Number 4
21	1	0	1	0	0	Station 5 (ITV/on)	Number 5	Number 5
22	1	0	1	0	1	Station 6 (VCR/on)	Number 6	Number 6
23	1	0	1	1	0	Station 7 (BBC1/on)	Number 7	Number 7
24	1	0	1	1	1	Station 8 (ITV/on)	Number 8	Number 8
25	1	1	0	0	0	Station 9 (BBC2/on)	Number 9	Number 9
26	1	1	0	0	1	Station 10 (BBC1/on)	Number 0	Number 0
27	1	1	0	1	0	Station 11 (ITV/on)	Full page	Full page
28	1	1	0	1	1	Station 12 (VCR/on)	Top	Top
29	1	1	1	0	0	–	Bottom	Bottom
30	1	1	1	0	1	Viewdata/on	Viewdata/on	Viewdata/on
31	1	1	1	1	0	Superimpose (mix)	Superimpose (mix)	Superimpose (mix)
32	1	1	1	1	1	Teletext/on	Teletext/on	Teletext/on

Note 1: RA indicates that these commands revert to analogue controls whenever the tv video is displayed

Note 2: The button 1* (No.7) allows up to 16 channels to be selected. To obtain station 14, for instance, press 1* followed by station 4.

Note 3: This command table applies to the production versions of our ICs. There are some differences to command tables applicable to development samples of the ICs.

Figure 11.7 Table of binary codes and remote control commands

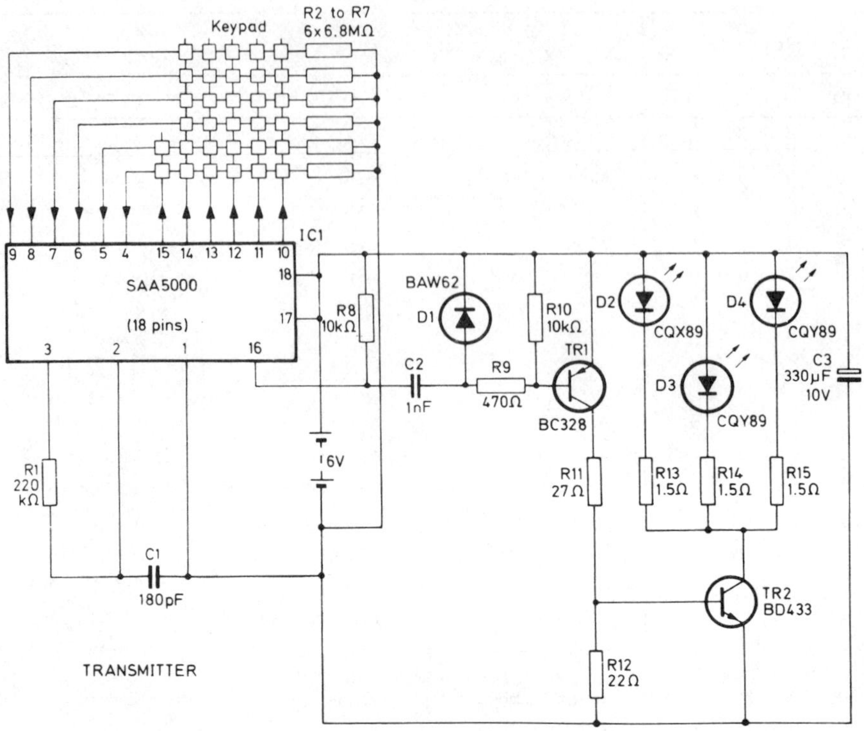

Figure 11.8 Infra-red transmitter

example if a scanning signal is returned on pin 4 the IC can 'calculate' that the lower right-hand switch contacts are closed. Alternatively if the scanning signal was returned on pin 9 then the IC would 'calculate' that the top right-hand contacts were closed.

Once the switch, and therefore the command requested, has been detected the SAA 5000 will generate the correct pulse sequence for the remote control command. This signal leaves the IC on pin 16 and is d.c. restored by C2 and D1 before being applied to the base of the driver transistor TR1. Under normal conditions this transistor is cut off, but the PCM output from IC1 will forward bias the transistor causing its output voltage across R12 to drive TR2 into saturation.

Three gallium arsenide infra-red LEDs, together with their series current limiting resistors, are connected in parallel to form the collector load of the output transistor. When a command is transmitted, the peak current required by this output stage is 4 A, and this large current is supplied by the electrolytic capacitor C3. When the output stage is idle the only current drawn from the 6 V supply battery will be the very small leakage current of C3 and the standby current of IC1.

The infra-red receiver

Figure 11.9 shows the circuit diagram of one of the remote control receivers that can be used with the transmitter of *Figure 11.8.* This circuit uses a combination of analogue and digital techniques to process the received command, and generate the

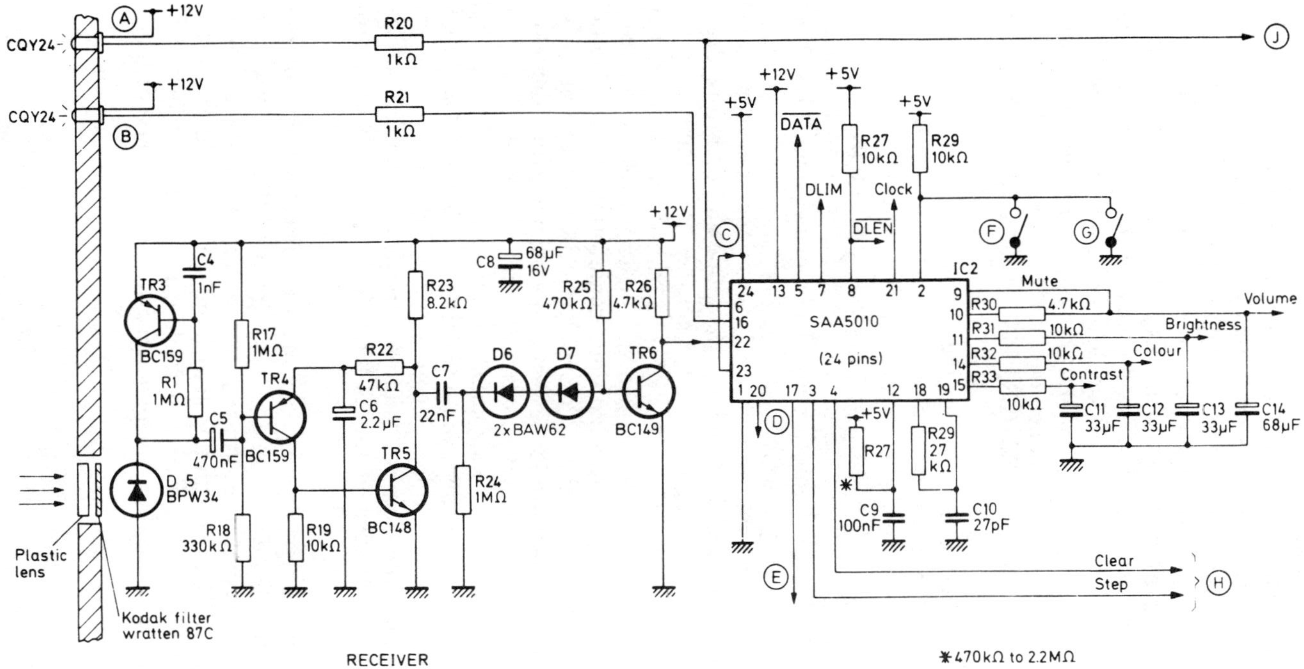

Figure 11.9 Circuit diagram for infra-red remote control receiver

A	On/standby indicator	F	Local reset switch
B	Message received indicator	G	Third mains switch contact pair
C	Ultrasonic/infrared selection	H	To touch tuner
D	Mode selection	J	On/standby control
E	'Picture on' input		

Resistors R_2 to R_7 are 3.3 kΩ for low-impedance switches

following outputs:

1. A serial digital signal which is applied to the Teletext decoder.
2. Four analogue outputs which are converted to d.c. voltages for volume, brightness, colour and contrast control.
3. Clear and step signals which are fed to the touch tuner unit.

The received infra-red light beam is focused by a clear plastic lens on to a photo-diode D5. As this diode is sensitive to light outside the infra-red spectrum a filter is included in the received light path to reduce the effects of unwanted wavelengths. This filtering process is by no means perfect and consequently D5 will respond to light radiated from tungsten and fluorescent lights as well as the light from the infra-red diodes. Before the voltage developed across D5 can be applied to the remote control decoder SAA 5010 the effects of this low frequency interference must be removed. This is achieved in several stages by the discrete component network comprising of TR3, TR4, TR5, and TR6 together with their associated components.

D5 and TR3 are connected in series across the 12 V supply, therefore TR3 forms the output load resistance for the receiver diode. The RC network R1 C4 has a long time constant when compared with the received infra-red command, but a short time constant when compared with low frequency light interference. Consequently the amplitude of the signal on the base of TR3 will change with frequency, this being low in amplitude for the remote control command signal, but higher in amplitude for the low frequency interference.

The interference voltage on the base of TR3 causes the transistor to conduct more heavily, reducing the load resistance for the diode and therefore decreasing the gain of the circuit at low frequencies. C5 couples the received signal to a two-stage amplifier formed by TR4 and TR5. As with the previous circuit, the frequency response of this amplifier is again shaped to provide a low frequency cut-off. Its output is coupled via C7 to a limiting circuit formed by D6 and D7. The anode of D7 is polarized towards the 12 V supply via R25 which normally holds TR6 into conduction. The amplitude of the output signal from the two stage amplifier TR4 and TR5 is developed across R24, and this must be sufficient to forward bias D6 and D7. The received command signal will be able to make these diodes conduct, but low level interference pulses will not. TR6 will therefore supply the remote decoder IC SAA 5010 with a clean digital signal, alternating between 0 and +12 V, which will be free from low and high frequency interference.

Once the SAA 5010 has received the 24-bit pulse sequence it carries out the following tasks:

1. It validates the received signal, and providing it is correct will ground pin 16 of the IC to indicate that a valid command has been correctly received. An LED can be used to indicate correct reception of a command, and in this example this has been achieved by LED (B) and its associated current limiting resistor R21.
2. Once the received command has been verified it is applied to a decoder which will determine which of the thirty-two commands has been received.

These commands can be broadly divided by function into three sections:

1. The analogue outputs

The four analogue outputs on pins 10,11, 14 and 15 produce a d.c. output voltage which is variable from 0–12 V in 62 steps. The CR network connected to pin 12 of the IC determines the rate of change of the analogue control, and the range available is between 36 and 200 mS per incremental output voltage step. The four outputs produced provide control over volume brightness colour and contrast.

2. The two outputs – Clear and Step

These are used to provide selection of the TV channel, and the outputs of these two signals is shown in *Figure 11.10*. When a channel change command has been received, verified, and decoded the clear output on pin 4 of the SAA 5010 falls to 0 V for 28 mS, and this will cause the tuner electronics to select channel one. 5.25 mS after the rising edge of the negative clear pulse a negative 1.7 mS pulse appears every 7 mS on the step output.

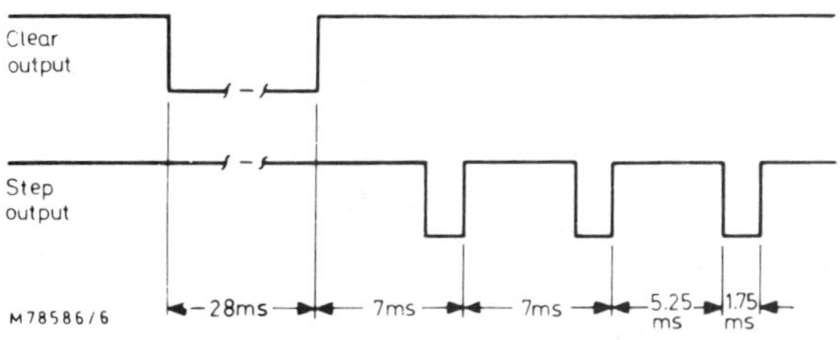

Figure 11.10 Outputs of 'Clear' and 'Step' signals

The number of pulses present on the step output depends on the channel number selected. If channel 1 was selected then only the clear pulse would appear. If channel 2 was selected then the clear pulse and 1 step pulse would appear. If channel 3 were selected the clear pulse and 2 step pulses would appear etc.

It can be seen that the procedure adopted to change channel in this remote control system is to reset the tuner to station 1, and then sequentially step it through the channels until the desired station has been reached.

3. Serial I bus signals

A large number of the integrated circuits which Mullard use in their Teletext decoders, Viewdata decoders and digital channel selection circuits require instructions from the remote control decoder IC. It has been seen from *Figure 11.7* that this remote control system supports a total of thirty-two separate commands, which produces a potentially difficult interfacing problem between the remote control decoder and the integrated circuits which require user instructions.

To simplify this interfacing problem Mullard have developed a communications system known as the I bus. This is a simple two-wire data highway which carries all the information required to control and interface a variety of integrated circuits with the remote control system. In the design of the I bus, Mullard have taken into account the need to maintain compatibility between existing and future integrated circuits. Since its introduction, a variety of sub-systems, such as the LUCY IC explained in chapter 10, have been added, and compatibility between integrated circuits has been maintained. The format of the I bus commands is briefly described in the following paragraphs.

A clock-and-enable signal designated DLIM consists of 14 clock pulses at a nominal 62.5 kHz rate which are generated whenever a command has been received. At the same time that the DLIM signal is generated, the 7-bit message code outlined in *Figure 11.7* is placed on the Bar Data line using the bit sequence illustrated in *Figure 11.11*.

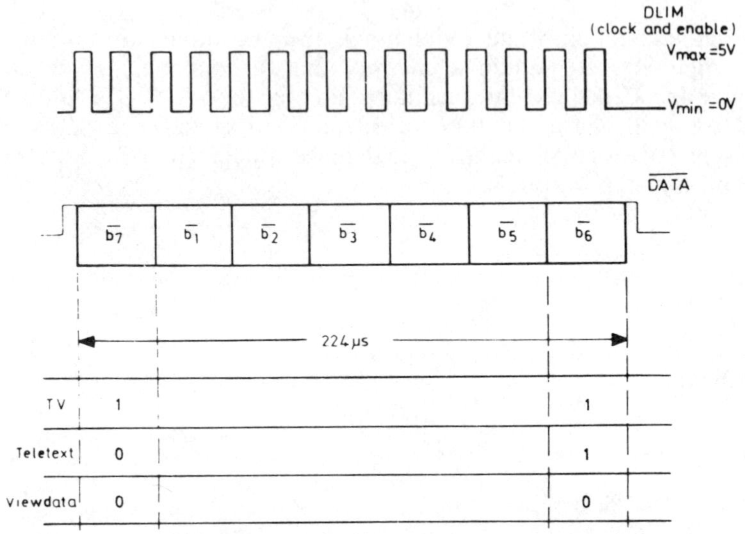

Figure 11.11 7-bit sequence

Each bit has a duration of two clock pulses causing the complete command to last for 244 microseconds. Bit 7 and bit 6 contain mode information which enables integrated circuits using the I bus to determine if the command being received carries information relevant to that chip. For example, if bit 6 is at logic '0' and bit 7 is at logic '1' then a Teletext command is being received, and is interrogated by the Teletext ICs only. Channel change circuitry will ignore the I bus command unless bit 6 and bit 7 are both at logic '0', while the Viewdata decoder circuits will only take action if the I bus data has bit 6 and bit 7 both set to a logic '1', see *Figure 11.7*.

The I bus was briefly introduced in concept only in chapter 9 where it was seen that the TAC, and TROM integrated circuits decoded information present on the I bus to determine the operational mode of the decoder. The I bus is also used for the range of integrated circuits developed by Mullard for digital tuning and Viewdata. Its application in these areas is described in the relevant chapters.

Chapter 12

Digital tuning systems

Variable capacitance tuners have been available to receiver manufacturers for many years, making the older mechanically operated versions obsolete. These electronic tuners employ a number of variable capacitance diodes as the tuning elements in the RF amplifier and oscillator/mixer stages. A single d.c. voltage, derived from a very stable supply, is used to control the tuning of the oscillator and RF amplifiers simultaneously.

The selection of the desired channel is achieved by manually adjusting a potentiometer so that the voltage produced on its slider causes the tuning elements of the RF amplifier and oscillator to resonate at the desired frequencies. This mechanically operated tuning system has several disadvantages, the most relevant being:

1. Mechanical switches and potentiometers are used to derive the tuning voltage, and these tend to be unreliable and deteriorate in performance with age. For example dirty switch contacts and worn carbon tracks, etc.
2. Mechanical components are expensive to manufacture, increasing the cost of the receiver.
3. Tuning the receiver is subject to human error, consequently the quality of the picture, and especially Teletext, may be impaired.
4. Accurate tuning can only be made when the desired TV channel is being transmitted. If the receiver is being installed from new then installation and final testing has to take place while all television channels are on the air simultaneously.
5. In receivers employing remote control, interfacing between the remote control receiver electronics and the varicap tuner can be unnecessarily complex. Interface circuits normally use discrete components which increase assembly costs and reduce reliability.

All of the above disadvantages can be overcome if the tuning voltage is derived from digital electronics, making the use of mechanical switches and potentiometers redundant. Although digital tuning systems tend to be complex in operation and circuit design, these two factors are heavily outweighed by the versatility and accuracy of the digital alternative. Before describing the techniques employed in such a tuning system it is necessary to quickly recap on conventional tuning arrangements so that the reader can appreciate the significant advances that digital tuning offers.

12.1 Variable capacitance tuning

A variable capacitance diode (Varicap Diode) is a simple *p-n* device which is always connected in the reverse bias mode. Under these conditions a depletion layer is formed at the *p-n* junction and a very small leakage current passes through the diode. It can be seen from *Figure 12.1* that a row of negative charges on the *p*-type side, and a row of positive charges on the *n*-type side produce a potential gradient across the depletion layer in a similar way to the potential produced across a charged capacitor. The reverse bias junction, with its very low leakage current, forms an insulating dialectric.

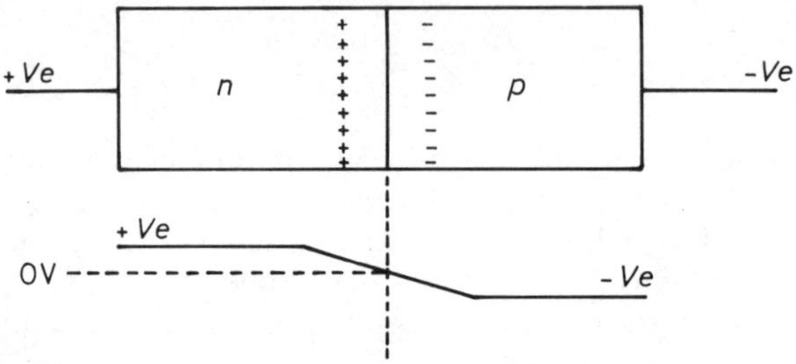

Figure 12.1 Simplified diagram of variable capacitance diode

If the reverse bias voltage across the *p-n* junction is increased two things will happen:

1. The depletion layer across the *p-n* junction becomes wider moving the positive and negative charges further apart.
2. The charge developed across the depletion layer also increases.

The effects of 1 and 2 above cause the capacitance to fall. If the reverse bias voltage is reduced the width of the depletion layer falls, as does the charge across the depletion layer. The capacitance therefore increases.

It should be remembered that the size of the capacitor plates, and the distance between these plates, is very small indeed because of the limited size of the *p-n* junction. Consequently practical varicap diodes have very small capacitance values, typical figures lie between 1.8 and 40 pF for the Signetics BB diode range.

Varicap diodes can be used in both series and parallel tuned circuits, not only in tuners but also in voltage controlled oscillators found in a variety of applications. A typical example being the subcarrier oscillator in the colour TV decoder.

The d.c. voltage used to control the resonant frequency of the tuned circuits in the TV tuner has to be very accurately controlled and completely free from noise and interference. The importance of this can be illustrated by examining the range of voltages required to tune the receiver through the UHF band.

The limits of the UHF band are from 471.25 MHz (Channel 21), up to 855.25 MHz (Channel 69), representing a frequency span of 384 MHz. This frequency range is covered by the tuner using a tuning voltage from 0–30 V. Hence a 1 V change in tuning supply will cause a 12.8 MHz change in tuner frequency, and a 100 mV change in tuner voltage will produce a 1.28 MHz change in tuner frequency. Relatively small changes in the tuner voltage will produce large, and very undesirable, changes in the tuner frequency.

To overcome these problems, dedicated voltage regulators have been developed which have heavily decoupled outputs. A typical example is shown in *Figure 12.2*. The TAA 550 is a custom-made shunt regulator specifically designed for tuner voltage applications. Its output load must ideally be fixed, and load variations caused by channel change must be kept to a minimum. This is achieved by placing a number of high value potentiometers across the stabilized 30 V supply. Voltages

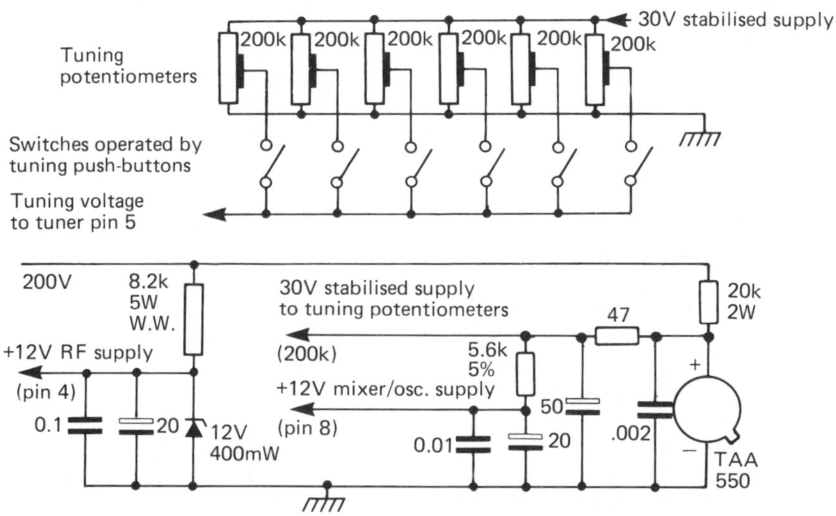

Figure 12.2 Dedicated voltage regulator and tuning potentiometers

produced at the slider of each tuning potentiometer are then switched by the channel selection buttons to the varicap tuner. Only one output voltage from one potentiometer slider will be used at a time, and the unused potentiometer slider circuits are open circuit when not selected. The load placed on the slider circuit in use causes very little current to flow, because all the varicap diodes in the tuner will be reversed biased, and the only current they will pass will be leakage current. Careful design of the tuned circuits themselves, and the use of high value series feed resistors, will ensure that the total current drawn from the tuning supply voltage is minimal.

A digital tuning system replaces the potentiometers and the mechanical switches used to select the desired channel. In their place is a system which uses a digital to analogue converter to create the tuning voltage for the channel or station requested. A closed loop frequency comparator is also included so that the digital tuning system can compare the oscillator frequency from the varicap tuner with the requested channel carrier frequency. As a result of this comparison process an error voltage is added to the nominal tuning voltage so that very precise control of oscillator frequency can be achieved.

12.2 A fundamental digital tuning system

A variety of digital tuning systems are available from a range of IC manufacturers. The example to be described is shown in block diagram form in *Figure 12.3*, and is based on the Mullard system. It is chosen as the basis for this chapter because it

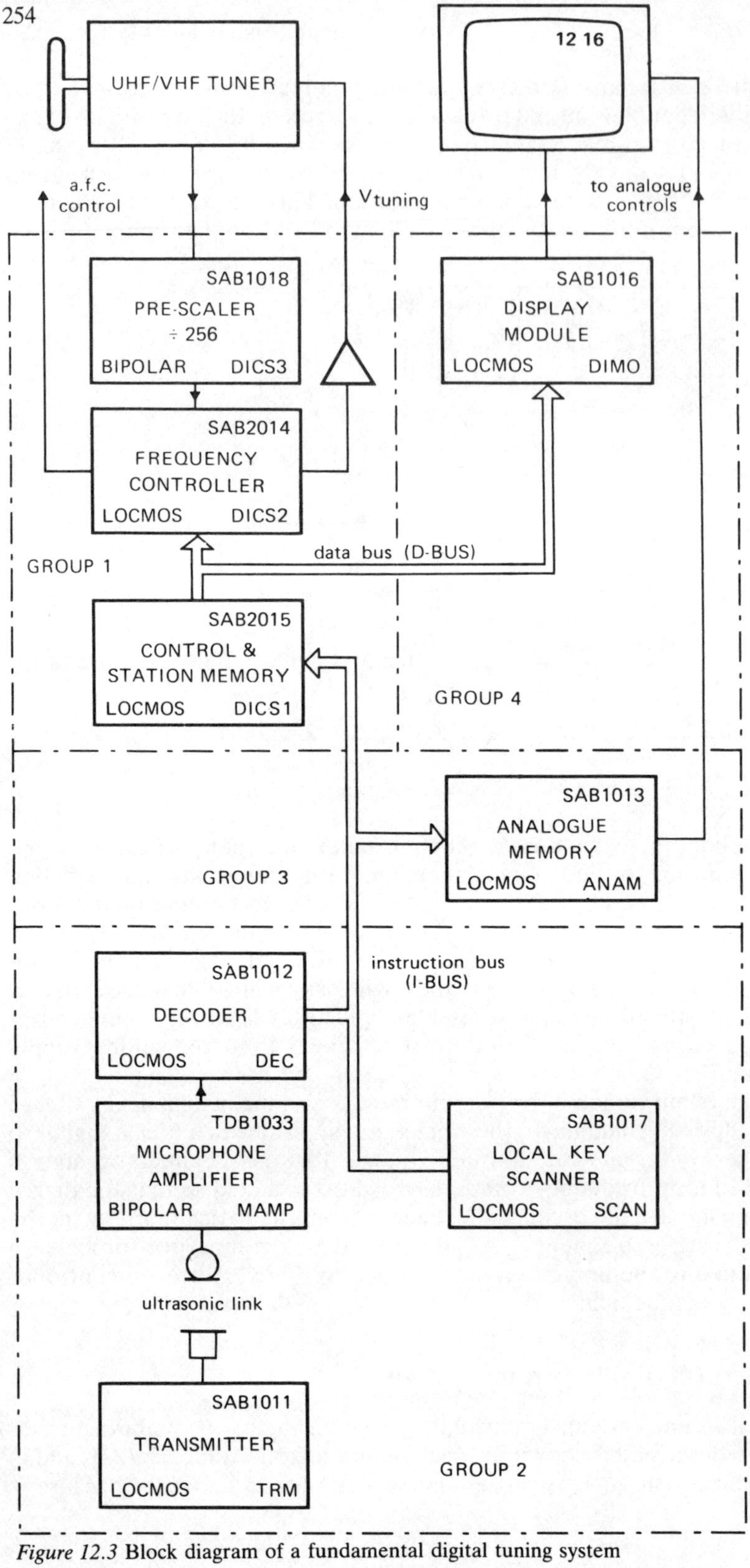

Figure 12.3 Block diagram of a fundamental digital tuning system

interfaces directly with other Mullard designed systems previously described in this book, and the complete system is available in varying degrees of complexity, and flexibility. The following facilities are provided:

1. Receiver tuning to all known CCIR (Comité Consultatif International Des Radio Communications), frequencies including B, G, I, and S channels.
2. 16 preselected stations can be stored in an electronic memory having battery backup.
3. The preselected station and/or channel number can be displayed on the receiver CRT in either the top left- or top right-hand corner.

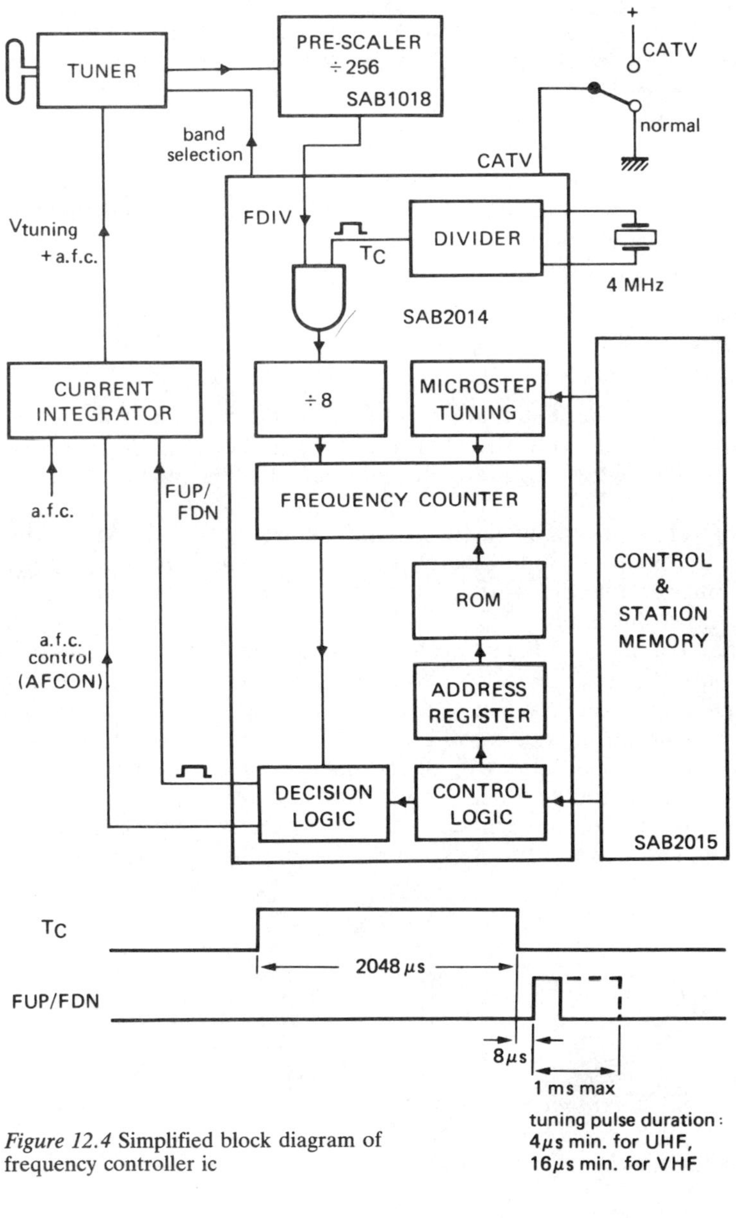

Figure 12.4 Simplified block diagram of frequency controller ic

4. Direct channel selection via remote control of any channel.
5. Up or down step tuning through all the CCIR channels.
6. Micro-step tuning to enable search of channels on unallocated frequencies.

The high degree of accuracy coupled with the versatility and flexibility of such a tuning system is reflected in its complexity. The full specification digital channel selection system from Mullard, including remote control, consists of a total of nine dedicated integrated circuits.

Figure 12.4 shows a simplified block diagram of the frequency controller IC (SAB 2014) together with its ancillary circuits. The local oscillator output from the tuner is divided by 256 by a pre-scaling IC SAB 1018. This divided oscillator frequency is designated FDIV (Frequency Divided), and provides an input to the frequency controller IC on pin 7. This frequency is AND gated with a 2048 microsecond timing pulse, (TC), derived by dividing the output of a 4 MHz crystal oscillator. The output of the AND gate will therefore be a 2048 microsecond burst of pulses which will be present when the timing pulse is at a logic '1'. This burst of pulses is further divided by eight before being fed to a frequency counter. Examination of the division ratios between the tuner and the input to the frequency counter are summarized below:

1. FDIV = f Osc/256.
2. TC = 2048 microseconds.
3. Freq. Counter input = f Osc/(256 × 8).
4. Freq. Counter input = f Osc/2048.

From the above listing it can be seen that the frequency counter input will consist of a 2048 microsecond period of (fc/2048). Therefore the number of pulses counted during the timing pulse TC will be equal to the number of MHz of the oscillator frequency.

The frequency counter is pre-loaded with a 10-bit word from a ROM, whose address is generated by decoding the channel number requested by the user. A listing of the ROM contents, together with its respective channel number and video carrier frequency is shown in *Figure 12.5*.

When the frequency counter has been pre-loaded with the ROM contents it is decremented by the output of the divide-by-8 counter for the duration of the timing pulse TC. If the oscillator frequency in the tuner is correct, the contents of the frequency counter after the timing pulse will be zero. If the oscillator frequency is not correct a positive or negative value will be left in the counter after the timing period. This remainder is used to produce a pulse whose width is proportional to the tuning error. The pulse is then integrated and used to modify the tuning voltage applied to the local oscillator in the tuner. A worked example is shown below.

Channel 21 frequency	= 471.25 MHz
Vision IF frequency	= 39.5 MHz
Tuner Oscillator frequency	= 510.75 MHz
Frequency counter input	= 510.75 × 10ˆ6/ (256 × 8) Hz
	= 2.494 × 1035 Hz
Periodic time	= 4.01 microseconds
Number of pulses in 2048 microseconds	= 2048/4.01
	= 510.7
ROM Contents for Channel 21	= 510

CCIR channel no.	video carrier freq. (MHz)	ROM contents	band switching info.
00	44,25	83	v.h.f. A
01	44,25	83	v.h.f. A
02	48,25	87	v.h.f. A
03	55,25	94	v.h.f. A
04	62,25	101	v.h.f. A
05	175,25	214	v.h.f. B
06	182,25	221	v.h.f. B
07	189,25	228	v.h.f. B
08	196,25	235	v.h.f. B
09	203,25	242	v.h.f. B
10	210,25	249	v.h.f. B
11	217,25	256	v.h.f. B
12	224,25	263	v.h.f. B
13	44,25	83	v.h.f. A
14	44,25	83	v.h.f. A
15	44,25	83	v.h.f. A
16	44,25	83	v.h.f. A
17	44,25	83	v.h.f. A
18	44,25	83	v.h.f. A
19	44,25	83	v.h.f. A
20	44,25	83	v.h.f. A
21	471,25	510	u.h.f.
22	479,25	518	u.h.f.
23	487,25	526	u.h.f.
24	495,25	534	u.h.f.
25	503,25	542	u.h.f.
26	511,25	550	u.h.f.
27	519,25	558	u.h.f.
28	527,25	566	u.h.f.
29	535,25	574	u.h.f.
30	543,25	582	u.h.f.
31	551,25	590	u.h.f.
32	559,25	598	u.h.f.
33	567,25	606	u.h.f.
34	575,25	614	u.h.f.
35	583,25	622	u.h.f.
36	591,25	630	u.h.f.
37	599,25	638	u.h.f.
38	607,25	646	u.h.f.
39	615,25	654	u.h.f.
40	623,25	662	u.h.f.
41	631,25	670	u.h.f.
42	639,25	678	u.h.f.
43	647,25	686	u.h.f.
44	655,25	694	u.h.f.
45	663,25	702	u.h.f.
46	671,25	710	u.h.f.
47	679,25	718	u.h.f.
48	687,25	726	u.h.f.
49	695,25	734	u.h.f.
50	703,25	742	u.h.f.
51	711,25	750	u.h.f.
52	719,25	758	u.h.f.
53	727,25	766	u.h.f.
54	735,25	774	u.h.f.
55	743,25	782	u.h.f.
56	751,25	790	u.h.f.
57	759,25	798	u.h.f.
58	767,25	806	u.h.f.
59	775,25	814	u.h.f.
60	783,25	822	u.h.f.
61	791,25	830	u.h.f.
62	799,25	838	u.h.f.
63	807,25	846	u.h.f.
64	815,25	854	u.h.f.
65	823,25	862	u.h.f.
66	831,25	870	u.h.f.
67	839,25	878	u.h.f.
68	847,25	886	u.h.f.
69	855,25	894	u.h.f.
70	44,25	83	v.h.f. A
71	44,25	83	v.h.f. A
72	44,25	83	v.h.f. A
73	44,25	83	v.h.f. A
74	44,25	83	v.h.f. A
75	44,25	83	v.h.f. A
76	44,25	83	v.h.f. A
77	69,25	108	v.h.f. A
78	76,25	115	v.h.f. A
79	83,25	122	v.h.f. A
80	105,25	144	v.h.f. B
81	112,25	151	v.h.f. B
82	119,25	158	v.h.f. B
83	126,25	165	v.h.f. B
84	133,25	172	v.h.f. B
85	140,25	179	v.h.f. B
86	147,25	186	v.h.f. B
87	154,25	193	v.h.f. B
88	161,25	200	v.h.f. B
89	168,25	207	v.h.f. B
90	231,25	270	v.h.f. B
91	238,25	277	v.h.f. B
92	245,25	284	v.h.f. B
93	252,25	291	v.h.f. B
94	259,25	298	v.h.f. B
95	266,25	305	v.h.f. B
96	273,25	312	v.h.f. B
97	280,25	319	v.h.f. B
98	287,25	326	v.h.f. B
99	294,25	333	v.h.f. B

Note: For channels 00 to 76, the CATV input to the SAB2014 is LOW: for channels 00 to 99, the CATV input is HIGH.

Figure 12.5 ROM listing with channel number and video carrier frequency

This system will select the desired local oscillator frequency accurately to within 1 MHz, and ensure that the received signal is within the capture range of the AFC. Once this has been achieved, AFC control (AFCON), is switched in to provide the final tuning voltage modification.

It should be noted that the tuning system operates by sampling the tuner oscillator frequency. Hence the ROM must contain the data which represents the incoming carrier frequency plus the IF frequency.

12.3 The tuning modes

There are three tuning modes for this system:

1. The station mode.
2. The channel mode.
3. Micro step tuning.

The station mode

This mode enables up to sixteen stations of the viewer's choice to be stored in a RAM contained in the control and station memory IC SAB 2015. A simplified block diagram of this integrated circuit together with the frequency controller and associated circuits is shown in *Figure 12.6*.

The SAB 2015 receives a serial digital signal representing the station requested by the user from the I bus. This signal is processed by the I bus receiver and station counter to generate an address for the RAM. Data stored in the addressed memory location is used to create a second serial signal representing the channel number and station requested, and this leaves the SAB 2015 on a three line data bus (D-BUS). This data bus is used to operate the mode of the frequency controller and the display module integrated circuits. The three signals used are:

1. The clock pulse, (CLCK), which carries a 62.5 kHz square wave derived from the 4 MHz crystal oscillator in the SAB 2014 frequency controller.
2. A Pulse Code Enable signal (PCEN) which carries a gating signal or 'Information Window', having a duration of 28 clock pulses.
3. A Pulse Coded Data signal (PCDA) which carries the station and channel number information in the form of a 13-bit serial binary word. Each bit of the PCDA word has a duration of two clock pulses, and each word is preceded by a leading '0'.

The format of the D-bus is shown in *Figure 12.7*, together with a timing diagram for the data highway. Note that the display mode bit is transmitted first and this is always preceeded by a '0'.

In order to prevent the integrated circuits which use the D-bus from operating intermittently a number of tests are carried out on the received data before the information is acted upon:

1. Presence of correct number of data bits.
2. Presence of a leading '0'.
3. Presence of noise on the data highway.

The data bus is used by both the display module IC (SAB 1016), and the frequency controller (SAB 2014). Information displayed to the user will include

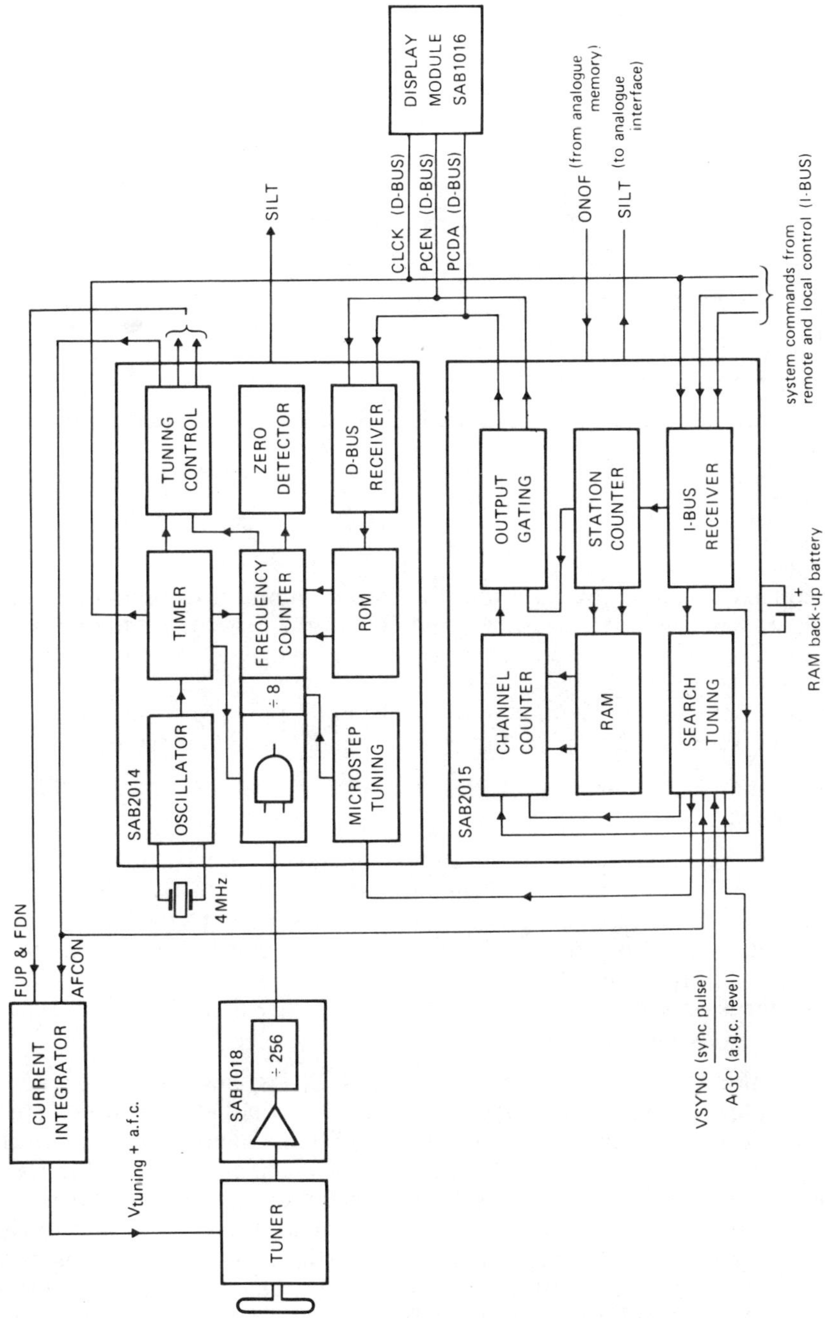

Figure 12.6 Control and station memory block schematic

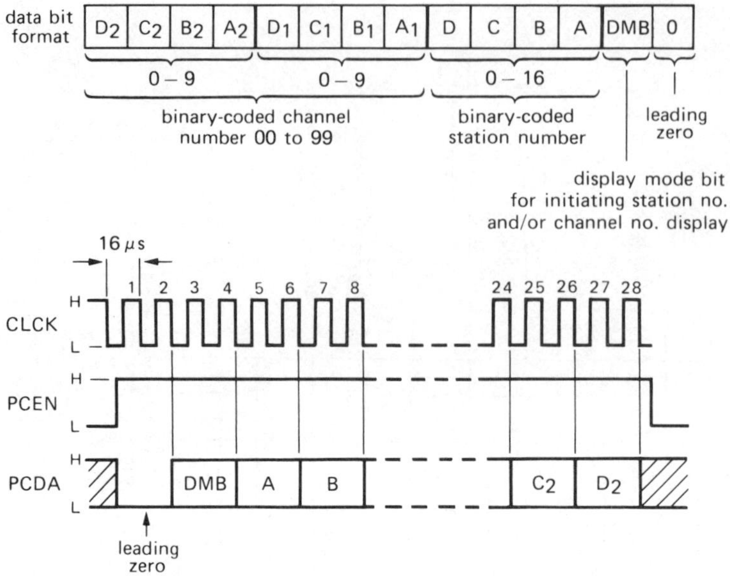

Figure 12.7 Format of D-bus

both the requested channel number and the station number. Consequently all of the data will be required by the display module. On the other hand the frequency controller only requires the CCIR channel number which is used to address its internal ROM.

Each of these integrated circuits contains a D-bus receiver, to de-serialize, check and process the received data from the control and station memory IC.

The channel mode

When the channel mode is selected, the circuits operate in exactly the same way as that described in the section on the tuning system earlier in this chapter. The channel requested by the user is again coded as a serial digital signal on the I bus, and is processed by the I bus receiver in the SAB 2015. Further processing in this integrated circuit converts the I bus signal to a Data bus signal before the coded channel number is passed on to the SAB 2014. There are three options which the user can select in the channel mode:

1. Direct selection of any channel stored in the ROM as shown in *Figure 12.5*.
2. Sequential step tuning, one channel at a time, in either direction. This is achieved by using the channel counter in the SAB 2015 and enables the user to manually step in either direction through the 100 channels stored in the ROM.
3. A search tuning mode which causes the system to tune slowly through all channels in either direction whilst continuously monitoring the AGC and video detector output signals for signs of a transmission. When a transmission is detected search will stop, displaying the received channel on the CRT. Once a channel has been found using this mode of operation the user has the option of storing its channel number in the SAB 2015 RAM so that it can be subsequently recalled.

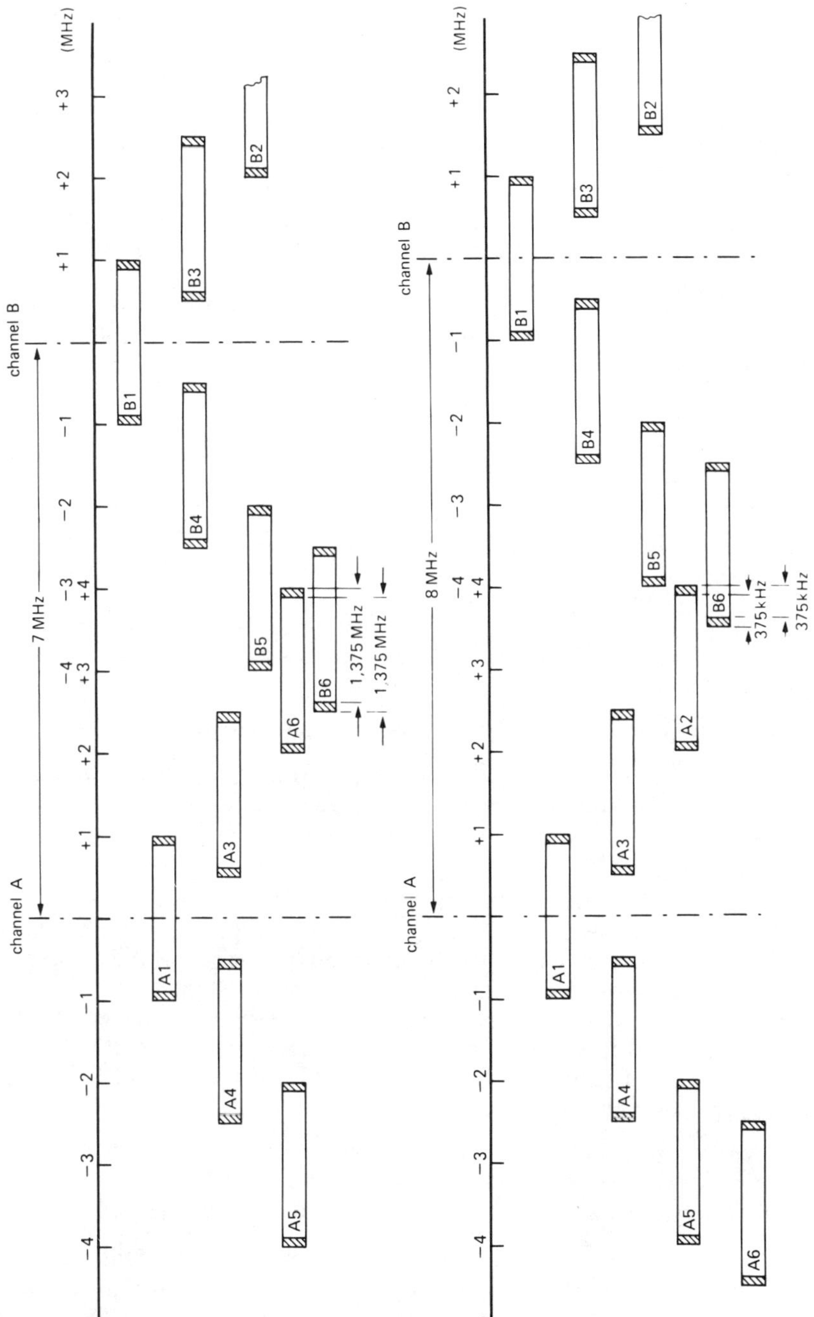

Figure 12.8 Microstep tuning on VHF band

Micro step tuning

Micro step tuning is automatically selected if the receiver selects a channel or station and finds that a transmission is not present. If the receiver AGC level indicates noise at the receiver input micro step tuning will de-tune the tuner oscillator frequency in five separate steps, checking the AGC level at each step for signs of a transmission.

This oscillator de-tuning is carried out in a set sequence, with two tuning steps above and three steps below the nominal oscillator frequency. Microstep spacing and the AFC pull-in range are so designed that the five steps cover the width of one channel. A schematic showing the coverage of microstep tuning on the UHF band is shown in *Figure 12.8*.

When a channel has been found, the DICS system carriers out a number of checks to ensure that the received transmission is of an acceptable quality for viewing. Referring to *Figure 12.9*, the first test made is to measure the level of the

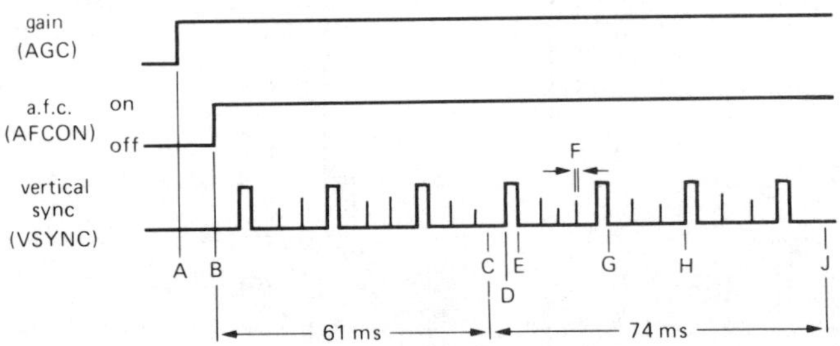

Figure 12.9 The DICS system checks

A: Check gain level
B: Wait for AFCON = HIGH, then test AGC = HIGH
B–C: 61 ms delay for a.f.c. action
D–E: Check HIGH duration of vertical sync pulses is between 32μs and 208 μs
F: Duration of spikes must not be more than 16 μs
G–H: Check LOW period between sync pulses is between 18,4 ms and 20,4 ms
J: End of test

AGC control voltage applied to the IF amplifier. If the AGC level is below a predetermined threshold point the test sequence is stopped, and if search tuning has been selected the system will advance to the next channel. If micro step tuning has been selected the system advances to the next micro step.

If test one passes, indicating that a transmission is being received, then AFC control switches on and the system pauses for 61 mS to allow the AFC circuits to settle. After this period of time an analysis is made of the received vertical sync pulses. Several parameters are tested:

1. Vertical sync width should be between 32 and 208 microseconds.
2. The duration of noise spikes must not be greater than 16 microseconds.
3. The interval between each field sync pulse should lie between 18.4 and 20.4 mS.

If all of these tests pass then the channel number of the received transmission is displayed on the CRT and the tuning sequence stops.

CRT display

The Mullard DICS system enables the CRT to display the station, and/or channel number either in the left- or right-hand top corners of the picture. The electronics used to achieve this is contained in a single integrated circuit (SAB 1016) known as the display module (DIMO). A block diagram of this IC is shown in *Figure 12.10*.

Four numeric characters are boxed in the TV picture for 2.5 seconds after channel/station selection. These digits use a dot matrix of 5 by 7 and employ character rounding.

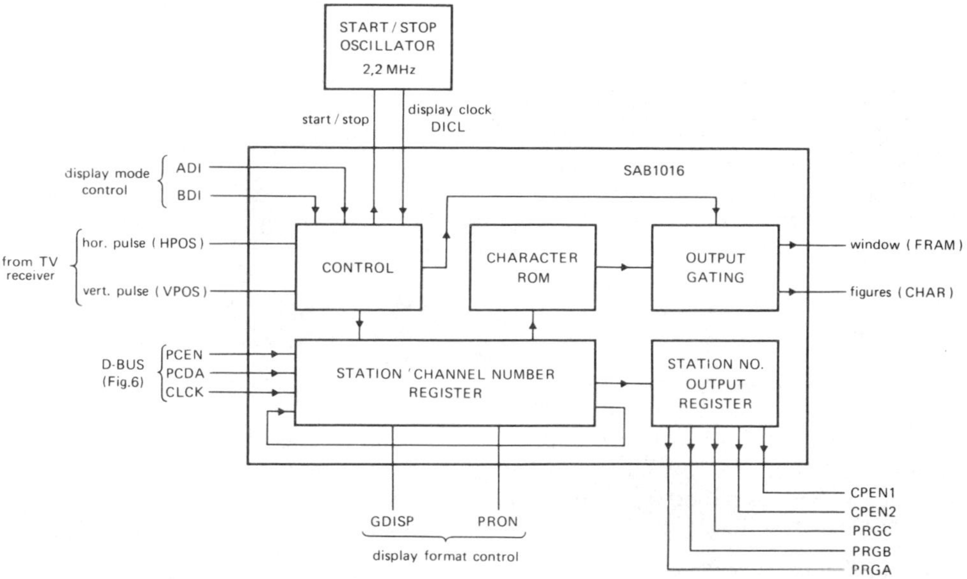

Figure 12.10 Block diagram of DIMO ic

The serial data signal (PCDA) on the D-bus is clocked into a station/channel number register by the D-bus clock signal during the presence of a PCEN window. Once the data is stored in the register it can be used to address the character ROM in order to produce the desired dot pattern for the output video waveform. The contents of the station/channel number register are also used by the station number output register. This register generates a five-bit digital signal which can be decoded by external circuits and be used for external switching of video cassette recorders, etc. A control circuit receives a variety of timing and control signals which it uses to select the display mode required. The horizontal (HPOS) and vertical (VPOS) synchronizing pulses derived from the receiver timebases provide the major timing references that determine the position of the boxed display.

Four display modes are available and these are summarized in *Figure 12.11*. Two display format control signals (GDISP) and (PRON) are used to select which of the four display modes are to be used. When GDISP is at a '0' the display is placed in the top left-hand corner and when GDISP is at a '1' the display is in the right-hand corner. The PRON input determines the length of the boxed display. When PRON is at a '0' the station number only is displayed, and if PRON is at a '1' a full four-digit display box is generated.

Reference to *Figure 12.3* will show that the four integrated circuits comprising of the pre-scaler, frequency controller, control and station memory, and display module, form the heart of the digital tuning system, and carry out the basic tuning and display operations. These integrated circuits are controlled by digitally-coded

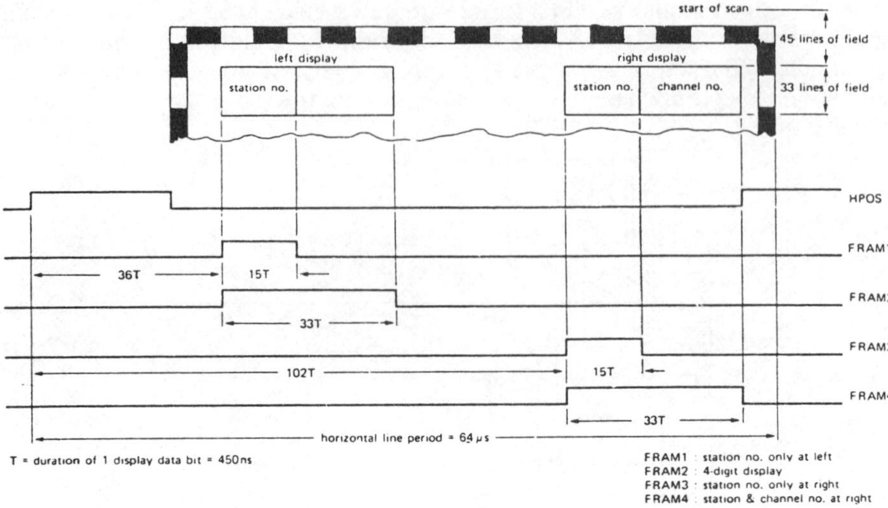

Figure 12.11 DIMO display modes

instructions on the I-bus from either a remote control system or a local keyboard scanner (SAB 1017). The techniques employed in these integrated circuits are very similar to those outlined in chapter 11.

DICS for Teletext receivers

The remote control system outline above was primarily designed for TV receivers without Teletext or Viewdata facilities. Consequently the remote control transmitter and receiver ICs are unable to cope with the added instructions required for Teletext and Viewdata. The design of a receiver incorporating DICS and Teletext/Viewdata creates several complex technical problems for design engineers, possibly the most significant of which is the number of remote control commands required to operate the receiver. Faced with these problems, Mullard developed a revised remote control and DICS chip set to cope with the additional load, and complexity of the combined system.

This new system (see *Figure 12.12*) was designated Video Tuning System (VTS). The transmitter is a single chip (SAB 3021) and its simplified block diagram is shown in *Figure 12.13*. The full specification transmitter incorporates a 64 push-button keypad and an additional two-position switch which enables the transmission of 128 different command codes. Each command is transmitted as a 6-bit code modulated on an infra-red beam, and can be used to control additional items of domestic equipment such as radio, hi-fi, etc.

The received infra-red beam is processed by a TDB 2033 to produce a clean serial digital signal representing the transmitted command (RSIG). This signal is then fed to the remote control decoder (SAB 3042) which is considerably more complex than the decoders described in chapter 11.

The decoder not only receives the RSIG serial data signal from the TDB 2033, but it also controls a local keypad mounted on the receiver front panel. This enables instructions for the decoder to be entered at the receiver or by remote control, the keyboard taking priority over remote control commands. Received

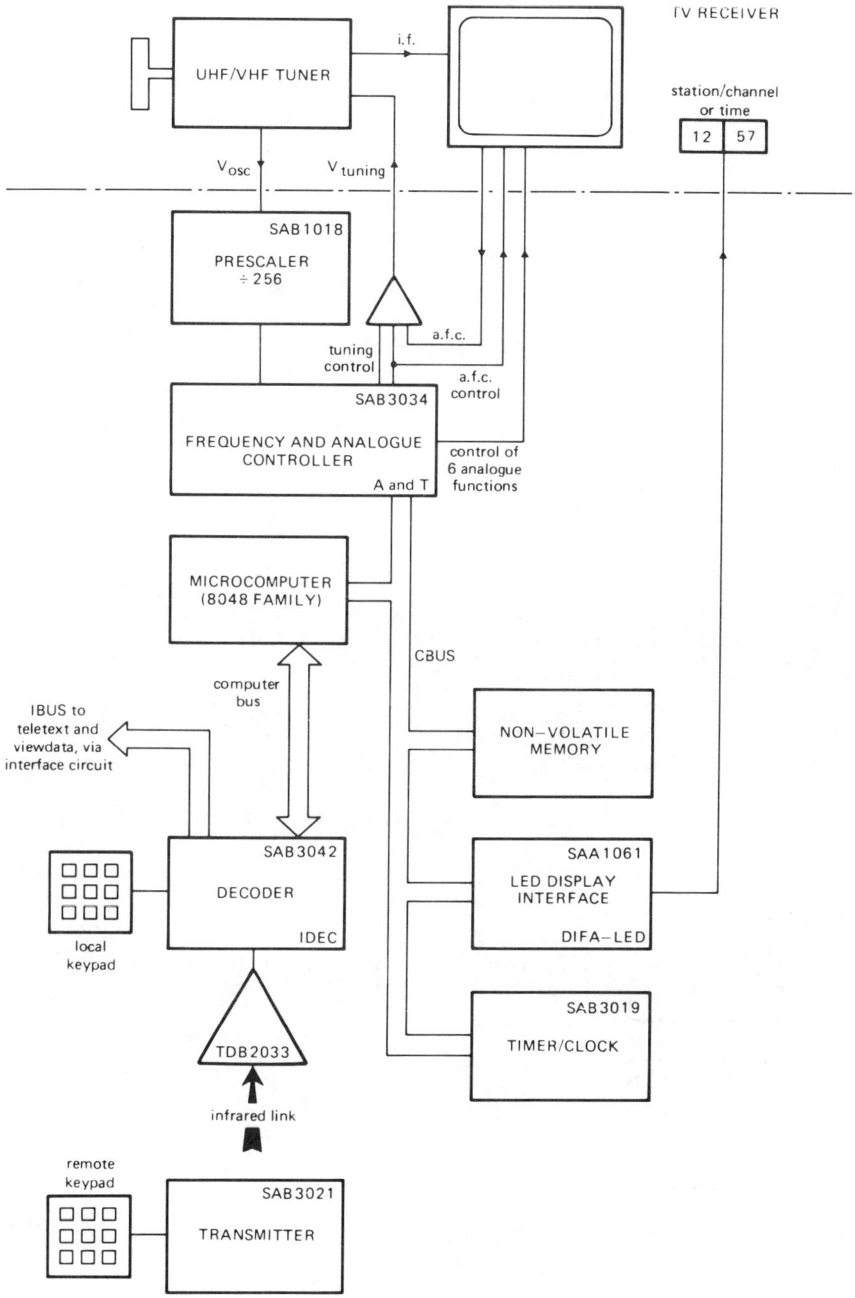

Figure 12.12 Video Tuning System (VTS)

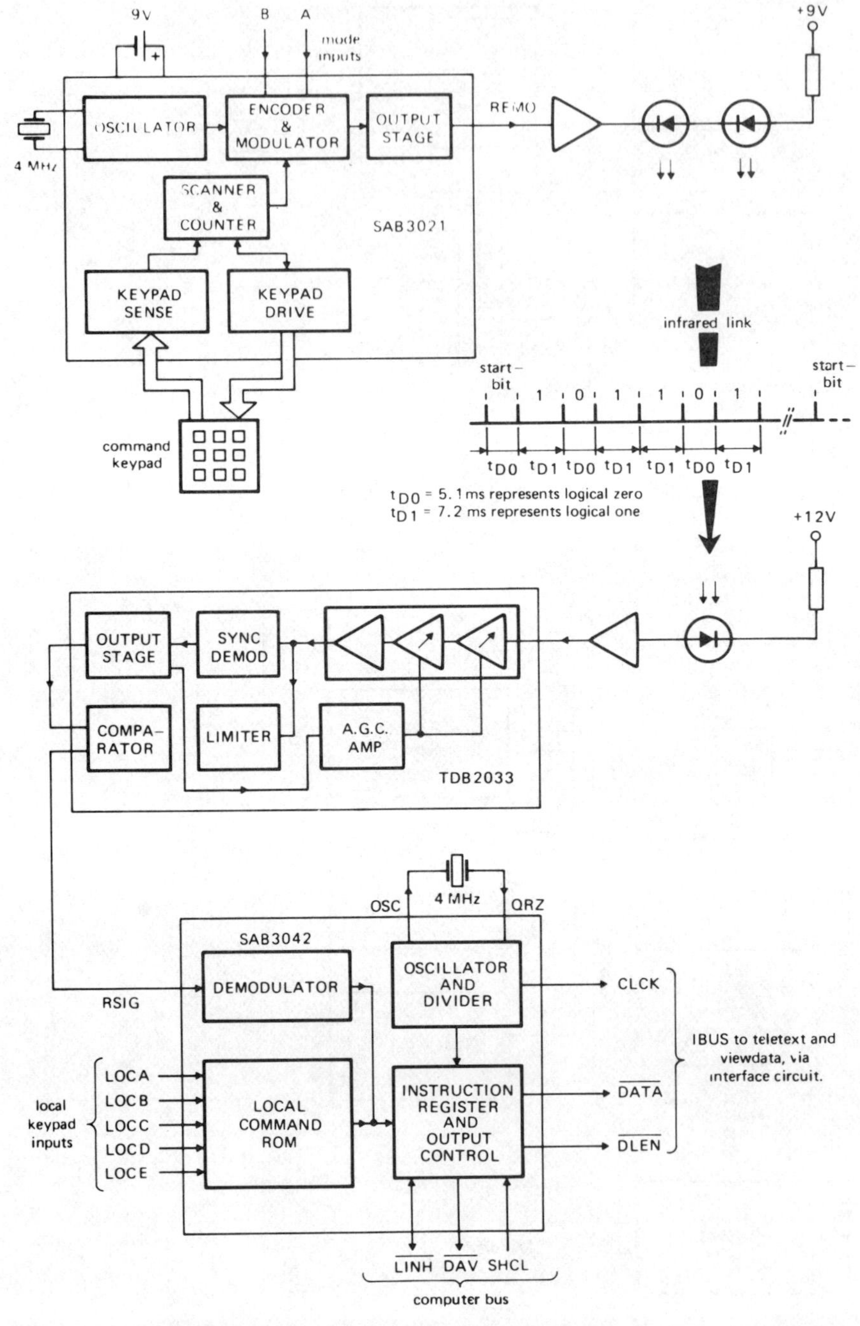

Figure 12.13 Block diagram for VTS

information from the local keypad or from the remote control system, causes the decoder to generate two bus outputs:

1. The I-bus which is used as previously described in earlier chapters for Teletext and Viewdata decoders. The I-bus leaving the SAB 3042 uses a 6-bit command, preceded by a start bit. We have seen that the Teletext and Viewdata decoders manufactured by Mullard use a 7-bit word, transmitted on a two signal I-bus, of which the first and last bits are reserved for mode selection. The I-bus generated by the SAB 3042 uses three signals designated DLEN (Data Line Enable), DATA and CLOCK. Consequently an interface circuit is required to convert the three signal I-bus from the remote control decoder to a two signal I-bus for Teletext and Viewdata decoders.
2. The computer bus is a three line system allowing the decoder to communicate with a single chip microcomputer in the 8048 family.

   This microcomputer is manufactured by the Intel Corporation and contains:

1 K by 8-bit ROM.	ALU.
64 by 8-bit RAM.	Accumulator.
Three 8-bit I/O ports.	Flag register.
A timer counter.	

Stored within the ROM of the microcomputer is a programme which controls the complete operation of the digital tuning system, and instructions from the microcomputer are sent to the other integrated circuits within the system via the C-bus.

The pre-scaler (SAB 1018) and the analogue controller (SAB 3034) use very similar techniques to those previously described, the major difference being in its interfacing techniques with the rest of the DICS system, and the absence of an internal ROM. Data that determines the initial loading of the presetable counter is now stored in the microcomputer ROM. A non-volatile memory is also connected to the C-bus and is used to store channels allocated to station numbers.

An SAA 1061 LED display interface has been devised to operate four 7-segment displays mounted on the receiver front panel, which are used to display channel and station information or the time of day. In the previous example of the DICS system this information was boxed in to the video display in either the top left- or the top right-hand corners. With Teletext however, this boxed information looks untidy and is confusing because it will block out the Teletext real-time clock if placed in the top right-hand corner, or the selected page number if located in the top left-hand corner. For these reasons it has been decided to display channel/station information separately.

Contained in the SAA 1061 is an interface with the C-bus which enables the relevant data to be received and stored. This latched data provides the inputs to a multiplexer which then scans the four 7-segment displays to produce the desired information. An SAB 3019 timer clock may be optionally included in the system so that the LED display can be switched between channel/station information or the time of day.

References

1. *Mullard Digital Tuning System for TV Receivers: Technical Information 46* Mullard Ltd, Mullard House, Torrington Place, London WC1E 7HD.

Index